D1088614

Molecular Biochemistry of Human Disease

Volume I

Author for Fundamentals of Clinical Diagnosis
George Feuer
Professor
Department of Clinical Biochemistry and Pharmacology
Banting Institute
University of Toronto
Toronto, Canada

Author for Pathological Supplements
Felix A. de la Iglesia
Director
Department of Pathology and Experimental Toxicology
Warner-Lambert/Parke-Davis Pharmaceutical Research
Ann Arbor, Michigan

CRC Press, Inc.
Boca Raton, Florida

Library of Congress Cataloging in Publication Data

Feuer, George, 1921-
Molecular biochemistry of human disease.

Bibliography: v. 1, p.
Includes index.
1. Metabolism--Disorders--Diagnosis. 2. Diagnosis, Laboratory. 3. Biological chemistry. 4. Pathology, Molecular. I. De la Iglesia, Felix A., 1939-
II. Title. [DNLM: 1. Biochemistry. 2. Molecular biology.
QU 4 F426m]
RC627.6.F48 1985 616.07'56 84-7156
ISBN 0-8493-6205-9 (vol. 1)

Direct all inquiries to CRC Press, Inc., 2000 Corporate Blvd., N.W., Boca Raton, Florida, 33431.

International Standard Book Number 0-8493-6205-9 (Volume I)

Library of Congress Card Number 84-7156
Printed in the United States

PREFACE

Medical education has undergone substantial changes during recent years, leading to several diverse forms of approaches. Due to many new facts our knowledge of the biochemistry of human disease is increasing. There is a trend, however, to decrease the basic science training given to medical students. As a consequence, it seems almost superfluous to say that compared to the present-day physicians, those of the future may be better trained but generally less informed about the accumulated biochemical knowledge and its uses in the treatment of disease. They may have as much or more absolute knowledge than now, but compared to the potential power of biochemistry at present or at that time their basic knowledge will be poorer.

The large and valuable store of facts learned in school and experienced in practice by the physician enables him to fit signs and symptoms into diagnostic categories and to predict probable consequences. The future physician will be, however, forced to rely more on biochemists working in clinical laboratories who may have an analogous store of facts about the biochemical aspect of diseases. The physician may consult a paramedical person who will be armed with biochemical knowledge to assess the relationship of the laboratory data to clinical signs, and who may be aware of the importance of drug therapy using a particular drug in a specific disease or can effectively help in selecting more specific therapeutic choices.

From the widening gap between the progress of clinical biochemistry and the patient-oriented curriculum of medical students a need arose to teach the basic aspects of biochemistry of human disease to nonmedical students. These thoughts prompted the initiation of a course a few years ago in which I became involved. The course was planned primarily for paramedical students, mainly pharmacy and nursing students in their last year, but through the years it has also been attended by students in biological sciences as well as taken as an elective by medical students. This book is largely based on the course offered at the University of Toronto aiming to provide some help in their studies.

The modest material of the original course has been greatly expanded. I undertook the enormous job of writing this book, which took close to three years. This attempt had the advantages of the necessity that a textbook should be homogeneous and should convey a uniform picture of the biochemical background and mechanism of various diseases. This is contrasted, however, by the disadvantages of my not being expert in some fields and by the increased time required to finish the job. To overcome these difficulties, I consulted many colleagues who provided help by pointing out lack of clarity in certain passages, inaccuracies, and errors by constructive criticism. It would not have been possible to conquer this task without their advice and thorough review. I owe them a great deal for their stimulation and encouragement and also for the sharp criticism which helped me to gain insight into many important problems. It is, however, inevitable that some errors, inadequacies, faulty interpretations or emphases have still remained in the book, for which I am solely responsible.

I received ample wise counsel and editorial comments from many colleagues and I wish to express my most sincere thanks and appreciation to the reviewers of the various chapters: Drs. David M. Goldberg, Thomas R. Csorba, James B. Cullen, John A. Kellen, Norbert A. Kerenyi, and Ronald A. Martin. My thanks go to the reviewers of the book arranged for by the publisher. I appreciate the help given by Dr. N. A. Kerenyi who provided microscopic slides of several disease entities which were kindly photographed by D. J. E. Fitzgerald. I am also grateful to my co-workers who helped in many organizational problems and who were the guinea pigs when I discussed with them several parts of the work: M.S. Ilyas Dhami, Rosa Drangova, Carlo DiFonzo, Rose Farkas, and Lidija Stuhne-Sekalec; to our librarian Mrs. S. Duda for helping with the compilation of references; to many secretaries who typed the various chapters of the manuscript: Marilyn Bailey, Fay Borysowich, Irene

Bowers, Mary Brenton, Nancy Gottschalk, Jackie Magee, Denise Zucchiatti; and my wife, who typed and corrected many parts of the book. Finally my deepest appreciation should go to my children who encouraged me and patiently missed my company and a partner in their games during many weekends and evenings while I was working on the manuscript.

Toronto, March 1983

George Feuer

THE AUTHORS

George Feuer, Ph.D., is a Professor of Clinical Biochemistry and Pharmacology (Toxicology) at the University of Toronto, Canada. Dr. Feuer received his B.Sc., M.Sc., and Ph.D. at the University of Szeged, Hungary from 1943 to 1944 and C. Med. Sc. at the Hungarian Academy of Sciences, Budapest, in 1952.

Dr. Feuer has worked in various fields of biochemistry. Originally, he was interested in the mechanism of muscular contraction and for eight years studied in this field. This area of study led to further investigations into the energy transfer process in relationship to acetylcholine synthesis in the brain. Interest in the metabolism of thyroid hormones commenced in 1955.

In 1960, Dr. Feuer was chairperson at a session of the Fourth International Goiter Conference in London, England. Between 1958 and 1963, he investigated the effects of endocrine glands on emotional behavior. From 1963 to 1968, he studied the effects of food additives on the liver. This area of study was extended to encompass the investigation into certain liver diseases such as neonatal hyperbilirubinemia and intrahepatic cholestasis.

Dr. Feuer's main interests, at present, are the investigations into the mechanisms of the development of intrahepatic cholestasis and hepatocarcinoma. He is also involved in the early detection of toxic side effects brought about by drugs and various other chemicals.

Dr. Feuer has written various books, reviews, and scientific papers including numerous publications on food and drug hepatic interactions and the biochemistry of liver diseases. Currently he is a member of the Editorial Boards of *Xenobiotica* and *Drug Metabolism and Drug Interactions.*

Felix A. de la Iglesia, M.D., is Vice President of Pathology and Experimental Toxicology at Warner-Lambert/Parke-Davis Pharmaceutical Research in Ann Arbor, Michigan, as well as Adjunct Professor of Toxicology at the School of Public Health at the University of Michigan, and Professor of Pathology, University of Toronto. He is a member of the Environmental Health Sciences Review Committee, External Consultant in Pathology and Toxicology to the National Institutes of Health, National Cancer Institute, and official expert in Pharmacology and Toxicology to the Ministries of Health of France and West Germany. In addition, he is Councilor, Society of Toxicologic Pathologists and President of the Michigan Chapter of the Society of Toxicology. Born in 1939 in Argentina, he received his M.D. degree in 1964 from the National University of Cordoba, Argentina, and completed postgraduate training in Experimental Pathology as Fellow of the Medical Research Council of Canada. In 1966, he joined the Warner-Lambert Research Institute of Canada in Ontario, Canada and developed various research and administrative positions until becoming Director of the Institute. After moving to the U.S. in 1977, he became Director of Pathology and Experimental Toxicology until assuming his present position. His research interests are in the area of toxicodynamics of subcellular organelle changes in drug-induced hepatic injury, toxicological aspects of novel anticancer chemotherapeutic agents, and safety assessment of novel chemical entities leading to the development of therapeutic agents. A Diplomate of the Academy of Toxicological Sciences, his publications include more than 250 articles, and currently he is the Editor of *Toxicologic Pathology,* and Editorial Board member of the journals *Toxicology and Applied Pharmacology* and *Drug Metabolism Reviews.*

To my teachers who initiated my intellectual development,
and to my family who have given the most
because of my passion for science.

INTRODUCTION

In the course of recent decades a fundamentally new view has emerged in the analysis of the complexity of disease processes and their relationship to the regulation of cellular metabolism. It is now firmly established that many derangements of the normal structure and function of the organism stem from some impairment of the biochemical organization. Defects in normal biochemical processes proved to be the reason for the primary anomaly, even though they do not always result in immediate pathological conditions. However, when such biochemical changes persist they become irreversible and thus normal cells are progressively transformed into abnormal ones, resulting in pathological changes and clinical symptoms. Today, we understand a disease only if the impairment can be clearly identified with alterations of normal biochemical processes as recognized by various laboratory tests. Correlations between these biochemical tests and clinical manifestations provide key information on the basic etiology of human illness. The interpretation of clinical biochemical data is available for the diagnosis and at the same time this knowledge aids in management of the disease and in understanding the rationale behind drug therapy.

Although in many instances these interrelations have not yet been established, we now can consider that the task of correlating clinical signs and symptoms of diseases with laboratory data is unfinished, awaiting what further research will reveal about the connections between the manifestations of disease and impaired biochemical function. Clinical biochemistry has been growing steadily as new steps are discovered in the mechanisms of well-known diseases and new insights are being gained into the background and origin of abnormal biochemical processes. New methods are introduced in diagnosis, and new drugs are tested for therapy. Generally, there has been improved experience of disease mechanisms. Further progress in investigations of the biochemistry of human disease will disclose additional correlations, allowing us more knowledge of the primary biochemical lesions, which manifest in diseases of presently unknown origin or mechanism.

Although the title of this book might imply a wider academic scope, this book considers only basic problems regarding the correlation between altered biochemical processes and illness. Its purpose is to describe the development of abnormalities in biochemical reactions, as well as in the underlying mechanisms, and to illustrate how the action of drugs fits into reversing the changes of the progressing disorder caused by the derangement in normal reactions. Disease processes are usually complex; in these situations several interrelated systems function in an integrated manner. Although the main arguments of the book are based on biochemical information, an attempt is made to incorporate various other aspects originating from pathological and pharmacological studies into a uniform view. Considering the sick man as a unit, this work conveys an integrated picture of the biochemical changes associated with disease.

The primary aim of the book is to help medical, pharmacy, and advanced students in science to understand the growing importance of continuously advancing biochemical concepts in human disease. It may serve as a *vade mecum* for clinical biochemists to review the basis of their practical experience. At the same time it may also help physicians to brush up the clinical biochemistry learned during their years in medical school. Several excellent texts on general biochemistry are available; hence basic information will not be given in detail. However, there is reason to believe that many students and physicians as well would welcome a book in which the fundamental biochemistry underlying the course of disease is presented in an extended and readily understandable form. Thorough discussions on the interrelationships between organ, cell, or cellular organelle and disease, and more space than usual are devoted to the interpretation of the biochemical nature of human disease.

Essential knowledge of physiology, biochemistry, pathology, and pharmacology is assumed. The basic features of biochemistry, well described in standard textbooks, are omitted in order to focus the interest upon important issues of the relationships between impaired processes and disease. However, where necessary, limited background information is given to provide the reader with an introduction to the basis of a multitude of diseases with their various and often interrelated manifestations. At the same time more complex associations are also described and the defects in the molecular organization of the diseased cell or cellular organelle are discussed in depth. Relevant interactions with pharmacologically active substances, either produced by the body or applied by drug therapy, which may influence biochemical processes and the progress of disease are briefly mentioned. Essentials of diagnostic methods and interpretations, are also presented.

In general, the basic philosophy of the book could be summarized as follows:

1. Most (if not all) diseases originate from an impairment of biochemical molecular mechanism of the organism.
2. Biochemical processes affected by the disease and manifested through pathological lesions, may be revealed by clinical biochemistry tests.
3. The aim of therapy based on this knowledge is to repair the damage.
4. The specific purpose of drug administration is to restore normal conditions; the action of drugs lies in reversing the clearly discernible changes which manifest in the biochemical mechanism.
5. Prolonged injury of the biochemical mechanism leads to irreversible alterations.

The subject matter includes:

1. General changes characteristic of cellular components
2. The mechanism whereby these changes alter homeostasis
3. Specialized changes occurring in individual diseases, associated with particular organs
4. Changes peculiar to unique situations, inborn errors, diseases of the newborn, and aging

The course of diseases is described as a continuous process similar to a flow sheet:

Normal cell ↗ composition ↘ biochemical processes →
Normal cell ↘ structure ↗

impairment → disease ↗ regeneration → recovery
impairment → disease ↘ degeneration → death

An understanding of the disease mechanism is essential for correct diagnosis and adequate therapy. This aim is presented in this book by summarizing our present knowledge on the molecular and cellular mechanisms of disease. This book is not comprehensive, because in several fields our knowledge is still fragmentary and no one could master all available information.

The various subjects have been arranged logically, starting from the participation of cellular elements in disease and continuing with disorders associated with a particular organ. The various topics may be read in sequence as they appear. Since, however, many biochemical findings accompanied by the progress of disease are not yet clearly understood, sometimes it may also be necessary to turn to earlier or later chapters. Hopefully, the background provided will be sufficiently clear to make it relatively easy to learn more about the various diseases from the general literature. A glossary of the essential terminology will be found in an Appendix. Each chapter is followed by references, plus suggestions for further reading. It should be understood that in order to avoid an encyclopedic aggregation it was necessary to limit the number of these references, including only basic illustrative examples, mainly from the latest available reports. Perhaps many important contributions have been omitted in the text. Individual references are mentioned only when they are fairly recent, and the references form our laboratory only indicate our interest in various areas. We hope that the references will provide the interested reader with a starting point for further enquiry.

BASIC REFERENCE BOOKS

BIOCHEMISTRY

Lehninger, A. L., *Biochemistry*, 2nd ed., Worth Publication, New York, 1975.

Montgomery, R., Dryer, R. L., Conway, T. W., and Spector, A. A., *Biochemistry: A Case-Oriented Approach*, Mosby Company, New York, 2nd ed., 1977.

CLINICAL BIOCHEMISTRY

Thompson, R. H. S. and Wootton, I. D. P., *Biochemical Disorders in Human Disease*, Academic Press, New York, 3rd ed., 1970.

Zilva, J. F. and Pannall, P. R., *Clinical Chemistry in Diagnosis and Treatment*, Year Book, Chicago, 2nd ed., 1975.

Cantarow, A. and Trumper, C., *Clinical Biochemistry*, W. B. Saunders, Philadelphia, 7th ed., 1975.

Gray, C. H. and Howarth, D., *Clinical Chemical Pathology*, Burroughs-Wellcome Foundation, 8th ed., 1977.

Gornall, A. G., *Applied Biochemistry of Clinical Disorders*, Harper and Row, New York, 1980.

CLINICAL CHEMISTRY

Tietz, N. W., *Fundamentals of Clinical Chemistry*, 2nd ed., W. B. Saunders, Philadelphia, 1976.

Varley, T. R., *Practical Clinical Biochemistry*, 5th ed., Heinemann, New York, 1976.

Henry, J. B., *Todd-Sanford-Davidsohn: Clinical Diagnosis and Management*, W. B. Saunders, Philadelphia, 16th ed., 1979.

Brown, S. S., Mitchell, F. L., and Young, D. S., *Chemical Diagnosis of Disease*, Elsevier/North Holland Biomedical Press, Amsterdam, 1979.

DISEASE

Walter, J. B., *An Introduction to the Principles of Disease*, W. B. Saunders Co., Philadelphia, 2nd ed., 1982.

Stanbury, F. B., Wyngaarden, F. B., and Fredrickson, D. S., *The Metabolic Basis of Inherited Disease*, McGraw-Hill, New York, 1978.

TABLE OF CONTENTS

Volume I

Chapter 1

BASIS OF ABNORMAL BIOCHEMICAL MECHANISMS

I. INTRODUCTION

The normal function of the living cell is based on several levels of organization. The first level is connected with the catalytic activity of simple or complex proteins. Cells may function because they possess enzymes capable of catalyzing many specific reactions. The high degree of specificity is the result of structural organization. Enzyme molecules bind substrates at the active site and accept only those molecules as substrates which fit into the site in an almost perfect fashion. Multienzyme complexes form a network and enzyme-catalyzed reactions are organized into consecutive sequences. They are controlled either by other part of the network or by the accumulation of an intermediate or a metabolite beyond a certain critical concentration acting through feedback mechanism. The activities of the cell are basically dependent on the transfer of genetic information. Any fault of the genetic system leads to some abnormalities of enzyme function or the failure of enzyme synthesis.

The second level is related to the structural complementation of the functional interrelationship existing between various enzyme activities.[25,78,80] The hundreds of enzyme-catalyzed biochemical reactions in the cell do not operate independently of each other. They are ordered into many sequences of consecutive reactions through common intermediates. Moreover, at the same time enzyme systems are separated from each other by compartmentalization. Subcellular structures are responsible for different cellular activities modulated by various reactions including energy transfer and mediators such as hormones.

Various organs and interrelated systems represent the third level in the organization of the living organism.[4,9] Cells are organized into more complicated multicellular tissues by the binding together of similar cells. These, in turn, relate to other cells, forming organs. Some of these organs have relatively simple structures and uncomplicated functional processes, such as the muscles; some of them are immensely complex both structurally and functionally, such as the nervous system; some of them contain several structures in association, such as the stomach wall with its muscular, epithelial, and connective tissues with fairly separate functions.

The three levels of organization of the living organism are interrelated. The process begins with simple precursor molecules obtained from the environment.[1,3] These are transformed by enzyme reactions into the building stones of the cell, some of them yielding macromolecules which are further assembled into subcellular organelles in a regular fashion.[2,36] The composition, size, shape, and surface characteristics of micromolecules and subcellular structures are important in their specific function and biological interactions, and also in the formation of the cell and whole organs. Living cells possess the ability to regulate the synthesis of this chain of structures — from the simplest to the most complicated ones. The cell can turn on and off the synthesis of any component from precursors. These self-regulating properties are fundamental to the maintenance of the steady state of the normal cell and in its reproduction.

The fundamental organization of living structures is based on the capacity of the cell to reproduce itself. This is associated with the nucleus and is compressed into the nucleotide sequence of very small amounts of DNA. DNA is constructed from four bases — adenine, guanine, thymine, and cytosine — and one sugar and phosphate. The genetic information stored in this molecule is very stable and is coded in the form of a specific sequence of four basic mononucleotide building stones producing the linear polymer DNA molecules. The one-dimensional information is translated into inherent three-dimensional design which synthesizes simple and complex proteins. Many different kinds of protein molecules are formed which contain a variety of 23 amino acids.[97] These proteins serve as the components of enzymes and building stones of membranes, ribosomes, and other subcellular organelles which are further assembled into cells and organs in a regular, organized fashion. Characteristic interactions and interrelationships in this organization constitute the molecular basis and regulation of the living state. The wide diversity of life forms differ from each other by the myriads of possible arrangements and rearrangements of basic structures.

The genetic make-up of our body and the sequence of synthesis of small molecules and complex structures are largely unaltered throughout life, although environmental effects exert constant changes. Our survival in this changing external environment is dependent upon the maintenance of a stable internal milieu, termed homeostasis. This is carried out by control mechanisms associated with specialized cells. Beyond environmental actions, in several diseases probably some genetic component are involved. This is dependent on the actions of many genes and chromosomes. The genetic component is present at birth and may be apparent in an immediate abnormality, but often it is not necessarily active throughout life, manifesting itself by an alteration to health only at appropriate times. Disturbance of the homeostasis by environmental or genetic effects represents the major factor in the etiology of many disease processes. Due to the adverse influence of these components, abnormalities in the organizing principles of our body lead to disease.

II. MOLECULAR ORGANIZATION OF THE CELL

Each cell contains a great number of small molecules, amino acids, fatty acids, purine and pyrimidine bases, and sugar derivatives; many of these derivatives are integrated into larger molecules called macromolecules. Amino acids are present in the cell alone, but are often linked to form polypeptides and proteins. Fatty acids, glycerol, phosphate, and some bases are combined to yield phospholipids.[7,11,55,87] Glucose units are aggregated to glycogen. Purine and pyrimidine bases are combined with ribose or deoxyribose sugars to form various nucleotides. Proteins play key roles as building stones of cellular structures necessary for the organization of cellular metabolism and also in forming the skeleton of enzymes.[19,23,39,44,87,89] Enzymes catalyze the conversion of one molecule to another one and thus are responsible for the processes of metabolism. Some peptides and proteins function as hormones. These molecules acting as messengers, deliver information from one organ to another and thus regulate the interrelationship of various metabolic processes.[98]

Lipids are also required for structural organization, associated with other macromolecules. Lipids form complex lipoproteins with proteins.[6,42,50,81,85,99] In combination with proteins and polymer carbohydrates, they aggregate to construct various kinds of membranes.[10,91] Cell membranes separate one cell from another and the cell itself from the extracellular space.[84,86] They serve many purposes: preventing the entry of undesirable compounds into the cell, excreting various metabolites needed by other cells, and eliminating waste products.[93] The membranes of cellular organelles provide separation and compartmentalization of many activities of the cell. These structures are capable of organizing manifold cellular processes, storing body constituents, and eliminating foreign elements by phagocytosis. Some lipids also play an essential part in the function of structure-bound enzymes and many of them are important in the intra- and extracellular transfer of lipid-soluble vitamins and other lipid-soluble foreign compounds.[56,57,76]

Nucleotides polymerize to ribonucleic and deoxyribonucleic acids. These two types of macromolecules are involved in the synthesis of proteins related to gene expression.[79] Ribonucleic, deoxyribonucleic acids, histones, and various other types of proteins together form the complex chromosomes. These are the sites of genetic information and work as programmers in duplicating a portion of the cellular genes and producing genotypes as well as specialized type of cells, phenotypes.

The presence of many different small molecules and macromolecular complexes in the intercellular space represents a microenvironment in which the cell operates. The constitution and composition of this microenvironment is dependent upon the function of the tissue, and show great variations from tissue to tissue in the amounts of electrolytes, metabolites, waste products, gases, and particulate materials. The background interstitial substance or matrix of the space contains varying amounts of proteins and lipids. Connecting tissue has greater amounts of proteins than lipids, whereas in the nervous system the protein content of the matrix is relatively low. Many factors operate in the microenvironment: (1) physical factors (space, surface, flow rates, shearing forces), viscosity and colloid characteristics of fluid suspension, colloid osmolality, charge distribution of cationic and anionic sites and osmolality; (2) microclimate (temperature) influencing chemical processes (oxidation/reduction potential, O_2 and CO_2 tension, pH); (3) essential nutrients (amino acids, lipids, glucose, electrolytes, trace elements, vitamins); (4) interactions (products of adjacent cells, inducers and specific activators and inhibitors of cell proliferation, antibodies, hormones); (5) and generally, cell density, membrane permeability, and blood and lymph flow. All these factors are well regulated and, in health, the microenvironment is homeostatically well controlled. In disease, the factors operating in the microenvironment are disturbed. During an inflammatory process, for instance, inflammatory cells occupy the intercellular space and fluid of differing composition is collected, leading to edema.

III. STRUCTURAL ORGANIZATION OF THE CELL

Within the individual cell, macromolecular complexes are organized into subcellular units.[71,90] The molecular structure of these cellular organelles is very complex, but it is probable that in the assembly of various units specific macromolecules represent different functional roles. These actions are partly modified when different macromolecules interact with each other. We do not yet know how these cellular organelles are assembled, and only little has been established about the structural arrangement of the various types of molecular complexes in these elements and about the forces that organize their architecture, special shape, and distribution within the cell and the structural-functional interrelationship between various organelles.

There are a number of intracellular organelles detected by electron microscopy within the cytoplasm in most cells.[62] These are the nucleus, nucleolus, chromosomes, endoplasmic reticulum, Golgi complex, mitochondrion, lysosomes, peroxisomes or microbodies, and microtubules and microfilaments. The function of many cellular organelles is known, although their role is not always clear — such as lysosomes in various tissues. There are some specialized organelles with known function representing transient structures, such as glycogen granules in hepatocytes as a storage form of glucose, or zymogen granules of the pancreas which are the production and storage site of enzymes. In diseases, the various subcellular entities are subject to structural disturbance and subsequently the structural impairment is associated with some pathological manifestations.[24] The abnormalities of these organelles will be described in more detail in a later chapter. Representative examples of subcellular organelles are seen in Plates 1 through 7.

A. Cell Membrane

The cytoplasm of the cell is separated from the extracellular environment by the cell membrane. Membranes are structurally complex, but their organization shows close similarity regardless of species.[3,13,15,49,54,92] The cell membrane regulates the entrance of molecules and interchanges between the interior and the exterior environment.[60,94] The membrane exhibits selective permeability; this function is connected with the structural arrangement of macromolecules, particularly that of proteins. Most structural membrane proteins have not been isolated and identified, yet we know some of their principal functions.

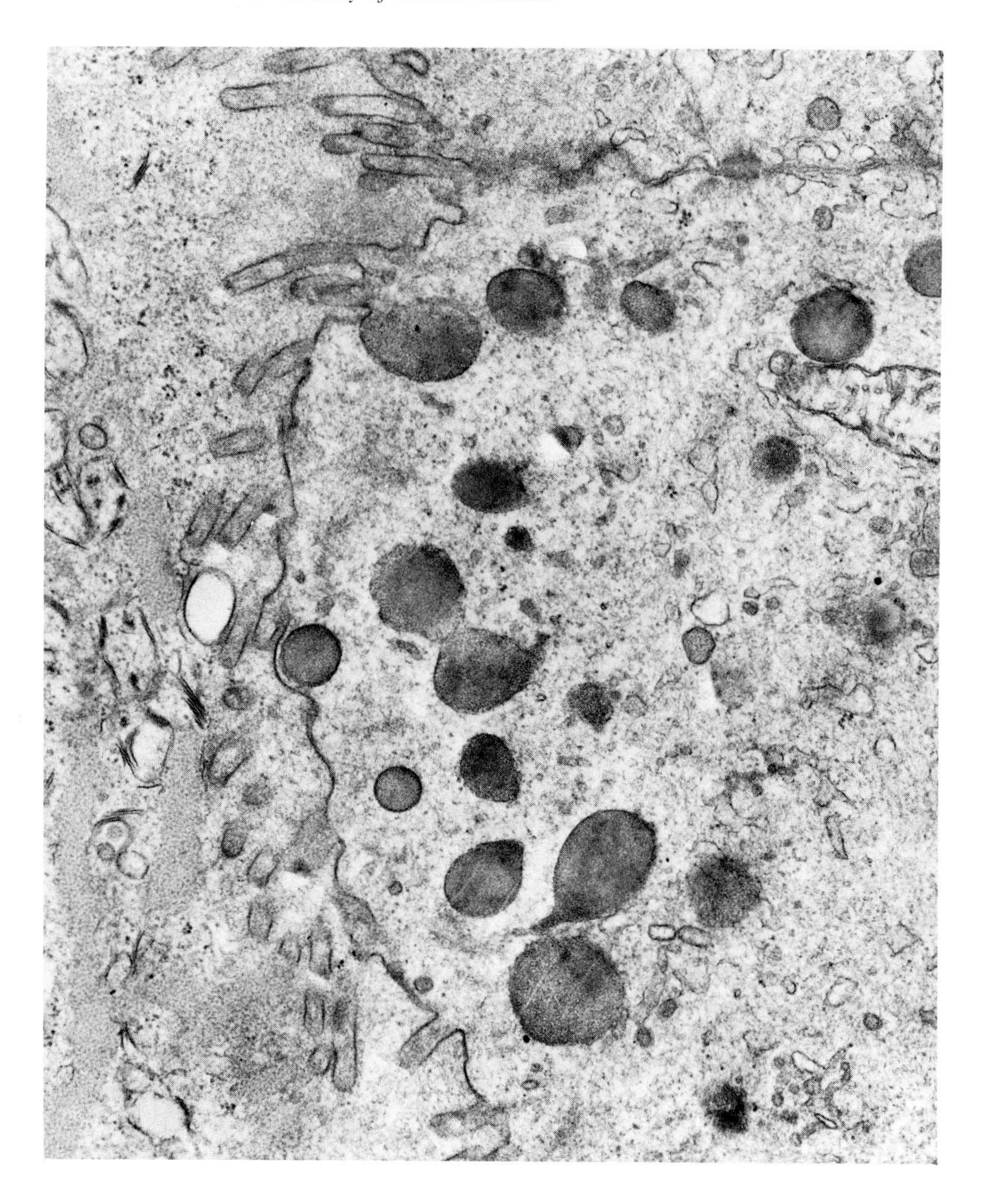

PLATE 1. Cell membrane. The cell membrane separates the cell from its surrounding fluid. It maintains the cell-to-cell communication. The cell membrane contains enzyme systems which can actively transfer substances from the environment into the cell, actively block substances from entering the cell, or actively eliminate substances from the cell. This EM illustration represents the biliary epithelium of the liver cell, which is a specialized area of the plasma membrance. In this biliary ductular cell, periodic structures constitute phospholiopoprotein material usually found as a constituent of bile. The villi protrude into the lumen to facilitate the excretory processes.

There are carrier proteins that transfer nutrients through the membrane, such as amino acids, glucose, and other small molecules. Glycophorin, found in the membrane of erythrocytes, is an important representative of these proteins.[74]

There are also messenger proteins that transfer information through the membrane, protein hormones that regulate metabolic processes, and antibodies imbedded into the surface of the cell that neutralize antigen actions. These special proteins, called receptors, play an

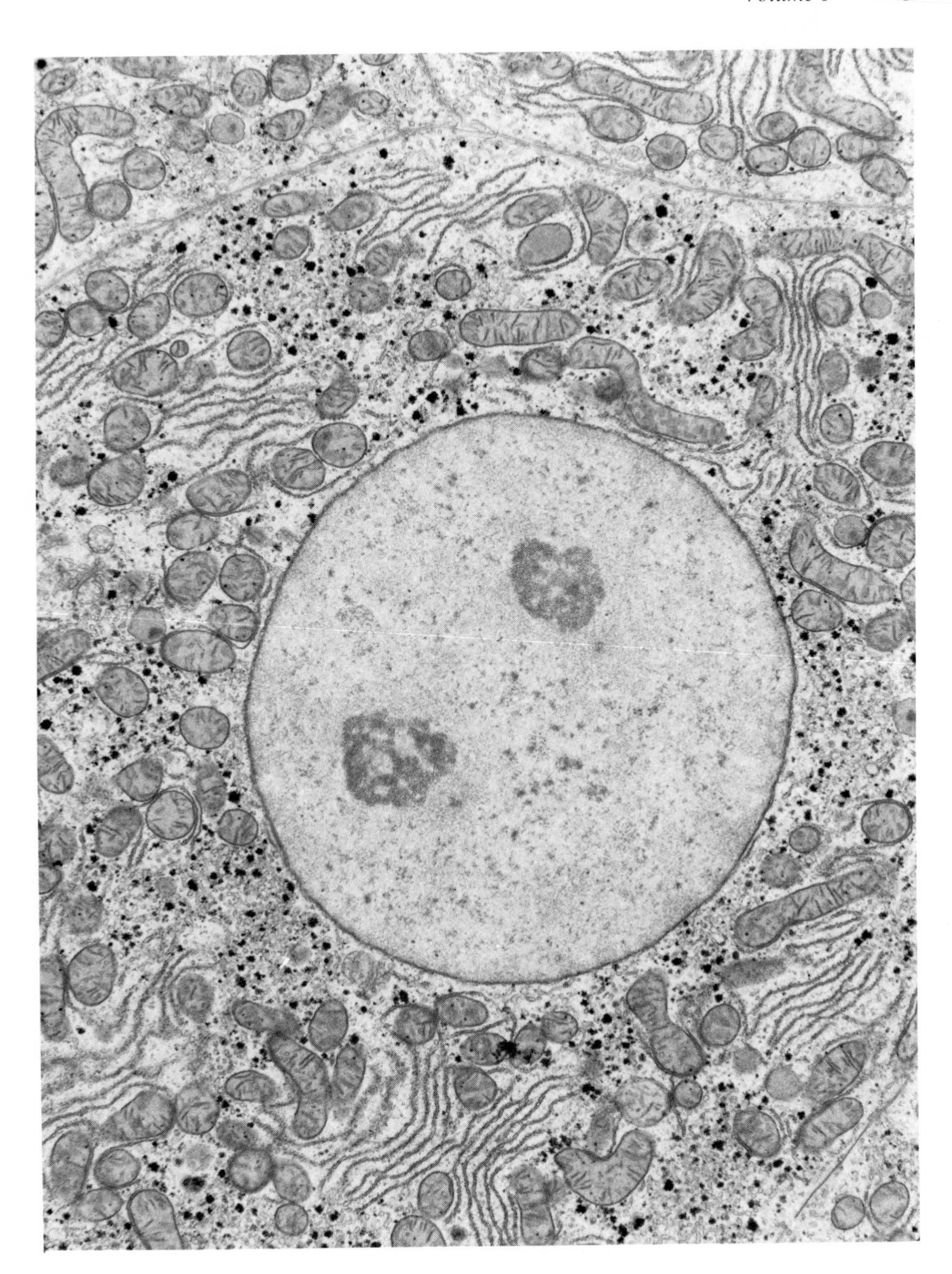

PLATE 2. EM of the nucleus of the cell. This is the largest and densest structure in the cell. The nuclear matrix is surrounded by the nuclear membrane, which regulates the exchange of substances between the nucleus and the cytoplasm. Within the nucleus the dark areas are nucleoli formed by a nucleolonema. The nucleus contains chromatin threads which carry genes, the units of hereditary control. Characteristic compounds of nuclei are deoxyribonucleic acid (DNA), where genetic information is stored. These structures are required for the maintenance of the whole cell. Different types of histones are contained in the nucleoplasm. Nuclei of all kinds of specialized cells have similar structure, with granules and clumps of chromatin and one or more nucleoli.

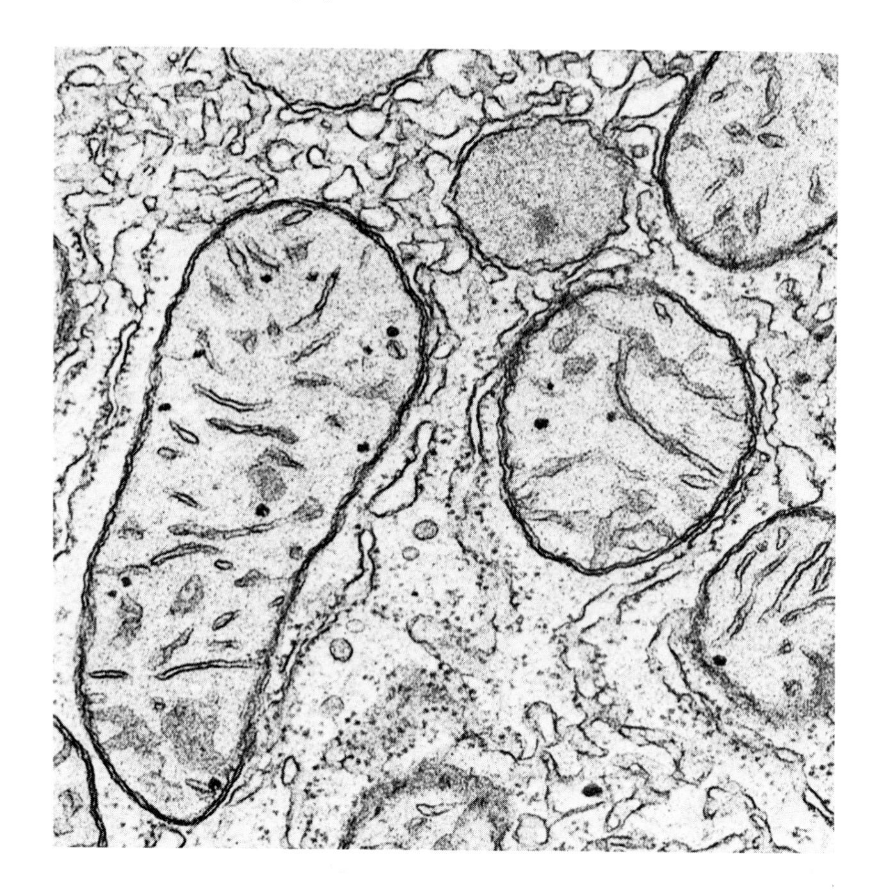

PLATE 3. Appearance of mitochondria under the electron microscope. These organelles possess cristae traversing the clear matrix. Each mitochondrion is bounded by a double membrane and divided into interconnecting compartments by partitions projecting inwards from the membrane. Respiratory enzymes are attached to the inner mitochondrial membrane, while the outer membrane is rich in monoaminooxidase. In mitochondria, molecular oxygen is used for the oxidation of fatty acids, and the released metabolic energy is used to synthesize adenosine triphosphate (ATP), which can supply energy to other regions of the cell. The mitochondria are the sites of oxidative phosphorylation, an important process in cellular metabolic activities. All enzymes of the mitochondria are organized in functional sequence.

important role in hormone action and in the cellular antigen-antibody reaction. Membrane-bound enzymes catalyze biochemical reactions, being involved in carrying sodium out and potassium into the cell against concentration gradients, such as Na^+- and K^+-activated adenosine triphosphatase. Many disorders arise from defects of substrate transport across cell membranes or from abnormalities of hormone transport or immunological response.

The cell membrane is involved in the elimination of unwanted products through some structural changes.[67,68,70] Interactions between the cell and the environment lead to transient alterations in the surface of the membrane known as phagocytosis and pinocytosis. In phagocytosis, particulate materials are engulfed by structures called pseudopods projecting from the cell membrane and the material is then incorporated within the cytoplasmic mass. In pinocytosis, microscopic fluid droplets are incorporated into invaginations of the cell membrane. In pathological processes both phagocytosis and pinocytosis play important roles. Disturbances of the structure or function of cell membranes bring about a deficient phagocytotic or pinocytotic process with retention of waste products, resulting in cellular impairment and deterioration.

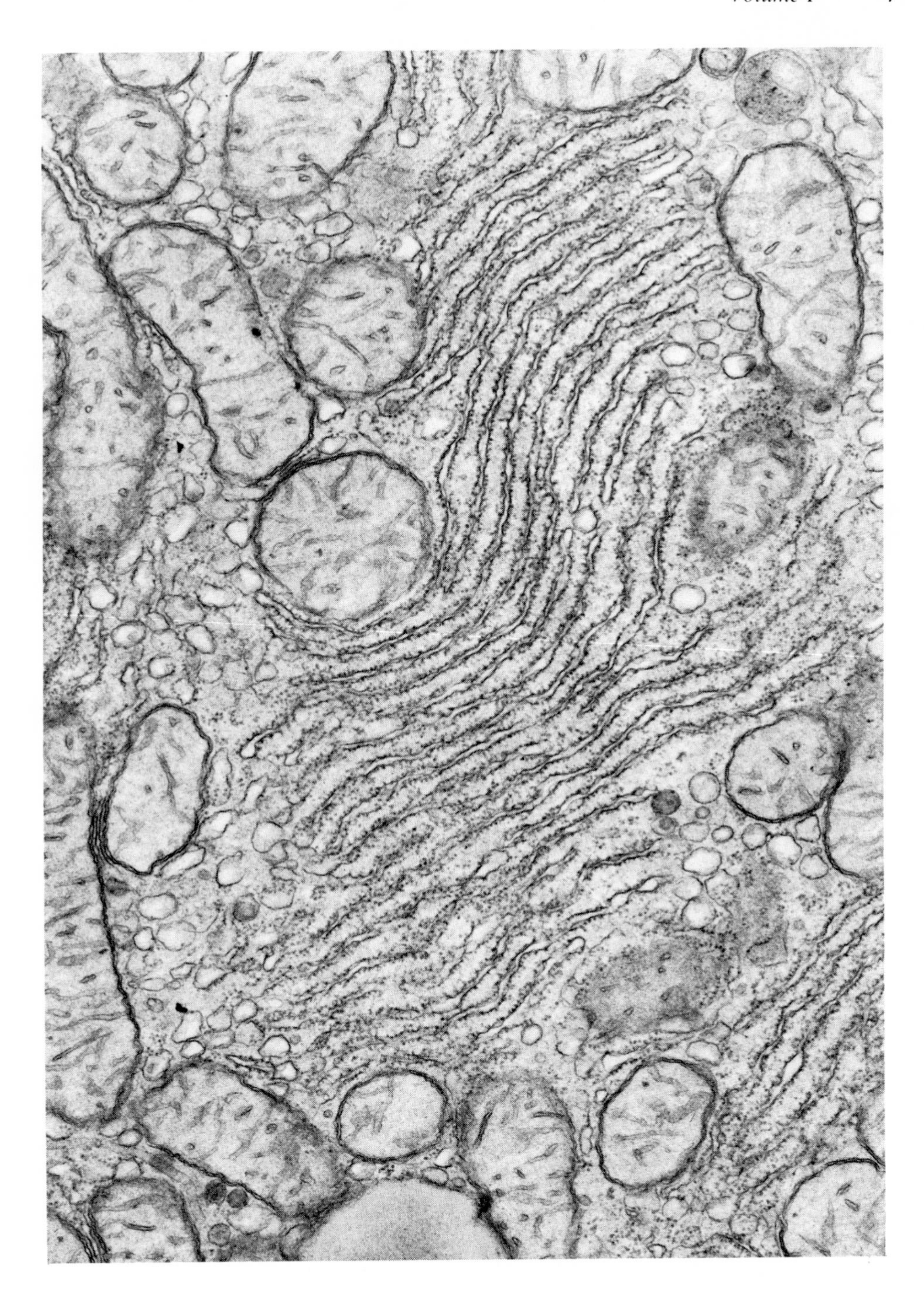

PLATE 4. Electron microscopic aspect of the rough endoplasmic reticulum. Parallel stacks of membranes are studded with ribosomes, thus representing the rough portion of the membranous apparatus. The large proportion of the membranes attached are involved in protein synthesis. Isolation of disrupted endoplasmic reticulum membranes results in microsomes. Microsomes are much smaller than mitochondria and can be sedimented only by high-speed centrifugation. The rough microsomes are rich in lipid and protein and contain more than half of total ribonucleic acid (RNA), concentrated in ribosomes. The cellular organelles regulate various processes of protein and phospholipid synthesis.

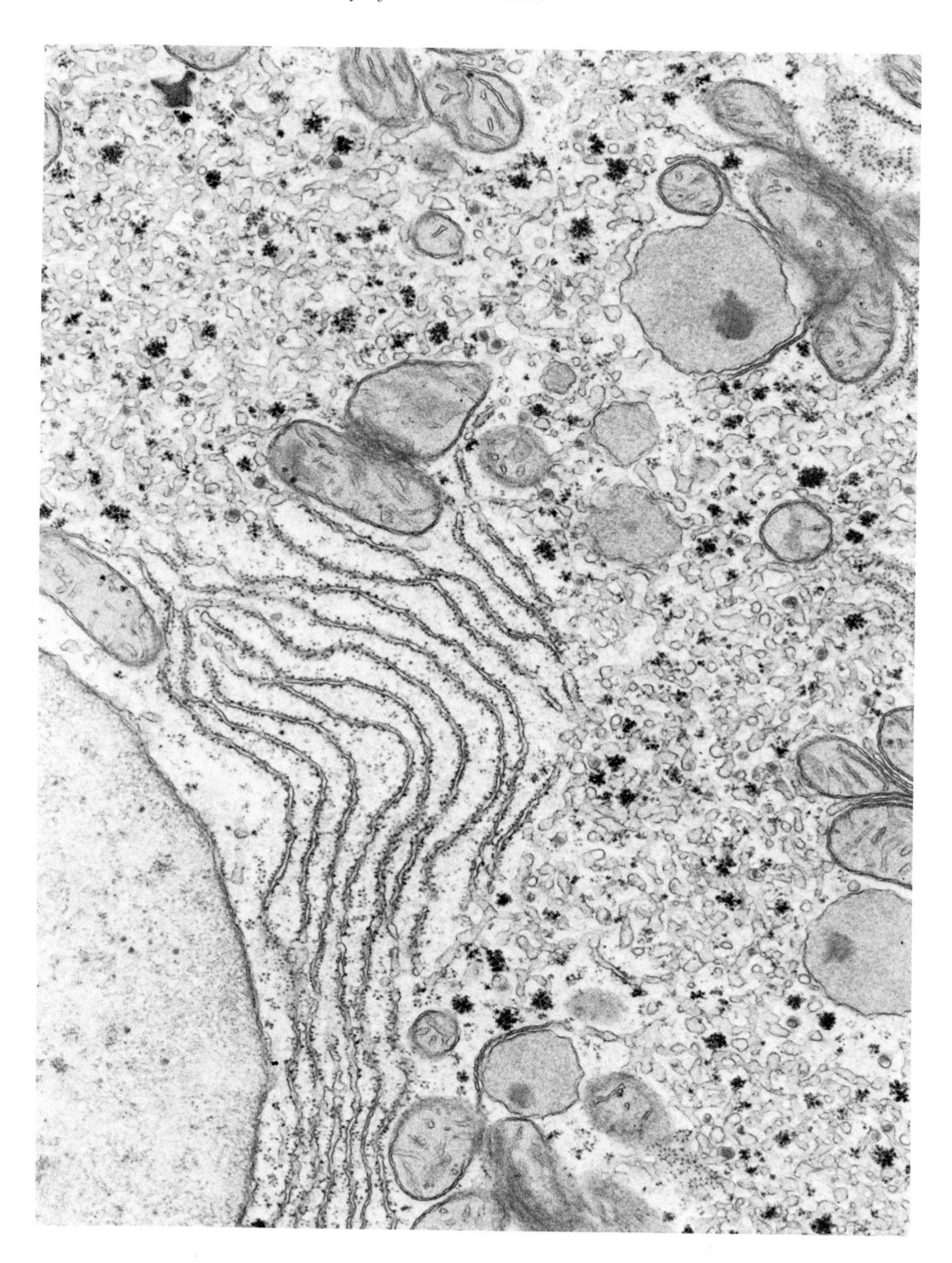

PLATE 5. Area of smooth endoplasmic reticulum membranes in the cytoplasm of a liver cell under the electron microscope. The membranes are denuded; therefore their "smooth" connotation. These membranes constitute the supporting framework for the enzymes involved in detoxification mechanisms primarily responsible for the metabolism of steroids and various xenobiotics. Connections with rough endoplasmic reticulum cisternae can be seen in the center of the illustration.

B. Nucleus, Nucleolus, and Chromosomes

The nucleus forms the center of the cell.[52,96] Usually, there is a single nucleus in each cell; there are, however, instances where the nucleus is lacking, as in the mature erythrocytes. In contrast, in some cells more than one nucleus may be present. The nucleus is separated

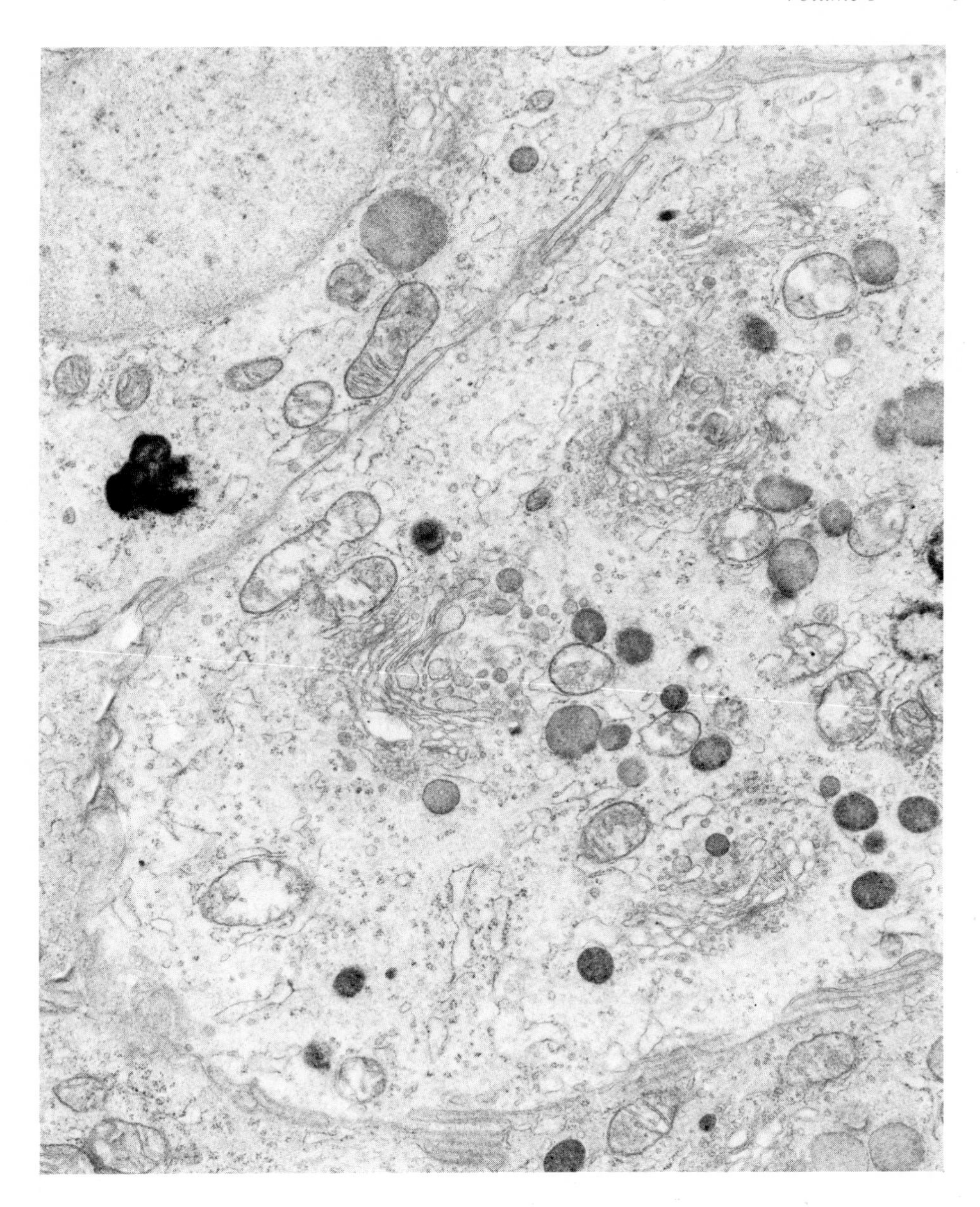

PLATE 6. The Golgi apparatus appears in fully differentiated functional cells. It represents an irregular network consisting of a group of large vacuoles, a series of flattened sacs, and clusters of small vesicles. This complex organelle is associated with lipoprotein synthesis and metabolism and elaboration of glycoprotein differentiation and secretion. It displays an array of concentric lamellae and saccules, and sometimes lipoproteins can be visualized in the lumen. The cell illustrated displays the multiple appearance of this organelle in biliary epithelial cells.

from the cytoplasm by the nuclear membrane, which resembles the cell membrane but has a unit structure and is perforated by pores. Thus, the nuclear sap and cytoplasm are in contact through these pores. The nucleoplasm represents a mixture of granules, filaments, and clumps and contains dispersed chromatin (deoxyribonucleic acid). The entire effect of the nucleus in cellular metabolism is not known, but its essential role is that it stores the chromosomes. In the nucleus of the human cell there are 46 chromosomes. These contain protein and enzyme units in an aggregated form called genes. Each pair of chromosomes is different, hence, they can be recognized by their shape.

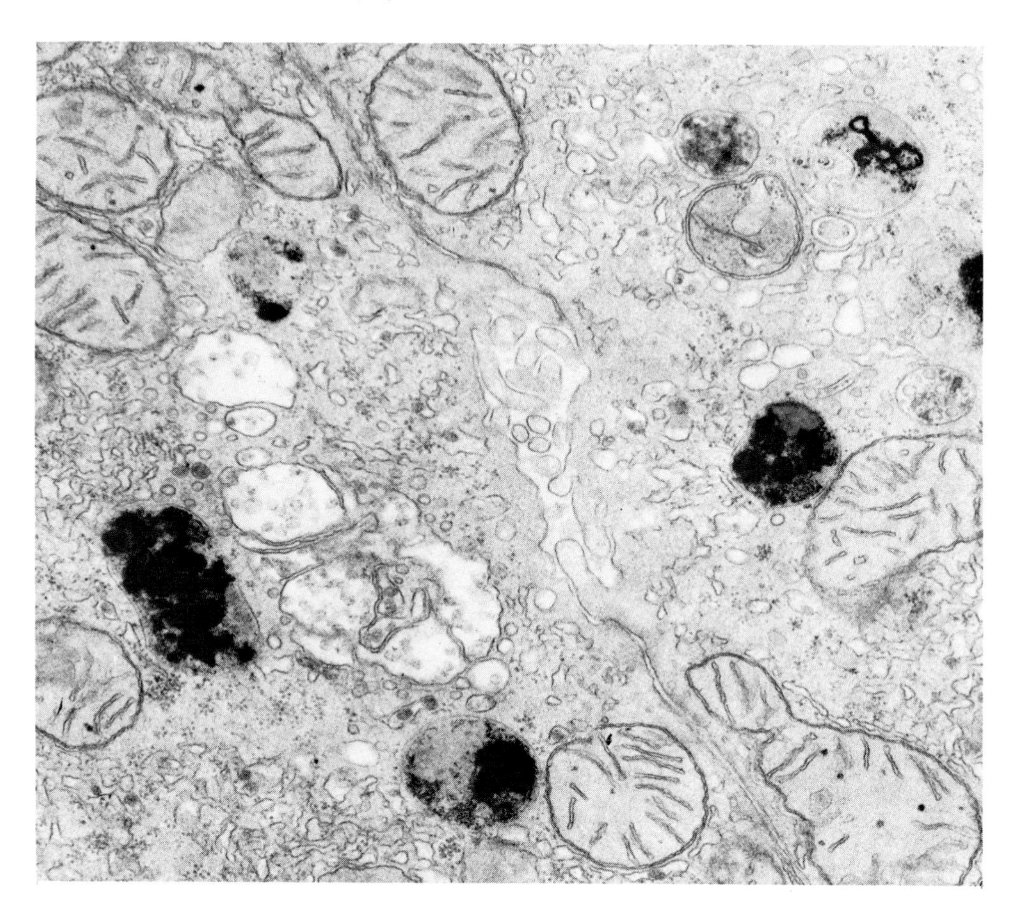

PLATE 7. Lysosomes in the vicinity of the bile canaliculus and Golgi apparatus, are shown in this figure. These highly heterogeneous particles contain a variety of hydrolases and other enzymes which break down large molecules into smaller components. Lysosomes are involved in the degradation of cellular byproducts which originate either from inside or outside the cell. In the dying cell, lysosomes quickly release their enzymes into the cytoplasm and bring about digestion of the cell. Thus, lysosomes are largely responsible for the autolysis and postmortem degeneration of tissues.

In the nucleus of every cell genes are responsible for the inheritance of traits that ultimately determine the body response in a health or diseased state. The gene contains genetic information that controls biochemical reactions. It consists of RNA and DNA aggregates. Most important is the presence of DNA, which is the true genetic material capable of genetic transmission. Whereas the DNA content of the cell is fairly constant, RNA acts as a transfer agent mainly located in the cytoplasm. Its amount varies with age and sex. DNA and RNA function together and complement each other to carry out all our biological functions.

The nucleolus is a distinct round structure in the nucleus, but is not really bordered by membrane. The nucleolus participates in the production of ribosomal RNA and is therefore important in protein synthesis. Thereby, this cellular organelle, together with the entire nucleus, is also an active site of genetic information. Any minor change is associated with pathological effects.

C. Cytoplasmic Organelles

Several smaller structural units exist in the cytoplasm in most cells. These include mitochondria, lysosomes, endoplasmic reticulum, Golgi complexes, microtubules, filaments, and centrioles. The roles of most organelles in cellular metabolism have been established.

There are specialized organelles, such as the zymogen granules of the pancreas and thyroglobulin-containing particles in the thyroid, with known functions. However, the roles of several organelles in special circumstances, such as the peroxisomes and lysosomes, have not been always classified. These structural entities show alterations in pathological conditions, although when a disease manifests at molecular level, structural alterations may not be apparent. Pathological manifestations and the participation of these intracellular organelles in disease processes will be described in more details in subsequent chapters.

1. Mitochondria

The mitochondria exist as a double membrane containing intracellular organelles.[8,40] Both the outer and inner membrane form in-folds inside the particles termed cristae.[43,66] The mitochondria are the sites of the electron transport chain containing many enzymes of the respiration and energy transfer. These enzymes are probably highly bound to the structure; the cytochrome oxidase system is attached to the inner membrane. Mitochondria are also responsible for fatty acid synthesis and oxidation, although the exact localization of these functions inside the mitochondrion is not well established. The exact pattern of the integration of enzymes involved in fatty acid oxidation within the mitochondrial membrane has not been established. These enzymes are probably not strongly bound to the structure.

Mitochondria are self-replicating organelles. They contain special enzymes, RNA, DNA and ribosomes, associated with the production of some proteins of the mitochondria. Several mitochondrial proteins are synthetized in the endoplasmic reticulum, such as cytochrome *c*. Defects developing in mitochondrial enzyme systems and energy transfer, mainly due to toxic actions of foreign compounds, are related to pathological effects.

2. Endoplasmic Reticulum

The endoplasmic reticulum is made of a more or less continuous system of vesicles containing tubules and cisterns, forming a network which may be continuous with both the cell and nuclear membrane. This cellular organelle is also related to other structures, altogether comprising the cytocavitary network.[56,57] In the three-dimensional endoplasmic reticulum structure, two types can be distinguished. Some membranes are studded with granules or ribonucleoprotein particles, ribosomes.[5] These form the rough endoplasmic reticulum; the smooth membrane is devoid of ribosomes. The endoplasmic reticulum is involved in numerous metabolic functions, the rough and smooth types having different roles.[20,21,45]

The major task of the endoplasmic reticulum is the synthesis of proteins and detoxication of foreign compounds.[2,3,5,69] Protein is synthetized by the rough endoplasmic reticulum, ribosomes playing a key role in this process.[58] On the other hand, smooth endoplasmic reticulum is the major site of detoxification processes. These membranes contain enzymes that detoxify endogenous and exogenous compounds which could exert a poisonous action on the cell. The enzymes involved in these processes are called mixed-function oxidases or drug-metabolizing enzymes. The action of this enzyme system is linked with a special electron transport chain which includes cytochrome P-450 and b_5.[28,69]

In view of the wide variety of functions of the endoplasmic reticulum, disturbances of cytoplasmic systems pertinent to any activity of these organelles may lead to disease.[27,35] The major causes are either inhibition of protein synthesis through blocking the production of ribosomal protein or toxic action.

3. Golgi Complex

The Golgi vesicles represent specialized smooth endoplasmic reticulum membranes usually located close to the nucleus.[41] This organelle plays an important role in the intracellular movement of metabolites and the modification of many substrates. Carbohydrates are synthesized in the Golgi; it is also involved in sulfation of polysaccharides and addition of sugar or lipid molecules to protein synthetized by the endoplasmic reticulum to form glyco- or lipoproteins. Toxic compounds and abnormalities affecting cytoplasmic enzymes may cause impairment in the function of the Golgi complex.

4. Lysosomes

Lysosomes are polymorphic cellular organelles; several morphologically different structures such as storage granules or primary lysosomes and autophagic and digestive vacuoles belong to these morphological units.[14,51,73] Lipofuscin pigment bodies are related or may be structurally identical to lysosomes. All these organelles are associated with (1) intracellular digestion through the processes of pinocytosis and phagocytosis, (2) the degradation of cytoplasmic structures in normal cells and tissues undergoing autolysis, (3) the disposition of secretory substances from glandular cells, and (4) the storage of insoluble lipid and pigment residues.[14] Many specialized events are also connected with lysosomal functions such as processes occurring in neurons, cellular interactions following differentiation, and shock or immunity.

The various lysosomes contain hydrolytic enzymes responsible for many intracytoplasmic digestion processes and for the autolytic destruction and elimination of dead cells.[46] These enzymes function at an acid pH optimum, for example, acid phosphatase. Normally, the lysosomal membrane provides a barrier which restricts the enzymes from leaking out from the organelles. Enzymatic degradation of many substances processed by lysosomes produce small molecules. These metabolites can be utilized within the lysosomes, retained in residual bodies or lipofuscin, or excreted into the cytoplasm.

In spite of the great variety of lysosomal structures, the sequence of morphological changes they undergo in each cell is relatively consistent. The structure of these membranes may be different relative to the substance processed and derivatives produced. Primary lysosomes or storage granules fuse directly with endogenous or exogenous particles, form a digestive vacuole, and release their content into phagocytic vacuole. Many tissues contain secondary lysosomes where the presence of primary lysosomes is not apparent. In this case, the engulfed material is accumulated within the vacuole which is usually located in the proximity of the Golgi complex. The fusion between secondary lysosomes and Golgi complex produces digestive vacuoles, known as protolysosomes. Another type of secondary lysosome is formed from the fragments of injured or degenerated intracellular particles such as mitochondria or endoplasmic reticulum. These are the autophagic vacuoles where enzymatic digestion is continuous, but concomitant swelling may also occur. Further changes produce dense bodies. When the lysosome contains nondigested debris, it is called residual body or lipofuscin granule.

In many inborn errors, genetic abnormalities are manifest in lysosomes. Due to the altered lysosomal enzyme activity, a metabolic block develops and various substances are deposited. The formation of secondary lysosomes is associated with several storage diseases such as lipidoses and polysaccharidoses. The retention of abnormal amounts of lipids or phospholipids may represent the signs of toxic action or lysosomal defects associated with aging.

IV. HOMEOSTASIS

In our external environment many factors show changes all the time; we can only survive in this changing environment if conditions within the body remain relatively unchanged. The parts of any organism function efficiently when supplies of oxygen, water, food, and subsequent energy and heat are maintained within certain limits. The maintenance of such a stable environment represents homeostasis; it has a primary importance to survival.

All living organisms have the ability to carry out many interrelated and diverse biochemical processes and also have the capacity to control these actions. The control mechanisms are built into the function of many specialized cells and these mechanisms serve to maintain the homeostasis within one organ and within the whole body. The regulation is associated with positive and negative controls. Every biological system exerts a positive and a negative control to turn on these processes under some conditions and to terminate them under other conditions. Briefly, in the organism there exists a highly coordinated and balanced mechanism. The function and operation of this mechanism ensures that both the composition of the *milieu interieur* and the operational processes are kept within well-defined narrow limits. The overall result of this mechanism is a finely tuned stable system representing homeostasis.[37,63]

Various organs are involved in the maintenance of homeostasis alone and in an integrated manner. From a functional point of view, the body may be regarded as being made up of several compartments. These are as follows: (1) the body fat, a storage depot in the general regulation of food intake and body weight; (2) body mass, dynamic functional soft tissues; (3) body fluids; and (4) metabolically relatively inert mineral residue, the skeleton. Homeostatic processes provide the regulatory mechanisms for these compartments.

Several major processes are essential for homeostatic control, namely, the maintenance of the intra- and extracellular environment and the regulation of metabolic reactions. There is an intrinsic relationship between these activities although, by and large, different organs are responsible for the maintenance of the intra- and extracellular environment, such as the kidney and lung, and for the regulation of body metabolic balance, such as the liver. The intracellular fluid represents the milieu where the metabolic processes are taking place and the extracellular fluid constitutes the environment of the cells within the body while the cells carry out all vital processes in this medium. Changes in the extracellular fluid reflect the effect of our external environment as well as the changes in the fluid existing within the cells, indicating the overall balance of cellular functions. Since the maintenance of this equilibrium, i.e., the maintenance of homeostasis, is an essential criterion of life, these functions require close control. The liver and the reticuloendothelial system are mainly responsible for the regulation of cellular activities which provide essential molecules for the hepatic, bone marrow, and spleen cells. Through the circulation various constituents reach all the organs, and their actions extend beyond the role of homeostatic regulators. It is also imperative to keep a relatively constant composition on tissue fluids, as is done by the kidneys which maintain the optimal chemical composition of the body fluids and the lungs which regulate tissue oxygen and carbon dioxide concentrations. These organs, therefore, do not merely function to remove metabolic wastes but actually perform considerable metabolic processes which represent part of the essential contributing factors in the maintenance of homeostatic functions.

The regulatory processes underlying homeostasis are numerous. Major processes include the regulation of body size and weight, fluid balance, metabolic pathways and interrelations, and normal ventilatory function. Disturbances of these internal organizations lead to impairments of homeostatic processes, shifting them beyond physiological limits to excessive pathological deviations, thereby leading to disease. Many diseases are thus associated with a disturbed homeostatis.

A. Regulation of Metabolism

This is related to the regulation of body weight and size. The control of overall body metabolism is primarily associated with food intake and ingestion. The volume of food intake is controlled by three mechanisms: the biometric, short-term, and long-term regulations.

The biometric regulation is varying and characteristic to each individual. The upper and lower limits of food intake is proportional to the energy requirement. The lower limit represents the energy need for basal metabolism; the upper limit is variable, being generally determined by stress situations such as demands required by exposure to low temperature, or any extreme physical exertion. The daily regulation of food intake is adjusted to energy losses. This depends on the turnover of carbohydrates, fats, and proteins. The liver is the major site of the closely integrated metabolic systems processing these body constituents. The homeostatic function of the liver is mainly responsible for the regulation of metabolism in other organs. For instance, it is associated with a control mechanism manifest in certain structures of the central nervous system. Ultimately, the liver is also responsible for the hormonal influence of the overall metabolism, hunger, satiation and appetite associated with insulin, glucogen, thyroid, and growth hormones. Variations of glucose levels in the extracellular fluid probably represent the basic stimulus turning this mechanism on and off.

The ventromedial area of the hypothalamus is the center of the regulation of the appetite and body weight. Lesions of this region, cerebral injuries, or tumors cause elimination of satiety, lead to overeating, and excess body weight. In contrast, damage to the lateral area of the hypothalamus brings about weight loss and eventual death from inanition.

The short-term or glucostatic mechanism represents an interrelationship between glucose levels and appetite. This is connected with the action of various hormones such as glucagon, insulin, growth hormone, and hypothalamic factors. The lateral hypothalamic feeding center is sensitive to changes in blood glucose level. When the glucose level falls, the activity of this center rises. In contrast, when the blood glucose rises, the hypothalamic activity is turned off, but the ventromedial hypothalamic satiety center is activated. The reciprocal activity of these centers is influenced by appetite, food intake, and associated blood glucose level. The alternate dominance of these centers mediates the activity of various hormones. Glucogen raises blood glucose, abolishes hunger contractions of the stomach, and achieves a transient sensation of satiety.

During fasting, glucose levels are maintained by the secretion of growth hormone from the anterior pituitary, which in turn is regulated by hypothalamic actions. Feeding diminishes this secretory mechanism and the rise of glucose in the blood triggers insulin secretion, increasing carbohydrate utilization and depot fat formation. Abnormalities of the short term glucostatic mechanism are seen in the hyperglycemia-obesity syndrome, in thyrotoxicosis, and in diabetes mellitus.

The long-term regulatory mechanism depends on lipostatic processes. This mechanism is represented by the daily metabolism of a constant portion of the body fat. This kind of homeostatic balance accounts for the relative stability of body weight under normal conditions: the increased fat content is compensated by enhanced fat mobilization. When this mechanism is impaired, the result is obesity. Hypothalamic lesions can cause obesity by abolishing satiety. Various psychological disturbances may lead to obesity as the result of overeating. Excessive food load impairs the efficacy of the regulatory mechanism. Other causes of obesity include pancreatic dysfunction with increased output of insulin, hyperglycemia, hyperlipemia, and an abnormally high rate of fat depot synthesis. Ovarian dysfunction in Stein-Leventhal syndrome also upsets this regulatory mechanism and causes a generalized somatic fat deposition.

In contrast, weight loss can be caused by lack of nourishment associated with starvation or inadequate food intake, thyrotoxicosis, malabsorption syndrome (gastric carcinoma, biliary obstruction, ulcerative colitis, Crohn's disease, idiopathic steatorrhea), obstructive disease of the esophagus, toxic states (tuberculosis, illness with high temperature), and cachexia in carcinomatosis.

B. Regulation of Fluid Balance

The total body water comprises 60 to 70% of body weight grossly divided into two compartments: (1) intracellular, about 50%, and (2) extracellular (vascular 5%, interstitial 15%). The overall water balance is connected with losses through the kidney as urine, by perspiration through the skin, as water vapor in the exhaled air from the lungs, and in the feces. Water is replenished by drink, water contained in food, and water produced by metabolic processes of the body. In the homeostatic regulation of the fluid balance, the primary role is played by the kidney.

Vascular and interstitial fluid compartments are separated by capillary membranes which are freely permeable to water and electrolytes. The volume of the vascular fluid space is constant, due mainly to the osmotic effect of plasma proteins. The hydrostatic pressure within the capillary represents the counter-force against the osmotic pressure of proteins, both being in a state of dynamic equilibrium. Disturbances of this equilibrium by accumulation of fluid in the interstitial space lead to edema. Edema occurs as the result of several impairments such as (1) decreased osmotic pressure by plasma proteins due to decreased production in hypoproteinemia or, especially, in hypoalbuminemia; (2) passive increase of the hydrostatic pressure in the capillaries as in congestive heart failure, or in circulatory insufficiency causing hypovolemia; and (3) enhanced permeability of the capillary membranes followed by exudation as in acute inflammation and certain allergic diseases leading to loss of protein, water, and electrolytes. Further details of edema formation will be discussed later.

Increased sodium intake into the extracellular fluid draws liquid from the intracellular compartment, leading to cellular dehydration. Depletion of the interstitial space is connected with the sensation of thirst, which in turn stimulates the drinking of water. Water is retained in the body through the action of an antidiuretic hormone; renal secretion is decreased, resulting in water conservation in the cell as a response to the initial osmolality differences. Any condition raising the osmolality of the extracellular fluid promotes a compensating mechanism — the secretion of antidiuretic hormone and retention of water. In case of excessive water intake, diuresis compensates for the increased fluid, which is a regulatory response of the normal kidney. If this is impaired, serious overhydration occurs with signs of water intoxication (vomiting, headache, muscular weakness, mental disorientation).

Excessive loss of salt, usually combined with water loss, leads to hypoosmolality of the extracellular space in relation to the intracellular compartment causing cellular hydration manifested by a shock-like state with falling blood pressure. Water deprivation occurs when normal water losses are not adequately replaced. The primary effect of this condition is a diminution of the extracellular fluid without a corresponding loss of electrolytes. Consequently, water comes out from the intracellular compartments with concomitant loss of potassium, which aggravates the situation by a further fall of intracellular osmolality. Antidiuretic hormone mediates the response of the kidney to conserve water. In salt conservation aldosterone secretion also plays part. More details of these pathological responses will be given later.

C. Respiratory Regulation

The respiratory system is associated with homeostatic regulation in two ways: (1) it carries oxygen gas into the body, and through hemoglobin binding it is distributed throughout the body, and (2) it eliminates volatile waste products, mainly carbon dioxide, and partly some other volatile materials. This mechanism is involved in the regulation of the acid-base balance. An increase in hydrogen ion concentration is usually neutralized by various buffer systems such as plasma proteins, hemoglobin, carbonate/bicarbonate and phosphate buffers. Uncompensated acidosis or alkalosis occurs if the capacity of these systems is exhausted. However, changes in hydrogen ions stimulate the medullary respiratory center and the increased hydrogen ions are rapidly compensated for by the elimination of carbonic acid in the form of carbon dioxide by the lungs. In alkalosis, respiration is retarded and carbon dioxide conserves hydrogen ions in the form of carbonic acid. Disturbances of the normal ventilatory action of the respiratory system lead to inadequate oxygenation of hemoglobin and cyanosis. Extensive pulmonary disease results in difficult breathing. Dyspnea manifests in acute pleurisy, bronchitis associated with influenza, or maybe in chest wall injury.

D. Cardiovascular Regulation

The homeostasis of the cardiovascular system is associated with (1) the transport of adequate amounts of oxygen and nutrients from the blood to the tissue and the removal of waste products from the tissues; (2) the regulation of cardiac output representing the distribution of larger or smaller volume of blood to different regions of the body in response to various physiological need; and (3) the maintenance of the blood pressure within physiological limits responsible for the blood flow and the efficiency of the transport into and out of tissues. Impairment of the homeostasis in the cardiovascular system is associated with cardiac failure. Abnormalities of blood pressure shed light on the type of the underlying disorders. The systolic pressure reflects the cardiac output, the diastolic pressure shows the peripheral arteriolar resistance. Elevations in systolic pressure alone may indicate minor stress conditions or represent symptoms of the onset of persistent hypertension. Elevations in the diastolic pressure are more indicative of a cardiovascular disease.

V. BIOCHEMICAL LESION

The concept of biochemical lesion presents the view that clinical symptoms or pathological conditions underlying disease derive from an impairment of the biochemical mechanism of the organism. Biochemical lesion is usually a reversible process; it initiates the development of disorder and is responsible for the primary anomaly. The biochemical lesion triggers the chain of events and the clinical symptoms are the direct expression of the altered biochemical processes; pathological cell damage may be seen or even remain undetected. Specific anatomical changes do not accompany acute or chronic poisoning, nutritional deficiencies, or the incorporation of abnormal substances into the metabolic chain. If the anomaly persists longer, however, it becomes irreversible and manifests itself in morphological changes. In many cases, the biochemical lesion or injury is genetically determined and associated with alterations of the germ cell. Such conditions are the inborn errors of metabolism, mutations, and perhaps various neoplastic diseases. The biochemical lesion leads definitely to pathological conditions in all these cases. Generally, the nature of biochemical lesion and the consequent disease condition are dependent on the cellular site where the impact occurs (Figure 1).

Biochemical lesions can be classified into two categories: (1) changes occurring directly in one or more cellular biochemical processes, or (2) modifications associated with genetically determined direct alteration of the germ plasma (Table 1). Several conditions belong to the first category, such as avitaminosis and other nutritional deficiencies, intoxication by various toxic compounds, metabolic antagonists, and lethal synthesis. The latter is associated with the often irreversible replacement of a normal constituent by a similar substance, leading to cessation of essential processes.

Ascorbic acid, thiamine, and nicotinic acid deficiencies are well-known examples of this type of biochemical lesion causing scurvy, neurological symptoms, and dermatological disorders due to the absence of protective mechanism or synthesis of coenzymes. The effects of toxic agents have been shown to act on proteins and modify enzyme activities. Trace heavy metals have a great affinity for sulfhydryl groups in proteins, enzymes, and coenzymes and inhibit their action by irreversible binding. Several organic compounds directly interfere with enzyme function by blocking essential groups or producing an enzyme-inhibitor complex. Among these are cholinesterase inhibitors and monoamine oxidase inhibitors. Some inhibitors compete with the substrate — metabolic antagonists are representative of these substances. Others are incorporated into the product of a biochemical system, replacing the naturally occurring constituents. Selenium containing amino acids can exchange sulfur amino acids and become part of proteins. If this protein is an important structural component, structural abnormalities may develop. Fluoroacetate is another important example, being

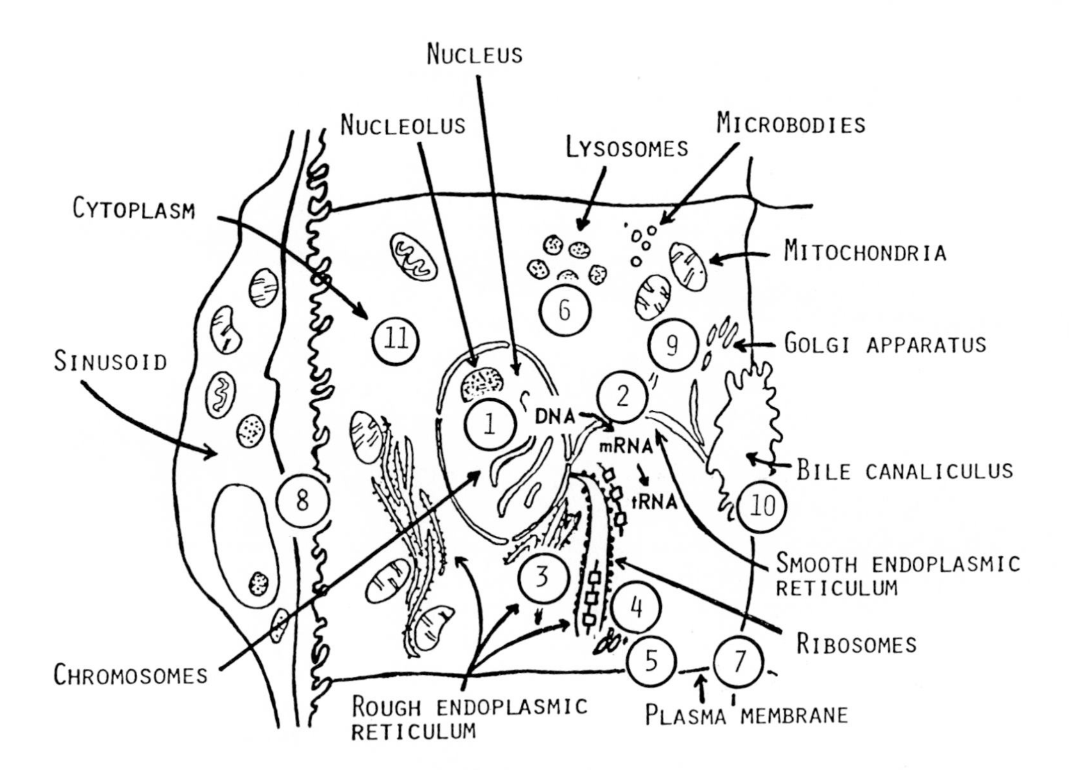

FIGURE 1. Sites of biochemical lesions in the cell. This is a representative liver cell but, generally, cells and cellular organelles of various organs reflect similar relationships between the site of lesion and disease processes. Defects in the double-strand structure of nuclear DNA related to gene function are associated with mutation and neoplasm (1). Inadequate DNA synthesis and subsequently faulty formation of complementary strand messenger RNA cause genetic inborn error. (2). Degranulation of ribosomes from the endoplasmic reticulum is connected with the action of toxic compounds and carcinogens (3). Toxic effect also causes dilatation of the membranes (4). Defects of the Golgi complex represent anomalies in the selection process, mainly abnormal complex protein synthesis (5). Lysosomal defects impair storage (6). Damage of the plasma membrane is accompanied by transport defects (7). Impairment of the cell membrane at the sinusoids is associated with abnormalities of cellular intake (8). Lesion of mitochondria is followed by swelling and impairment of cellular oxidation processes (9). Defect in the function of bile canaliculus impairs the elimination processes (10). Anomalies of cytoplasmic enzymes are connected with errors of metabolic and transport processes (11).

metabolized to fluorocitrate which blocks the Krebs cycle at the level of aconitase, and thus citrate accumulates in various tissues. Citrate cannot be excreted by the kidney but, in excess, binds ionized calcium in body fluids and cells, causing ossification abnormalities. If the failure of the tricarboxylic acid cycle is prolonged, this leads to inadequate pyruvate oxidation and symptoms of general intoxication such as convulsions, heart failure, and eventually to death. Many organofluorine compounds are toxic because they are activated by the organism and metabolites are incorporated into normal processes through lethal syntheses.

Many analogues of important body constituents cause biochemical lesions. Fluorocitrate is an analogue of citrate; selenocysteine is an analogue of cysteine; ethionine is an analogue of methionine. Up to a certain point these analogues, abnormal amino acids, and other constituents are handled in the same way as the normal components. Ethionine is activated and combines with adenosine triphosphate, as does methionine. However, activated ethionine blocks further methionine utilization. Sulfonamide is an analogue of *p*-aminobenzoic acid which is an important part of the growth factor pteroylglutamic acid. In the presence of adequate sulfonamide concentration bacterial growth can be blocked. However, this drug, in excess, may influence the normal process of animal cells. Some amines may represent

Table 1
BIOCHEMICAL LESIONS

Cellular changes	Mitotic changes
Nutritional deficiency	Inborn errors
Lack of essential amino acids	Amino acid metabolism
Lack of essential fatty acids	Urea cycle
Lack of essential metals	Plasma protein synthesis
Lack of vitamins	Protein assembly
Choline deficiency	Carbohydrate metabolism
	Lipid metabolism
	Purine and pyrimidine metabolism
Release of feedback control	Molecular transport
Drug interaction	Mutation
Increased or decreased metabolism	Structural changes
Receptor site or structural binding	Functional changes
Competition with endogenous substrate	
	Teratogenic effects
Toxication	
Mycotoxins	
Bacterial toxins	Tumorigenic effects
Drugs and other chemicals	
Lethal synthesis	
Incorporation of Se-cystein	
Incorporation of F-citrate	

analogues of catecholamines which can bind firmly to specific receptors in the nervous system, thus bringing about disorders of neural function associated with these transmitter molecules.

Modification of the germ cells also cause biochemical lesions. These include the incorporation of purine or pyrimidine bases into nucleic acid synthesis which not only alter the physicochemical and biological properties of these molecules, but leads to mutation and disease. Mutations affect the genetic apparatus of the cell and its regulation, thereby allowing an interaction with any changes occurring in the environment. The response at molecular level is in association with enzyme-proteins actually produced by the genetic apparatus. In response to the environmental changes the induction of an enzyme is modified. The adaptation does not regularly represent basic change since the induction of a new enzyme does not alter the organism basically. However, if these changes fundamentally modify the composition of chromosomal DNA, mutations are produced. Mutations may exert no essential influence on the life of the cell and may be beneficial by bringing about evolutionary changes in the organism. On the other hand, mutations can interfere with the synthesis of essential proteins that may elicit profound effects on the organism.

Mutations at the DNA level include (1) loss of a base-pair in the sequence of the DNA strands, (2) replacement of one base-pair of DNA; this can be a substitution of pyrimidine base with another pyrimidine base or a purine for a purine: transition, or a purine replace a pyrimidine or vice versa: transversion; (3) reversal of the normal order of the base-pair sequence along the strand, and (4) insertion of a base-pair normally not present. Many derivatives of these bases will induce mutations, such as 5-bromo-uracil. Alkylating agents such as nitrogen mustard produce mutations by alkylation of nitrogen of the guanidine ring at the 7-position, resulting in a detachment of the base from the strand, and this space is subsequently replaced by another base. Mutations are connected with the action of many chemicals: mutagens or chemical carcinogens. Some heterocyclic compounds such as phenazines or acridines may disturb the regular architecture of the chain and may be bound between bases, leading to errors in replication. Radiations, such as ultraviolet light, X-rays, or γ-rays, may elicit potent mutagenic action. Mutagenesis often leads to abnormal development of enzyme systems and to marked reduction or lack of enzyme activity, resulting in conditions known as inborn errors of metabolism.

VI. MOLECULAR BASIS OF INHERITANCE

The human body is made up of various cells. In the development of these cells the interaction between genetic inheritance and environmental influences plays an important role. The genetic make-up shows great diversity in all human beings and remains essentially unaltered throughout life, whereas environmental effects are constantly changing. There are some genetic components in almost every disease process although environmental alterations or cell mutations may change these factors. These factors are dependent on the actions of genes and chromosomes. They are often latent, but manifestations may occur and normal health state turns to clinical and even pathological conditions.

A. Cytogenetic Abnormalities

During cell division a number of small bodies known as chromosomes, which are composed of DNA on a protein framework, are formed in the nucleus. DNA segments are organized in genes, which are the units of heredity. It is believed that about 100,000 genes exist, being arranged in various chromosomes. When the cell divides, the chromosomes replicate yielding pairs formed in a very precise fashion so that each daughter cell possesses an adequate complement of the chromosomal material.

Each human cell contains 23 pairs of chromosomes; one member of each chromosome pair is derived from the father and one from the mother of the individual. Among these chromosomes, 22 pairs are alike in both sexes and are called autosomes, while the remaining pair, the sex chromosomes, differ between males and females and thus determine the sex. In females, the sex chromosomes are identical and are known as X chromosomes; in males they are different, one is like the female chromosome and is called X chromosome, the other is known as the Y chromosome and is characteristic of the male.

Cell divisions follow two patterns. Somatic or body cells have two of each type of chromosome and replication is associated with the division of the chromosomal material into identical parts through mitosis; these are present in the daughter cells. Germ cells or gametes have only one of each type of chromosome; these are the haploids. During cell division, called meiosis, the genetic material is partitioned precisely so that on the union of two germ cells, an ovum and spermatozoon, the diploid somatic cell complement is restored.

The mechanism of cell division, either by somatic or germ cells, may be influenced by environmental changes at various stages of the process. These factors may cause death of the cell, but there is some biochemical adaptation which may counteract the alterations brought about by environmental effects. However, chromosomal abnormalities can cause impairment in both mitosis and meiosis. In many cases it is difficult to diagnose these errors early since in most inherited disease there is no morphologically detectable difference from normal in the number or size and shape of the cellular chromosomes. There are, however, some diseases which usually occur as a consequence of deficient cell divisions accompanied by chromosomal changes. Few patients with these diseases are capable of reproduction, so the effects of these cytogenetic disorders are not transmitted to the offspring.

In general, in cytogenetic diseases either a chromosome is missing or extra chromosomes are present. There are diseases where more than two chromosomes assemble as a pair and one or two possess fewer chromosomes. Many genetic errors are associated with numerical aberrations, these are the polyploidies; in particular, triple chromosomal units are the most common, known as trisomy. This occurs when one part of a chromosome is translocated to another, thus resulting in a change of shape. These disorders are often not directly inherited, but some inherited characteristics may predispose to this condition. In the affected patient it is probable that there is an abnormality in the mitosis of the gametes shortly after fertilization. It may be that the chromosomes are ruptured, or inadequately separated, or did not recombine properly. One of the fragments of the ruptured chromosomes may become linked with another pair and hence becomes the third chromosome, leading to trisomies. The imperfect meiosis may be stimulated by environmental effects. In such cases, all the daughter cells of the gametes probably have distorted chromosome patterns, causing abnormalities of the whole organism that grow from it.

More frequently, the parents of afflicted patients with trisomy or any other cytogenetic abnormality are often older than parents of normal children. The mitotic disorder may, therefore, be related to the reduction of endocrine activity or senile changes in the parental gametes. Abnormal meiosis leading to trisomies involve either changes in sex chromosomes or autosomes. Theoretically, trisomy could occur at any chromosome pair or involve any part of the chromosomal structure and arrangement, but there are certain chromosome pairs which are more susceptible to mitotic accidents.

In the process of mitosis several sources of error may cause abnormal chromosome arrangements. One example of an alteration of the normal mechanism of cell reproduction is known as mosaicism. Normally, reproduction leads to a genetically identical cell line which is called a colony or clone. If an error occurs in reproduction, a mutant cell is produced which forms its own clone and consequently two kinds of cells exist within the same organism. In mosaicism, not one but two or more of the same cell lines are present in the body. This occurs in certain diseases of the blood, when two immunologically distinct lines of erythrocytes or leukocytes may coexist. Mosaicism is produced temporarily at blood transfusion or tissue transplantation. Another example for the autosomal trisomies is characterized by the syndrome of mongolism. There are several types of mongolism or Down's disease — the most frequent ones being associated with the trisomy of autosome number 21.[75] Occasional mongolism occurs from the defect of other chromosomes, or sometimes no chromosomal abnormality is apparent but the disease is due to abnormal translocations. Trisomy in autosome number 18 also causes a fairly specific syndrome. It is known as floppy baby syndrome. The affected children are hypotonic, ossification is defective, hands and feet deformities are frequent, and they usually develop small jaws.

In several chromosomal abnormalities sex chromosomes are involved. Two of these are the Turner's syndrome and Klinefelter's syndrome. In Turner's syndrome, although the external genitalia indicate a normal infantile female, no ovaries are present; this disorder is compatible with life; some deficiency may be apparent. In these patients instead of the genotype XX sex chromosome, one gene is absent in the chromosomal locus. Their single X chromosome is enough to the assumption of a female phenotype, but it is not adequate to maintain a full gonadal development. In Klinefelter's syndrome, the chromosome abnormality is a mirror image of Turner's. These individuals have a Y chromosome, but also two or even three X chromosomes. This condition represents a male phenotype, but the appearance is eunochoid or feminine. The presence of an increased number of the feminine X chromosomes prevents complete masculine maturation, probably by stimulating estrogen synthesis.

Since trisomy can occur at an alteration of any chromosome pair, subsequently many such children are born with severe deformities. Several different trisomies are not compatible with life and many spontaneous abortions or premature deaths are probably due to defects of mitosis. Cytogenetic diseases are generally fairly uncommon, but they are important because they represent exaggerated examples of much more frequently occurring minor mutations. These minor mutations may be associated with the spontaneous onset of inheritable disease, tissue changes occurring during aging, and be responsible for some types of cancer. The development of cytogenetic syndromes are influenced by the environment, especially by the intrauterine environment. Events occurring in the womb may represent the most important contributing factors to the production of congenital malformations, which is considered the major cause of perinatal and neonatal death.

B. Transfer of Genetic Information

In cell structure and metabolism, proteins and nucleic acids are the most essential macromolecules. There is an interrrelationship between the synthesis of proteins and nucleic acids and both are together essential in understanding the main processes of cellular synthesis. At a higher level the course of cell division, differentiation, and growth are functions of the genes and chromosomes which are finally responsible for protein synthesis. Any error in nucleic acid or protein formation leads to interference with the macromolecular arrangement of the cell and ultimately to a disease process.

Protein is synthesized in the rough endoplasmic reticulum. The information required for the synthesis of a specific protein with specific amino acid sequence is stored in a DNA structural gene. The linear arrangement of purine and pyrimidine bases in a segment of DNA contains the information necessary to organize the specific assembly of the various amino acids in the protein chain. The nuclear DNA controls the synthesis of most proteins in the cell by producing another template on itself. The primary product of the structural gene is constructed from ribonucleic acid known as messenger RNA. The messenger RNA carries the code for a specific polypeptide which is formed when protein synthesis is required. On being detached from DNA, it leaves the nucleus and in combination with ribosomes attached to the endoplasmic reticulum the process of protein synthesis is organized. The priming of ribosomes by the messenger RNA, and probably with small amounts of cytoplasmic DNA, enables these structures to assemble peptide chains. The assembly proceeds from the amino end of the polypeptide and new amino acids are added sequentially to the carboxyl group. The sequence of these chains is determined by the template of the messenger RNA. Three or four ribosomes are often aggregated to construct polysomes. Ribosomes only become fully functional if they are attached to endoplasmic reticulum and form the rough endoplasmic reticulum. Certain toxic compounds and carcinogens cause degranulation of the ribosomes from the membrane, which leads to dysfunction of these cellular organelles and thus either to inhibition of protein synthesis or to an aberration of normal processes. Dissociation between ribosomes and the endoplasmic reticulum in the liver cell is considered to be one of the essential phases in the initiation of the formation of hepatic neoplasms by hepatocarcinogens. There are circumstances when proliferation within the cell leads to the increased production of smooth endoplasmic reticulum. The induction of these membranes occurs when drugs or other foreign compounds are admininstered in adequate doses. Smooth endoplasmic reticulum does not contain ribosomes and its main function is associated with the elimination of the inducer itself by detoxification processes.

The structural genes are essential in the primary formation and proper conformation of specific polypeptides. Other genes, known as control genes, are responsible for the expression of the structural genes. There are two types of control genes; operator and regulator genes. The operator genes are associated with certain regions of the DNA strand and initiate the messenger RNA formation. It is probable that a single operator controls the transcription of adjacent structural genes on the same chromosome. The activity of a group of genes is coordinated by the operon. This only operates on the same chromosome and elicits no effect on the other member of a chromosome pair.

The role of the regulator gene is to control the function of the operator gene. This role is carried out through the production of a specific substance, called a repressor. The repressor combines reversibly with the operator, blocks the initiation of the transcription process, and inhibits the synthesis of the specific proteins which are governed by structural genes associated with the specific operons. The effect of the repressor on the operon is always negative, preventing its action. In certain circumstances, the repressor is inactivated by combining with small molecules such as drugs and other exogenous or endogenous substances. This process represents the mechanism of enzyme induction (Volume II, chapter 3). In contrast, regulator gene may be inactivated by the end product of a metabolic sequence, resulting in an arrest in the synthesis of specific proteins. This process represents the mechanism of enzyme repression.

The first step in the formation of the peptide chain is the activation of the amino acid with adenosine triphosphate and the coupling with transfer RNA. Transfer RNAs are different, corresponding to the different amino acids. Each transfer RNA is coded to carry the particular activated amino acid to the messenger RNA onto a specific place on the template. When this amino acid is properly oriented on the surface, it is linked to the preceding amino acid and the peptide chain grows by the one amino acid unit. After the new amino acid is bound in the peptide the transfer RNA is set free and moves back to the cytoplasm to bind another activated amino acid and the process is repeated. When the whole peptide chain is completed, it is released from the ribosomal surface. If several peptides join to form a larger protein molecule, this linkage usually occurs while the peptide is still attached to the ribosomes and the peptide chains remain oriented. After detachment from the ribosome, secondary and tertiary structures are formed. In several cases, further association with another polypeptide or protein occurs and thus functional proteins are produced. The production of a complete molecule is a fairly rapid process since the messenger RNA is often unstable and does not remain attached to the ribosomal surface for a long time. It is probable, therefore, that in most cases only a few peptides are formed and bound simultaneously by each molecule of messenger RNA. It has been suggested that if the synthesis of more peptides is required, during the process of peptide synthesis the nuclear DNA produces new messenger RNA or long-acting messenger RNA. It has not yet been established how the DNA-induced synthesis of messenger RNA is regulated: how much DNA is required and what special fragment takes part in the program, what is the stimulus that triggers off the process and when does it stop, or when the quantity of messenger RNA is enough and the amount of peptide produced satisfies cellular needs. It is certain, however, that RNA functions in a very well-organized manner under a selective regulatory mechanism which initiates and terminates the whole procedure. The environment containing the embryonic organizers and also the cellular constituents produced previously are important parts of this system.

VII. INBORN ERRORS OF METABOLISM

All inherited diseases are ultimately connected with changes either in the quality or quantity of one or more specific proteins.[18] Many proteins function as enzymes which catalyze essential biochemical reactions necessary for the proper maintenance of life. When essential enzymes are not formed, their lack may lead to an abnormal function. Decreases or absolute failure in the synthesis of these enzymes do not always cause disease. Some enzymes are not vital, and if they are absent alternative pathways will replace their role. Other enzymes are more important or absolutely essential and the alternative pathways cannot replace their function. Accordingly, the lack of these latter enzymes leads to disease. When such important enzymes are deficient or not formed at all, fuctional and even associated morphological defects often occur. These enzyme defects characterize the development of pathological conditions called inborn errors of metabolism.

Inborn errors of metabolism are generally associated with mutations of several types. The point mutation is correlated with a substitution of a single purine or pyrimidine base in the messenger RNA for a different base. Subsequently, the deficient base alters the code responsible for the order of specific amino acid combination. This may happen under the control of different gene and the altered messenger RNA inserts the amino acids into the polypeptide chain in a different order. Changes in DNA segments may result in the presence or absence of an extra base in the RNA, which also produces modified protein or no protein at all.

Whereas most inborn errors of metabolism are not manifested in disease, several are extremely severe and lead to early death. Many inborn errors are, however, probably not diagnosed since they do not have any special symptoms. Their occurrence varies between incidences greater than 1 per 1000 births to less than 1 per 100,000 births. All human beings have a difference of biochemical make-up just as they have variations in fingerprints or facial features. It may be that some individual differences bring quite a few people closer to the borderline of inborn errors. Individual differences in the quality and quantity of enzymes are probably related to variations in susceptibility to disease.

Many inherited metabolic disorders only represent normal variations and are symptomless, such as bitter taste recognition of thiourea derivatives, pentosuria, most cases of fructosuria, and β-aminoisobutyric aciduria. Some others are also symptomless, but become symptomatic by some environmental circumstances. Inherited deficiencies of blood coagulation factors have no consequences unless a trauma triggers the disease. Acute intermittent hepatic porphyria or some hemolytic anemias are induced by the exposure to certain drugs. Other inborn errors cause mild to moderate problems throughout life, but they are still compatible with long life. Gout, alkaptonuria, and familial bilirubinemia do not interfere with a reasonably normal existence unless secondary problems such as biliary stones develop. There are further conditions when the error may have lethal consequences. Several diseases interfere with normal neurological functions leading to irreversible changes and mental deficiency such as uncontrolled phenylketonuria, branched chain aminoaciduria, and lipid storage diseases; some of them are often fatal in infancy. It is probable that many cases of miscarriage, spontaneous abortion, or stillbirth are also connected with inborn errors of metabolism in the embryo.

A. Metabolic Abnormalities

1. Metabolic Basis

Most proteins function in the cell as enzymes. The enzyme activity depends on various factors: (1) inherent properties of the enzyme, (2) amount of enzyme, (3) intracellular transport of substrates, (4) intracellular substrate concentration, (5) enzyme activators or inhibitors, (6) availability of cofactors, (7) presence of activatory or inhibitory metals, (8) architecture of the cell, and (9) participation of cellular organelles.

The entrance of substrates into the cell is under active control. This may be exerted through carrier proteins which influence the extracellular transport as well as the passive diffusion of substrate into the cell. The influx may be regulated by enzymes found in cellular membranes which affect the active transport into the cell or through other enzymes which influence the transport out of the cell and produce a diffusion/concentration gradient. Deficiency in carrier proteins has not so far been described. Membrane transport defects have been associated with disorders of renal tubular transport and related to some toxic symptoms occurring in some aminoacidemias.

One limiting factor in the velocity of enzyme reactions is the substrate concentration. Several enzymes function at the maximum rate, when the substrate level is high and the enzyme is always saturated. In this case the reaction rate is dependent on the amount of available enzyme. Any changes in enzyme concentration play a role in regulation of the reaction rate. The occurrence of Type I glycogen storage disease is due to the lack of glucose 6-phosphatase. There are some suggestions that in primary gout glutaminase activity is deficient and the excess glutamine is converted to purines.

Enzyme activators and inhibitors regulate enzyme actions. Many enzymes need cofactors or essential metal constituents for their activity. Some cofactors are synthetized in the body, others are supplied by the diet entirely or in precursor form. Pyridoxal phosphate is an essential cofactor in many enzymatic processes involving transformations of amino acids. Isonicotinic acid hydrazide (isoniazid) binds directly with this cofactor and during prolonged administration toxic side effects may develop due to the decreased activity of several enzymes such as amino acid aminotransferases or decarboxylases. The reduction of aminotransferase by cefazolin may be due to an effect on the production of pyridoxal phosphate from pyridoxine by the phosphokinase enzyme. Many dehydrogenases need pyridine nucleotide coenzymes synthesized from nicotinic acid. This is partly supplemented in the diet and partly synthesized from tryptophan. Membrane transport defects occur in pellagra and Hartnup disease, where the clinical symptoms of these diseases are connected with the limited tryptophan absorption.

Some trace metals are essential constituents of enzymes while others are toxic. Even essential trace metals show impaired action when present in excessive amounts. These toxic effects are related to binding of enzyme sulfhydryl groups required for their function.

2. *Mechanism of Abnormalities*

It is apparent from what has been said before that mutations affecting structural genes lead to the production of proteins which contain different amino acid sequences as compared to normal protein. Due to the mutation, basic alteration occurs in the protein structure and consequently these proteins have different physical and chemical properties, resulting in various disturbances in their function. Differences in amino acid sequence at the active site of an enzyme alter or may even completely eliminate its function, the presence of a variation in amino acid composition causes the formation of abnormal hemoglobin, such as hemoglobin S. This heme protein shows chemical and physiological characteristics very different from the normal hemoglobin A.

If during mutation a nonfunctional messenger RNA is formed as the result of some difference in the composition of purine and pyrimidine bases, protein is not synthesized at all. On the other hand, the mutation of regulatory gene has an effect on the quantity of the protein produced. The amount of product can be greatly decreased or elevated, depending on the nature of mutation. In some cases it is difficult to distinguish whether the defect is caused by a structural or regulatory gene. The consequence of a genetic defect and alteration in the quality or quantity of the protein depends on the importance of the role this protein plays in normal circumstances. The disabilities of the cell show variations as to whether the change involves a transport protein, a structural protein, or an enzyme. Defects in transport proteins cause disorders in specific functions. Disturbances in circulating proteins such as plasma proteins, alterations in hemoglobin structure, and defects in membrane transport belong to this category. Defects of structural protein synthesis are associated with many genetically determined disorders. The majority of metabolic abnormalities known until now is, however, associated with consequences of abnormalities in enzyme functions.

Genetic alterations in the synthesis of an enzyme can cause a metabolic block. The potential effect of a block may disrupt metabolic sequences (Figure 2). If in a reaction sequence governing substrate-product conversion from S_1 to S_4 by $E_1$to E_3 enzymes a defect occurs at the regulator gene R_3, this might be due to a mutant protein replacing E_3. The replacement may have various consequences dependent upon the properties of this enzyme even if E_3 is not formed at all. The consequences of this change may be various:

1. S_4 is not formed, or is only produced in greatly reduced amounts. If S_4 represents an important metabolic product, clinical symptoms may be manifest. Such conditions occur in the failure of cortisol or thyroxine synthesis associated wtih adrenogenital syndrome or goitrous cretinism, respectively.
2. If the reaction sequence $S_3 \rightarrow S_4$ is reversible, due to the block the reaction is shifted backward by end product inhibition and S_3 will accumulate. This change is followed

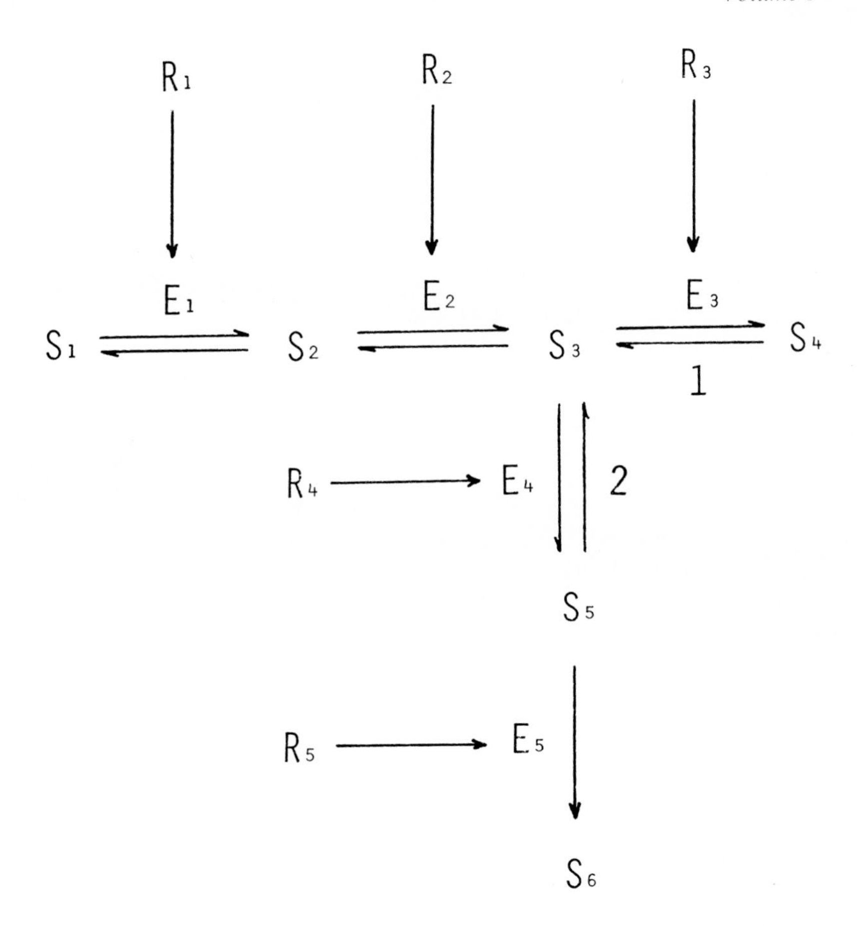

FIGURE 2. Mechanism of genetic effects on the synthesis of enzyme systems. A reaction sequence may be regulated by different enzymes and each enzyme is formed by genes acting independently. If an enzyme defect occurs at the regulatory gene R_3 (1) enzyme E_3 is not formed and the metabolic block disrupts the sequences. Consequences of this block the lack of formation of S_4 or some other products derived from S_4, and previous precursor S_3 or other precursors S_2 and S_1 are accumulated. The block may augment minor pathways (2) and otherwise minor metabolites S_5 and S_6 are produced in greater quantity.

by the accumulation of many metabolites which are characteristic of disorders such as lipidosis, glycogen storage disease, or jaundice due to bilirubin accumulation.

3. The whole metabolic sequence is interrupted and intermediates are also accumulating in excessive amounts. This occurs in the absence of hepatic glucose 6-phosphatase when not only glucose 6-phosphate, but remote precursors such as glycogen are also present in greater amounts in the liver or lactic acid in the blood. Furthermore, the abnormal accumulation of metabolites or precursors in the reaction sequence preceding the block may lead to deposition in tissues. This interferes with the function or has a possible disruptive effect such as accumulation of galactose-1-phosphate in galactosemia, limit-dextrins or glycogen in glycogenosis, or iron in hemochromatosis. When there is a widespread storage, complicated systemic effects occur.
4. Alternate collateral pathways $S_3 \rightarrow S_5 \rightarrow S_6$ are increased and the bypass often results in the formation and excretion of unusual metabolites or normal metabolites in unusual amounts. This clinical condition occurs in various porphyrias or in phenylketonuria when the production of phenylketo acids are greatly enhanced.
5. Finally, when the collateral pathway is blocked the main pathway is increased and the availability of more substrates raises product formation and accumulation.

Table 2
INBORN ERRORS OF AMINO ACID METABOLISM

Disease	Biochemical manifestation or enzyme defect	Clinical symptoms
Hyperglycinemia	Glycine metabolism	Mental retardation, osteoporosis, neutropenia, ketosis
Hypervalinemia	Valine transaminase	Mental retardation
Hyperprolinemia	Proline oxidase δ-Pyrroline 5-carboxylate dehydrogenase	Mental retardation, growth retardation
Histidinemia	Histidase	Mental retardation
Maple syrup disease	Branched chain ketoacid decarboxylase	Excretion of ketoacids from leucine, isoleucine, and valine gives maple sugar smell in the urine
Phenylketonuria	Phenylalanine hydroxylase	High plasma phenylalanine, phenylketones in urine, pigment disorder, mental retardation
Alkaptonuria	Homogentisic acid oxidase	Homogentisic acid in urine, ochronosis of cartilage
Albinism	Tyrosinase	Impairment of melanin production
Tyrosinosis		Tyrosine accumulation in tissues, liver disease, myasthenia
Homocystinuria	Cystathionine synthetase	Osteoporosis, hypercoagulation of blood, ectopia lentis, mental retardation
Cystathionuria	Cystathionase	Mental retardation
Cystinosis		Cystine accumulation in viscera, kidney stones, mental retardation

B. Defects of Amino Acid and Protein Metabolism

1. Defects of Amino Acid Metabolism

In this section, certain inborn metabolic errors of amino acid metabolism occurring in man will be discussed in general.[37,47,82,98] More details will be given subsequently in various other chapters relevant to the major target organ of any particular defect. Although most of these diseases are infrequent, their study provides important data for the elucidation of amino acid metabolism in normal human subjects. These disorders are detected in infancy, are often fatal at an early age, and often associated with irreversible brain damage and mental deficiency if left untreated. Their early detection and appropriate treatment is absolutely essential. Since several enzymes involved in these disorders are detectable in aminiotic fluid cell cultures, this procedure may provide a possibility in making a diagnosis. The most widely occurring inborn errors of amino acid metabolism and diseases of urea synthesis are given in Tables 2 and 3. Some disorders are connected with transport defects of amino acids and other small molecules by the renal proximal tubules (Table 4).

These metabolic disorders originate from a mutation of the genetic code followed by a modification of the protein primary structure. Depending on the nature of the primary alteration, other changes in the secondary or tertiary protein structure may be manifest. Some modification in the primary structure may have little or no effect on the enzyme activity, but other changes may markedly affect the three-dimensional structure of the regulatory or catalytic sites and activity will be profoundly altered. The modified enzyme may exert changed catalytic activity or ability to bind to prosthetic groups or allosteric regulators. Most protein contains 100 or more amino acids, thus many modifications may occur in the primary structure. Since a large number of enzymes take part in the metabolism of amino acids, a great number of disorders may exist. The same error in amino acid metabolism may be due to several simultaneous enzyme defects.

Table 3
INBORN ERRORS OF THE UREA CYCLE

Disease	Biochemical manifestation or enzyme defect	Clinical symptoms
Citrullinemia	Arginosuccinic acid synthetase	Ammonia intoxication
Arginosuccinic aciduria	Arginosuccinase	High blood ammonia, cerebral intoxication
Hyperammonemia	Carbamylphosphate synthetase	High blood ammonia
	Creatinine transcarbamylase	

Table 4
INBORN ERRORS OF SMALL-MOLECULE TRANSPORT OF THE RENAL PROXIMAL TUBULES

Disease	Biochemical manifestation or enzyme defect	Clinical symptoms
Renal tubular acidosis	Potassium and bicarbonate loss in the kidney	Systemic acidosis, dehydration
Hypophosphatasia	Serum alkaline phosphatase reduction	
Cystinuria	Loss of basic amino acids	Mental retardation
	Loss of cystine, arginine, ornithine, lysine	
DeToni-Fanconi syndrome	Amino acid, phosphate glucose loss in the urine	
Hartnup disease	Neutral amino acid loss in the urine	Mental retardation

Disorders of amino acid metabolism involving glycine include glycinuria and hyperoxaluria. The former is probably attributable to a defect in the renal tubular transport of glycine. In the latter disorder, high urinary excretion of oxalate can lead to bilateral calcium oxalate urolithiasis and nephrocalcinosis. Similarly, abnormalities of sulfur-containing amino acids such as cystinuria are probably associated with a transport defect. In cystine storage disease, various renal functions are seriously impaired.

Although the major site of urea synthesis is the liver, disorders involving the urea cycle mainly cause pathological changes in the brain. These will be mentioned among the diseases of the nervous system (Chapter 13). Other amino acid metabolic disorders such as phenylketonuria, branched chain ketoaciduria, homocystinuria and cystathionuria, histidinemia and hyperglycinemia will also be included in that chapter. Many inborn errors have an effect primarily on the liver, such as tyrosyluria and tyrosinemia.

Defects may occur in the formation of fatty acid derivatives, some of them serving as precursors of amino acids by the aminotransferase reaction. This group comprises disorders of propionate and methylmalonate metabolism. There is a metabolic relationship between these compounds and between vitamin B_{12}. The consequence of these abnormalities is a profound metabolic acidosis. Propionyl CoA is synthesized from isoleucine and methionine from fatty acids containing odd number of carbon atoms and from the side chain of cholesterol. The conversion of propionyl CoA to intermediates is associated with the carboxylation to methylmalonyl CoA, which requires biotin as a coenzyme. Methylmalonyl CoA is formed directly from valine, then is converted to succinyl CoA by isomerization. This reaction is dependent on 5′-deoxyadenosylcobalamin coenzyme, a vitamin B_{12} derivative.

Methylmalonic aciduria is characterized by the excretion of large amounts of methylmalonate in the urine. This condition disappears if sufficient quantities of vitamin B_{12} are

Table 5
INBORN ERRORS OF PLASMA PROTEIN SYNTHESIS

Disease	Biochemical manifestation or enzyme defect	Clinical symtoms
Pseudocholinesterase deficiency	Pseudocholinesterase	Sensitivity to general anesthetics
Agammaglobulinemia	Gamma globulin production	Susceptibility to infection
Analbuminemia	Albumin production	Edema
Hemophilia A	Clotting factor VIII production	Excessive bleeding
Hemophilia B	Clotting factor IX production	Bleeding
Parahemophilia	Clotting factor V production	
Hagerman factor deficiency	Factor XII production	Slight bleeding
Hypoprothrombinemia	Prothrombin production	Bleeding
Afibrinogenemia	Fibrinogen production	
Wilson's disease	Ceruloplasmin production	Copper deposition in tissues, hepatic and lenticular lesions, cirrhosis

given. It has been shown that the defect in this disorder lies in the inability of the liver to synthetize 5′-deoxyadenosylcobalamin from normal levels of vitamin B_{12}. There is another form of methylmalonic aciduria which does not respond to the administration of vitamin B_{12} In this disease, the enzyme methylmalonyl CoA isomerase is not functioning.

Propionic aciduria is due to propionyl CoA carboxylase deficiency. The site of the defect is in the leukocytes; propionic acid catabolism is defective and, consequently, high serum propionic acid levels develop.

2. *Defects of Plasma Protein Synthesis*

The synthesis of all plasma proteins is under genetic control. Some variations occur in health, but excessive changes may lead to disease (Table 5). These abnormalities may be connected with some general alteration of protein synthesis and metabolism as will be discussed in Chapter 2. There are, however, numerous inborn errors of the plasma protein production and many inherited diseases are associated with the proteins involved in coagulation.

The plasma contains a complex mixture of simple proteins, glycoproteins, lipoproteins and metalloproteins; their concentration total normally is about 7 g/dℓ. Some proteins exist in combination with other proteins. Several proteins bind nonprotein molecules such as various ions, including heavy metal contaminants. Human plasma contains about 100 different proteins, 40 of which make up 90% of the total plasma. Some of the remaining 60 proteins are present only in microquantities. The molecular size of these proteins varies from 45×10^3 to 1.5×10^6 dalton. Many plasma proteins have special roles; some proteins possess enzyme activity as derived from various tissues, and their presence in the blood stream is due to leakage caused by cell destruction. Others function as carriers of substances or only exhibit the physicochemical characteristics of large colloid particles.

Plasma proteins are responsible for many of the biological and physiochemical properties of blood. The viscosity of plasma proteins is important in maintaining the suspension of erythrocytes and in regulating the distribution of water between the blood and interstitial tissue through their osmotic effect. Plasma proteins are carriers of many substances — lipids, lipid-soluble vitamins, organic compounds synthesized within the body such as bilirubin, various hormones, drugs, and their metabolites — and transport them into many organs where they are utilized or further metabolized.

The site of most plasma protein synthesis is the liver. Almost all albumin, globulin, particularly γ-globulin, and fibrinogen circulating in the blood are formed in the liver and influenced by a variety of physiological and pathological conditions. Extensive loss of blood increases globulin biosynthesis, infectious diseases enhance γ-globulin formation, decreased blood pressure accelerates albumin production; inflammatory processes increase fibrinogen biosynthesis. In the production of α- and β-globulins, both the liver and the reticulo-endothelial system take part. Plasma protein level is lowest at birth, gradually increases to a peak at the age of 5 years, then slowly decreases until the age of 25 years when the normal adult value is reached.

γ-Globulin represents an electrophoretically homogeneous fraction of the serum, but includes at least 20 different components demonstrated by immunological techniques. Separate antibodies are produced against influenza, measles, mumps, diptheria, and other infections. Several inborn errors are connected with the absence of γ-globulins.

3. Globulins

Some mutations are known which alter the synthesis of serum globulins. Primary agammaglobulinemia is probably associated with deficiency of all the immunoglobulins. In this disorder all the three major components, IgG, IgM, and IgA are involved. In some cases small amounts of IgG are found. The clinical symptoms appear only in later life. Genetic factors may be involved in the development of primary acquired agammaglobulinemia. The lack of antibody production in agammaglobulinemia is connected with increased susceptibility to infections. If γ-globulin is not administered to these patients, otitis, bronchitis, pneumonia, and sinusitis frequently occur after each other. Gastrointestinal problems and chronic diarrhea are also common, but the urinary tract is rarely infected. Complications, including septicemia, may follow meningitis or osteomyelitis with fatal consequences.

A special type of agammaglobulinemia exists where antibody deficiency is combined with lymphopenia. All primordial lymphoid tissue is absent, particularly in thymus and tonsils. Familial lymphopenia occurs in early infancy; it shortens the life span to only a few months. Another type of antibody deficiency is connected with normal or slightly reduced levels of IgG. This disorder is known as dysgammaglobulinemia. Immunoelectrophoresis indicated the lack of IgM alone or IgM and IgA together. IgA can be missing in normal individuals.

Wilson's disease is associated with the lack of ceruloplasmin, a copper carrier protein. Various apparently unrelated clinical symptoms characterize this disease, such as cirrhosis and hepatolenticular degeneration. Striking neurological symptoms are often associated with mental deterioration.

4. Albumin

Several mutations have been described which influence serum albumin synthesis. The absence of detectable albumin, or very low levels, characterize analbuminemia. The symptoms of this inborn error are relatively mild; only slight edema and fatigue are apparent. The lack of albumin causes a decrease of the blood colloid pressure. This is partly compensated by the increase of various globulins, a slight decrease of blood pressure, and marked reduction of the capillary pressure.

Bisalbuminemia represents another abnormality, but it is also present in normal people. The existence of the two albumins in the blood has been detected by electrophoresis. One component corresponds to normal serum albumin, migrating with the same speed as the normal protein. Both components are present in approximately equal amounts and the total of the two is equivalent to the total albumin concentration of a normal individual. The immunological reactions are the same, but the slower-migrating anomalous albumin shows a structural mutation of the peptide chain with a replacement of only one amino acid residue.

5. Transferrin

Transferrin is a glycoprotein; it carries iron in the serum in bound form. Each molecule of transferrin binds two atoms of iron in ferric form. The bond is ionic and the interaction

Table 6
INBORN ERRORS OF BLOOD FORMATION

Disease	Biochemical manifestation or enzyme defect	Clinical symptoms
Hemoglobinopathies	Hemoglobin production	
Thalassemia	Hemoglobin production	
Methemoglobinemia	Methemoglobin production	Congenital cyanosis
Congenital methemoglobinemia	Methemoglobin production	
Drug-induced anemia	Glucose 6-phosphate dehydrogenase in erythrocytes	
Nonspherocytic hemolytic anemia	Pyruvate kinase in erythrocytes	
Congenital spherocytosis	Round red cells	Hemolysis, splenomegaly
Elliptocytosis	Ellipsoid red cells	No symptoms

between binding sites and ferric ions is weak. The dissociation is pH dependent; on lowering the pH from 7 to 6 about 50% of bound iron is dissociated from transferrin. The affinity of iron to the carrier protein is considerably greater than to synthetic chelating substances.

In normal plasma only 30 to 35% of the transferrin is saturated with iron. In various individuals, 17 molecular species of transferrins have been identified by electrophoretic separation and the presence of each component is genetically determined. Unique transferrins occur in a particular population, but the genetic variations are symptomless. In contrast, the hereditary absence of transferrin causes retarded growth and increased sensitivity to infection.

C. Diseases of Blood Cell Formation

1. Hemoglobinopathies

Many inherited diseases affect the synthesis of various blood components (Table 6). Increased hemolytic processes may be associated with an intrinsic defect in erythrocytes.[53] In hemophilia, a congenital absence of a protein required for normal blood clotting is the cause of the disease. Inborn errors may affect enzyme activities of red cells, such as pyruvate kinase and glucose 6-phosphate dehydrogenase deficiency; some are connected with the synthesis of abnormal hemoglobins such as hemoglobinopathies, methemoglobinemia, and thalassemia. In this chapter only diseases directly connected with abnormalities of the erythrocytes will be discussed.

The principal protein responsible for oxygen transport in the body is hemoglobin. This complex protein is a component of the red blood cell and contains iron which forms a ferrotetrapyrrol complex, ferroprotoporphyrin IX or heme. Four of these units are linked to a carrier protein, globin. The heme prosthetic group is embedded into the protein part of the molecule by binding two histidine residues. When oxygen is transported, one histidine bond is replaced by molecular oxygen (Figure 3). Furthermore, the heme prosthetic group also occurs in cytochromes and vitamin B_{12}. When the iron-prophyrin part of the molecule is removed the protein moiety, globin remains. It has a relatively constant molecular weight at about 66,700 daltons and consists of four peptide chains, each binding one heme group.

2. Adult and Fetal Hemoglobin

The hemoglobin species present in the adult is hemoglobin A. It contains two α and β chains, which are all bound to one another. Another hemoglobin, hemoglobin F is present in infants at birth in concentration varying between 60 to 90%. The fetal hemoglobin contains two α and two γ chains. It gradually disappears from the circulation and usually only traces are found by the age of 4 months. Hemoglobin A is not a homogeneous protein; by electrophoresis it can be separated into at least two fractions, hemoglobin A_1 amounting to about 98%, and hemoglobin A_2 which contains two α and two γ chains, representing about 2%.

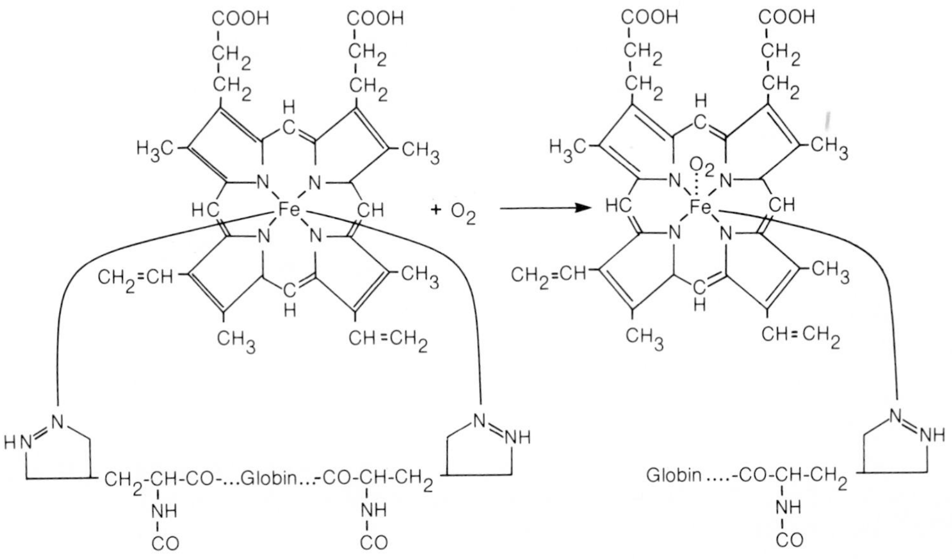

FIGURE 3. Binding of heme to globin. Four identical heme groups (ferroprotoporphyrin IX) are attached to one molecule of globin through histidine residues embedded into the protein. When oxygen is transported, one histidine residue is replaced by the molecular oxygen by freely dissociable secondary bondage.

These two adult hemoglobins are produced and inherited independently from each other, and the amino acid composition of their peptide chains is distinctly different from each other. Variations in the hemoglobin A_2 component have been seen in some hemoglobinopathies.

There are numerous differences between adult and fetal hemoglobins including oxygen affinity, antigenic activity, solubility, ultraviolet spectrum, and resistance to denaturation by alkali. Considering that the half-life of the red cell is only 120 days, hemoglobin F is not present in the blood of normal children in appreciable amounts beyond 4 months. If the production of hemoglobin F persists, it represents various pathological conditions such as infantile anemia. It is also found in many other syndromes such as sickle cell anemia, nutritional anemias, thalassemia, leukemia, and spherocytic jaundice.

3. Sickle Cell Anemia

This disease was discovered in anemic patients, when in low oxygen tension the formation of sickle-shaped erythrocytes was observed. The red cells resumed their normal shape when exposed to sufficient oxygen. Sickling of the cells and the sickle cell anemia is due to an abnormal hemoglobin, hemoglobin S, which is the most prevalent form of abnormal hemoglobins. Exposure of hemoglobin S to carbon dioxide leads to the formation of sickle-shaped crystals. In erythrocytes, excess carbon dioxide causes the whole red cell to sickle, thereby losing its normal ellipsoid flat appearance. Although the oxygen carrying capacity of the red cells remains unaffected, the mechanics of blood flow are altered. The sickle cell does not pass easily through small capillaries and the abnormal shape enhances the production of clumps which may obstruct capillaries and sinusoids. Clogging of the small blood vessels may lead to their closure and thus to deprivation of normal blood supply to the neighboring tissues. Such crises of sickle cell anemia lead to ischemia, which may cause death to these cells.

Hemoglobin S differs from the normal adult hemoglobin in some physical properties, too, such as changed migration on electrophoresis. The altered behavior is caused by only one amino acid (Table 7); in hemoglobin S, valine replaces glutamine at position 6 in one of the beta peptide chains. The sickle cell anemia is therefore caused by a change of only one

Table 7
DIFFERENCES IN HEMOGLOBIN STRUCTURE IN VARIOUS DISEASES

Disease	Hemoglobin	Structure	Substitution
Normal	A	$alpha_2{}^{A}beta_2{}^{A}$	None
	A_2	$alpha_2\ delta_2$	
Sickle cell anemia	S	$alpha_2{}^{A}beta_2{}^{6val}$	glu → val
	C	$alpha_2{}^{A}beta_2{}^{6lys}$	glu → lys
	E	$alpha_2{}^{A}beta_2{}^{26lys}$	glu → lys
Thalassemia	F	$alpha_2\ gamma_2$	Only 71% identical with beta chain
Methemoglobinuria	M_{Boston}	$alpha_2{}^{58tyr}beta_2{}^{A}$	his → tyr
	$M_{Milwaukee}$	$alpha_2\ beta_2{}^{67glu}$	val → glu
	$M_{Saskatoon}$	$alpha_2\ beta_2{}^{63tyr}$	his → tyr
	M_{Ewate}	$alpha_2{}^{87tyr}beta_2$	his → tyr

nucleotide base which alters the relevant gene responsible for the determination of the hemoglobin type.

Sickle cell anemia is mainly hereditary in blacks. The onset of anemia is the consequence of the increased fragility of sickle cells. They break down and hemolyze easier than normal cells thus reducing their number in circulation. Due to a compensatory mechanism, the bone marrow of a sickle cell patient becomes hyperplastic. However, the increased production of new cells replacing the hemolyzed ones never offsets the rate of destruction. The permanent decrease of red cells leads to a reduced oxygen-carrying capacity of the blood, thereby increasing cardiac output, which may in turn result in an increased total blood volume. This may finally cause tissue congestion and edema.

Sickle cell anemia is associated with acute crises, like other hemolytic anemias. These crises can be related with either increased hemolysis or to a failure of the bone marrow to produce new cells at the maximum required rate. During the critical period, a severe decrease of circulating erythrocytes and hemoglobin develops.[47] The hemoglobin is metabolized to bilirubin which is, however, not eliminated at an increased rate; consequently, the patient becomes jaundiced.

4. Atypical Hemolytic Anemia

Several abnormal hemoglobins were found to be associated with atypical hemolytic disease, such as hemoglobin C, D, and E. The presence of hemoglobin C is also connected with sickle cell anemia. These patients are heterozygotes; their erythrocytes contain hemoglobin S as well as hemoglobin C and they probably have two abnormal hemoglobin genes. Their blood contains a high level of hemoglobin S. However, the severity of anemia is modified mainly due to the synthesis of smaller total amounts of hemoglobin S in the cell than that produced in homozygous sickle cell anemia. Hemoglobin C alone causes no symptoms, yet the presence of hemoglobin C implies disease. It is often associated with an enlargement of the spleen. Under favorable conditions the hemoglobinopathy is latent, but the decreased half-life of the red cells represents an additional stress to the patient.

Hemoglobin D shows electrophoretic similarities to hemoglobin S. Both hemoglobin D and E differ from the normal adult hemoglobin in only one amino acid (Table 7). Since hemoglobin S, D, and E show a difference from hemoglobin A in the β chain, it is likely that separate genes control the formation of the α and β chains. Hemoglobin D occurs in 1% of some populations in southern India, and in 2% in Punjab. Hemoglobin E also exist in high frequency in non-Negroids.

5. Thalassemia

In normal adults, hemoglobin F is present only in negligible quantities[17] while larger amounts are found in patients with sickle cell anemia.[95] However, this abnormal hemoglobin has a major role in thalassemia. Beginning at early life, thalassemia represents an erythroblastic anemia and, if severe, is associated with spleno- and hepatomegaly and characteristic bone changes. Thalassemia is a hereditary disease in people of Mediterranean origin and occurs mostly in Italians and Greeks. It appears to be due to abnormalities of the genes which control globin synthesis, reflected in a failure to produce erythrocytes with normal hemoglobin content. The decreased hemoglobin production leads to reduced erythrocyte volume. Sometimes there is increased hemolysis as well. The synthesis of red cells by the bone marrow may be increased but the simultaneous destruction, in most cases, is also accelerated. Anemia may develop, depending on the equilibrium which is established in a patient between the formation and destruction of erythrocytes. This delicate balance is disturbed when additional demands are raised, such as during pregnancy or following infectious diseases.

There are different thalassemias, some are severe and may be fatal, others are asymptomatic. Severe and mild forms of the disease are described as thalassemia major and minor, respectively. Thalassemia major patients have a marked reduction in the number of erythrocytes, and cells with smaller size are produced, representing microcytosis. Moreover, the variability of the shape and size is great. Some cells are characteristic, with hemoglobin spots in the center, called target cells. In this disease, episodes of hemolysis and small tissue infarcts are frequent. The destruction of the red cells leads to jaundice and the excess iron is deposited in the reticuloendothelial system, causing hemosiderosis. The accumulation of iron in the Kupffer cells is often followed by cirrhosis or scarring of the liver. Hemolytic crises and infections frequently cause early death. Heterozygous persons with thalassemia minor have varying symptoms — some are even symptomless, some show characteristic target cells but otherwise remain asymptomatic. Mild anemia may occur, but no jaundice develops. Occasional hemolytic episodes are usually mild.

The biochemical background of the disease is probably derived from the mode of inheritance regulating the synthesis of various peptide chains of hemoglobin. The location of the genes which determine whether the β, γ, or δ chain is produced is closely linked and controlled by an operator-suppressor system (Figure 4). In the fetus, the operator gene for the α-chain functions at maximum from the onset of hematopoiesis. However, the production of the β- and δ-chains by the genes is suppressed and only γ-chains are synthesized. The α-chains can readily combine with any of the other three chains, but since only γ is produced, the fetal hemoglobin F will be formed by the combination of $\alpha_2\gamma_2$. After birth, the genes for β- and δ-chains start operating. The production of the β-chain is more efficient than γ- or δ-chain formation, resulting in normal adult hemoglobin A with an $\alpha_2\beta_2$ chain combination, small amounts of hemoglobin A_2 with an $\alpha_2\delta_2$ combination, and hemoglobin F with an $\alpha_2\gamma_2$ combination. In β-thalassemia, the defect occurs in the production of the β-chain; the α-chain is formed normally so these patients have relatively less hemoglobin A and more A_2 and F, depending on whether γ- or δ-chain production is also suppressed. In α-thalassemia, the synthesis of the α-chain is reduced. Since β-, γ-, and δ-chains are produced normally, in α-thalassemia an abnormal hemoglobin tetramer is formed, e.g., containing all β molecules (β_4, hemoglobin H) — or in β-thalassemia all γ-molecules (γ_4 hemoglobin Barts). The hemoglobinopathy may not be the only reason for the hemolytic episodes in thalassemia; a defect in the synthesis of the red cell membrane may be also connected with the increased fragility. An interrelationship between different thalassemia genes is probably responsible for the genetic discrepancies observed in thalassemic families and for the wide range of clinical symptoms.

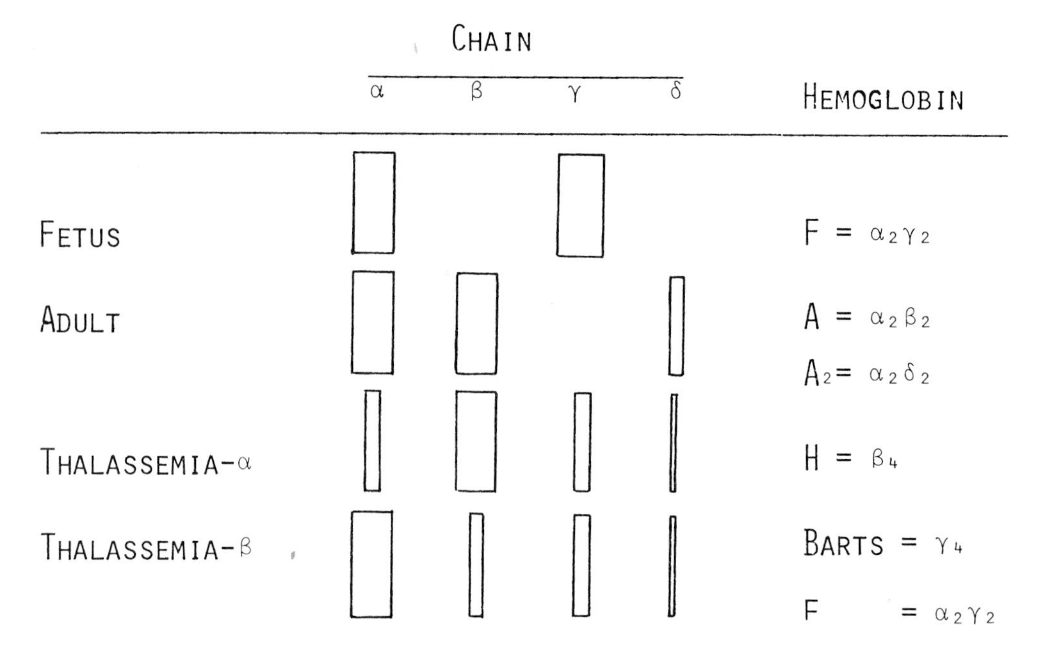

FIGURE 4. Mechanism of hemoglobin variant synthesis in normals and in thalassemia. Independent genetic control determines the formation of the various chains. In the normal fetus, α and γ chains are synthesized resulting in the production of hemoglobin F. In the normal adult, this is changed to α and β chains and to a minor extent δ chains, producing hemoglobin A and A_2. In α-thalassemia the formation of α chain, in β-thalassemia β-chain is reduced γ-chain is present, consequently hemoglobin H and hemoglobin Barts and F are produced in these disease conditions, respectively. Other abnormal hemoglobins are probably dependent on variations of the genetic code which determine the nucleotide sequence of RNA along with the amino acids assembled on the ribosomes. This is responsible for the synthesis of the amino acid sequence of the polypeptide chain.

6. *Methemoglobinemia*

When hemoglobin is oxidized, the bivalent ferrous iron of the heme group is oxidized to the trivalent ferric state, producing methemoglobin. In methemoglobin, the iron is no longer capable of binding oxygen and cannot serve the properties of an oxygen carrier. Methemoglobins derive from all known human hemoglobin and therefore several methemoglobins exist. Methemoglobin can be reduced to hemoglobin, which can form oxyhemoglobin with oxygen, and methemoglobin can bind various substances, particularly nitrogen-containing molecules.

In the presence of air when reducing substances are absent, hemoglobin is slowly oxidized to methemoglobin. In the erythrocytes permanent oxidation of hemoglobin does not take place, some processes reverse methemoglobin to hemoglobin as soon as it is formed. Normally, therefore, only 1% methemoglobin is present in the red blood cell. The presence of greater amounts indicates an impaired balance between reductive and oxidative reactions. Congenital forms of methemoglobinemia are associated with the absence of an enzyme which is able to reduce methemoglobin, or with an abnormal hemoglobin containing different globin molecules which, when transformed to methemoglobin become resistant to methemoglobin reductase. The reduction process is accelerated when glucose or lactate is present.

The acquired form of methemoglobinemia has considerable clinical importance in young children. Moderate methemoglobinemia causes cyanosis, dyspnea, and compensatory polycythemia. In hereditary methemoglobinemia, chronic cyanosis is associated with increased amounts of methemoglobins within the red cells.

The activation of the methemoglobin reducing process by glucose indicated that in enzymatic hereditary methemoglobinemia the deficiency is related to the NADH-diaphorase

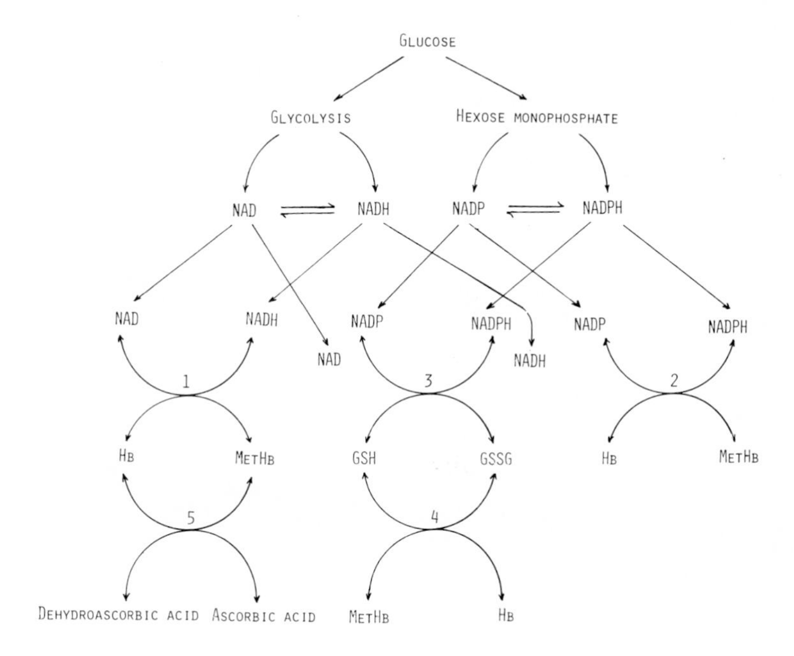

FIGURE 5. Mechanisms associated with the production of methemoglobin and its reduction to hemoglobin. Normal red cells contain a small but definite amount of methemoglobin. Reversible reduction of methemoglobin is connected with three enzyme systems and one nonenzymatic pathway. NADH-diaphorase (1), NADPH-diaphorase (2), and glutathione reductase (3) represent the enzymatic regulation. Glutathione reductase activity is connected with the control of reduced and oxidized glutathione and the actual ratio of these components nonenzymatically (4) regulates the hemoglobin:methemoglobin ratio. The ascorbate:dehydroascorbate system (5) also reduces methemoglobin in the red cell at a slow rate. The relative rates of methemoglobin reduction by these systems are NADH-diaphorase:NADPH-diaphorase:glutathione reductase:ascorbate = 1:12:1.5:2. (Modified from Scott, E., Duncan, I., and Ekstand, V., *Fed. Proc.*, 22, 467, 1963.)

system (Figure 5) which is greatly reduced or sometimes completely missing. Cytochemical investigations revealed that NADH reductase deficiency is not uniform in all red cells, but is distributed unevenly, that is, the older cells contain more methemoglobin than younger ones. It is assumed that in younger cells a secondary mechanism partially compensates for the defect. This secondary defect is probably associated with ascorbic acid or glutathione and the hexose monophosphate shunt. Methemoglobin may also be reduced by NADPH alone as generated by glucose 6-phosphate dehydrogenase from the oxidation of glucose 6-phosphate. The hydrogen is not transferred directly to methemoglobin, but a specific enzyme NADPH-dependent methemoglobin reductase is required.

There are methemoglobins, or hemoglobin M, where the molecular abnormality lies in the globin part. In the blood of the affected patients usually up to 40% hemoglobin M is formed. Mutation may occur in the amino acid sequence in both the α-chain and the β-chain. To date, more than ten molecular variants of hemoglobin M have been described. The abnormal structure of the polypeptide chain has been established in some cases (Table 7). These molecular changes facilitate the oxidation of ferrous to ferric ion.

Recently a new form of methemoglobinemia has been described. In this case the blood contained large amounts of benzoquinoneacetic acid. The origin of this compound is not known, but is may represent a block in tyrosine metabolism.

D. Defects of Carbohydrate Absorption

These defects are related to the absence or deficiency of essential enzymes such as lactase, sucrase, and disaccharidase.

Intolerance to lactose, present mainly in milk causes the syndrome of lactase deficiency.[61] There are three different forms of the disease: primary low-lactase activity, secondary low-lactase activity, and inherited lactase deficiency. In all three forms, the signs and consequences of lactose intolerance are the same, including diarrhea, flatulence, and abdominal cramps caused by lactose accumulation in the intestines. The osmotically active sugar molecule holds water within the lumen and the fermentative action of intestinal bacteria on lactose produces gases and other products which exert intestinal irritation.

Primary low-lactase activity is a relatively common disorder. The lactose intolerance does not occur in early life, but the symptoms appear in adulthood suggesting that the development of this disease is attributable to a gradual reduction of the lactase activity in susceptible individuals as they grow older.

Secondary low-lactase activity is the consequence of several intestinal diseases. Even when normal human lactose digestion is low, intolerance to milk often develops. Many gastrointestinal diseases are associated with secondary low-lactase activity such as colitis, gastroenteritis, kwashiorkor, and celiac sprue. These diseases are prevalent in tropical countries, but also occur frequently in nontropical countries. Loss of lactase activity is observed in patients after surgery for peptic ulcer.

Inherited lactase deficiency is a relatively rare disease. The prominent feature of the disease is the presence of lactose in the urine. The various biochemical abnormalities are due to the toxic action of lactose on the intestines. The symptoms of the disease are intolerance to milk, wasting, diarrhea, incidence of fluid and electrolyte disturbances, and reduced growth due to inadequate nutrition. This syndrome develops very early after birth. Withdrawal of milk and a lactose-free diet eliminates the symptoms and normal development is restored.

Sucrase deficiency is an inherited disease. The lack of this enzyme occurs in the mucosa of the small intestine. The symptoms of this disorder are apparent in early infancy following the sugar uptake. Patients with this disorder should have a modified sugar-free diet. In adulthood, most patients eventually overcome the difficulty of sucrose intolerance and small amounts of sugar can be consumed without deteriorating consequences.

Disaccharidase deficiency is diagnosed by the increased urinary excretion of disaccharides. The intestines are virtually impermeable to disaccharides and, practically speaking, they are not metabolized. Intestinal damage may promote the accumulation of disaccharides in the intestines and excretion may exceed 300 mg daily.

E. Defects of Carbohydrate Metabolism

Inborn metabolic errors of carbohydrate metabolism involve the deficiency or absence of an enzyme along the pathway of glycolysis[38] in the processes of glycogen breakdown or synthesis, the hexose monophosphate shunt, or interconversion of sugars (Table 8).[16,26,30,31,33,65] Some of these inborn errors cause only minor symptoms such as pentosuria or fructosuria; other defects are connected with pathogenic conditions such as glucose 6-phosphate deficiency, glycogen storage disease, or galactosemia.[77,100] The disturbance of glucose 6-phosphate metabolism mostly affects the red blood cells, including drug-induced hemolytic anemia. This inborn error of metabolism will, therefore, be discussed in another chapter.

1. Pentosuria

This disorder is benign and can be divided into essential pentosuria or chronic essential pentosuria and alimentary pentosuria. The former is present from birth, associated with a constant urinary excretion of L-xylulose resulting from an impairment of glucuronic acid

Table 8
INBORN ERRORS OF CARBOHYDRATE METABOLISM OR RELATED CONDITIONS

Disease	Biochemical manifestation or enzyme defect	Clinical symptoms
Diabetes mellitus	Several, related to glucose metabolism	Hyperglycemia, glucosuria, angiopathy
Galactosemia	Galactose 1-phosphate uridyl transferase	Jaundice, mental retardation, cataract
Fructosuria	Hepatic fructokinase	No symptoms
Pentosuria	L-Xylitol dehydrogenase	No symptoms
Lactase deficiency	Intestinal lactase	Malabsorption syndrome
Glucose 6-phosphate de-hydrogenase deficiency	Erythrocyte dehydrogenase	Hemolysis
Glycogen deposition[a]	Various enzymes of glycogen metabolism	Affected organs: liver, kidney, spleen, heart, skeletal muscle
Oxalosis	Glyoxylic glutamic transaminase Glyoxylic dehydrogenase	Kidney stones

[a] Further details in Table 9.

metabolism. In alimentary pentosuria, small quantities of xylose or arabinose appear in the urine following the ingestion of large amounts of fruits. Small amounts of D-ribose are excreted by normal individuals but larger quantities are found in the urine of patients with muscular distrophy, probably due to the excessive breakdown of nucleotides containing ribose in degenerating muscular tissue.

In essential pentosuria the oxidation of glucuronic acid is defective; in pentosuric individuals this sugar derivative is incompletely metabolized. Along this pathway glucose 6-phosphate is converted to glucuronic acid which is further oxidized through 3-ketogluconic acid to xylulose. Further transformation of this compound is blocked (Figure 6). The importance of this pathway in human metabolism is relatively small since no ill effect is found in people suffering from pentosuria. Life expectancy is not decreased as compared with normal individuals.

2. Fructosuria and Fructose Intolerance

Two different inherited disorders exist in the metabolism of fructose — essential fructosuria and hereditary fructose intolerance.[30,31,72] In essential fructosuria, following the ingestion of fructose an abnormally high fructose level occurs in the blood, however, the lactate content is not increased. In hereditary fructose intolerance, the administration of fructose causes severe hypoglycemia and vomiting.

Fructosuria is present only after the digestion of foods containing glucose. It is harmless and asymptomatic. Glucose and galactose metabolism is normal in these individuals and insulin treatment has no effect on fructosuria. In this disturbance the primary enzyme defect is probably associated with the deficiency of hepatic fructokinase (Figure 7).

Patients suffering from hereditary fructose intolerance show no signs of disease provided they do not ingest any food which contains fructose. However, various symptoms develop on the intake of fructose as high blood fructose levels, accompanied by a sudden fall of blood glucose, causes severe hypoglycemia and elevated serum inorganic phosphate. Further clinical manifestations of the disease are protracted vomiting, nausea, liver and spleen enlargement, jaundice, aminoaciduria, and albuminuria.

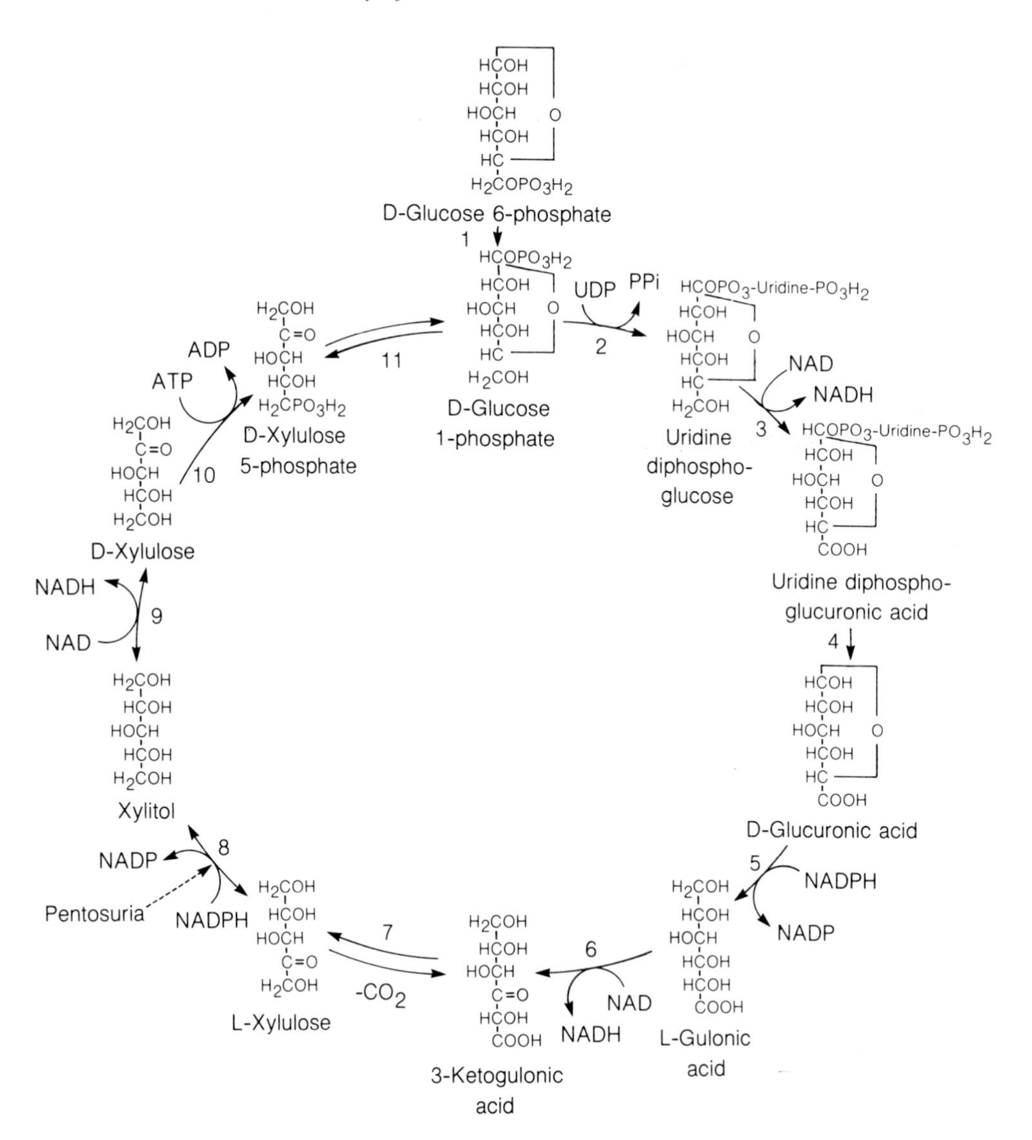

FIGURE 6. Mechanism of glucuronic acid oxidation and its relationship with pentosuria. D-Glucuronic acid is produced by glucose 6-phosphate via glucose 1-phosphate and uridine diphosphoglucose by phosphoglucomutase (1) and UDP-glucose pyrophosphorylase (2) and further converted to uridine diphosphoglucuronic acid by UDP-glucose dehydrogenase (3). Hydrolysis (4) leads to glucuronic acid, which is reduced to gulonic acid by glucuronic acid dehydrogenase (5). L-Gulonic acid is the precursor of ascorbic acid in some animals which can synthesize this compound. In man, gulonic acid is further oxidized (6) to 3-ketogulonic acid, which is carboxylated (7) to L-xylulose. L-Xylulose is reduced (8) to xylitol, which is oxidized (9) to D-xylulose by NAD/NADH-dependent dehydrogenases. D-Xylulose is transformed to D-xylulose phosphate by an ATP-dependent kinase (10) reaction. D-Xylulose phosphate is the precursor of glucose through the transketolase-transaldolase pathway (11) representing the hexose monophosphate shunt. In pentosuria, L-xylulose excretion is greatly enhanced indicating a metabolic block in the reversible xylulose-xylitol reaction found in liver, kidney, and spleen. (Modified from Burns, J. J., in *Metabolic Pathways*, Vol. 1, Greenberg, D. M., Ed., Academic Press, New York, 1967, 394.)

The primary enzyme defect is the virtual absence of fructose 1-phosphate aldolase in the liver (Figure 7). Secondary effects include blocks of fructose 1,6-diphosphate aldolase and fructokinase activities directly related to fructose 1-phosphate synthesis. Other secondary alterations include a defect in the phosphorylation of glycogen to glucose 1-phosphate and

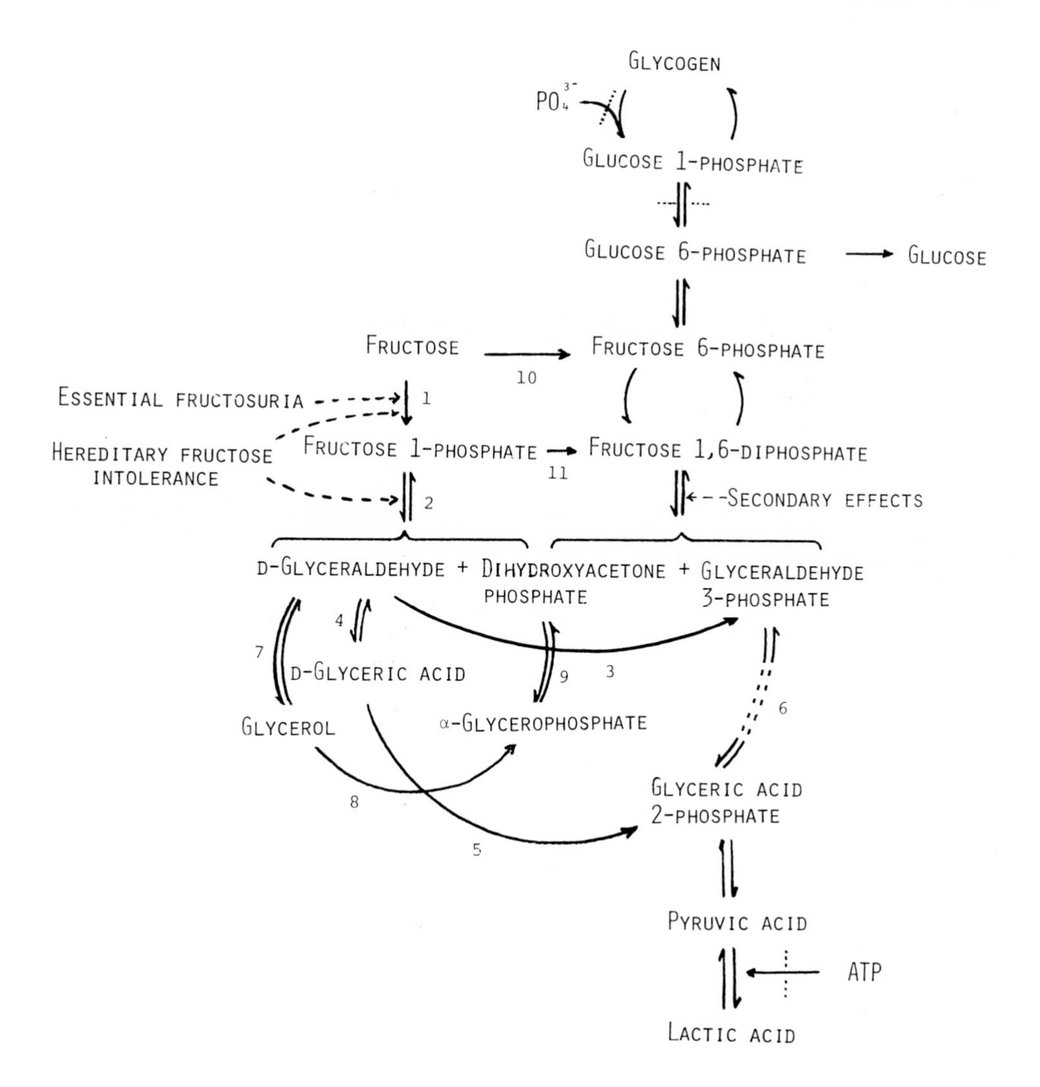

FIGURE 7. Mechanism of fructose metabolism and its relationship to essential fructosuria and hereditary fructose intolerance. Fructose is phosphorylated by an ATP-dependent and Mg^{++} activated specific fructokinase enzyme (1). Fructose 1-phosphate is cleared into D-glyceraldehyde and dihydroxyacetone phosphate by 1-phosphofructoaldolase (2). D-Glyceraldehyde is metabolized along three routes. It is converted to glyceraldehyde 3-phosphate catalyzed by triosekinase (3), oxidized to glyceric acid by glyceraldehyde dehydrogenase (4) and further converted to 2-phosphoglyceric acid by D-glycerate kinase (5). 2-Phosphoglyceric acid is also formed from glyceraldehyde 3-phosphate in a complex way (6). The minor pathway of glyceraldehyde metabolism leads to glycerol by glycerol dehydrogenase (7) and by subsequent phosphorylation by glycerokinase (8) to α-glycerophosphate. This compound is converted to dihydroxyacetone phosphate by reversible reduction (9). Another relatively major pathway of fructose metabolism is its phosphorylation to fructose 6-phosphate by hexokinase (10). The conversion of fructose 1-phosphate to fructose 1,6-diphosphate by 1-phosphofructokinase (11) in presence of ATP and Mg^{2+} represents a minor metabolic pathway. Through various reactions, fructose metabolism is connected with glyconeogenesis and the formation of pyruvate. Hereditary fructose intolerance is associated with a defect in fructokinase (1) activity and lack of fructose 1-phosphate splitting 1-phosphoaldolase (2) enzyme. (Data from Hers, H. G. and Joassin, G., *Enzymol. Biol. Clin.*, 1, 4, 1961; Froesch, E. R., Wolf, H. P., Baitsch, H. P., Baitsch, H., Prader, A., and Labhart, A., *Am. J. Med.*, 34, 151, 1963.) Secondary alterations manifest at other different sites (-----). In essential fructose intolerance fructokinase (1) is deficient. (Data from Schapira, F., Schapira, G., and Dreyfus, J. C., *Enzymol. Biol. Clin.*, 1, 170, 1962.) All these changes are associated with the lack of hepatic production of the defective enzymes.

the intracellular depletion of ATP. These are secondary to the rapid phosphorylation of fructose caused by inhibited aldolase activity. The clinical signs of the disease following oral intake of fructose — vomiting and nausea — may indicate that intestinal metabolism is also disturbed. The development of aminoaciduria and proteinuria are connected with impaired renal function brought about by a severe derangement of the tubular metabolism of fructose 1-phosphate.

3. Galactosemia

In this rare disease the ability of patients to convert galactose to glucose is impaired with varying degrees of severity.[83] Manifestations may occur either in the neonatal period or even several years after childhood. The administration of galactose gives rise to the clinical symptoms, which may become fatal if the ingestion continues. If, however, galactose-containing foods are removed from the diet, the disease quickly regresses and might even completely disappear. The affected infants appear normal after birth, but the symptoms develop shortly and are related to galactose production from the lactose present in the milk. The intake of milk for several days or weeks triggers the characteristic vomiting, diarrhea and associated dehydration. Due to subsequent poor nutrition, in the severe form impaired growth, spleen and liver enlargement, jaundice and cirrhosis, and mental retardation develop.

In this disease, the characteristic finding is an increase of the blood galactose level since the patients are unable to utilize this hexose which is thus retained in the blood. The galactose tolerance curve is elevated. The biochemical effect in galactosemia is connected with a block in the transformation of galactose 1-phosphate to glucose 1-phosphate associated with the deficiency or lack of erythrocyte galactose 1-phosphate uridyl transferase (Figure 8). The liver of these patients is deficient in this enzyme. Minor transformation of galactose 1-phosphate in tissues may be associated with the compensatory action of galactose pyrophosphorylase. Through this auxiliary mechanism, or by an another pathway, there is some utilization of galactose occurring in galactosemic patients. If galactosemic children are maintained on galactose-free diet until maturity, they gradually acquire galactose and uridine diphosphogalactose pyrophosphorylase, which produces uridine diphosphogalactose directly. Once this compound is formed the patient may convert it to glucose for further utilization. As these children grow older they can tolerate increasing galactose load without any considerable side effects.

Possible factors responsible for toxic action have been suggested. It is probable that the accumulation of galactose 1-phosphate inhibits glucose 6-phosphate dehydrogenase activity. Increased galactose-1-phosphate levels may damage various tissues. Cataract is common in galactosemia, as caused by the accumulation of this substance in the lens. Elevated galactose 1-phosphate also causes a reduction in the adenosine triphosphate level of the cells. Galactose 1-phosphate is stereochemically similar to glucose 1-phosphate. It is utilized by phosphoglucomutase, and prevents the enzyme from converting glucose 1-phosphate to glucose 6-phosphate. When the metabolism of glucose 1-phosphate is inhibited, glycogen breakdown to glucose is delayed and the blood glucose level remains lower than in normal individuals. Galactose 6-phosphate produced by phosphoglucomutase from galactose 1-phosphate competes with glucose 6-phosphate and causes a reduction of glucose 6-phosphatase activity, which again influences blood glucose levels. The reduced tissue glucose level alters the production of uridine diphosphoglucose which is not only required for the metabolism of galactose, but is essential in many biochemical processes. Uridine diphosphogalactose is used to synthetize cerebrosides in the white matter of the brain. The reduction of uridine diphosphogalactose and consequent hypoglycemia, due to abnormal glucose metabolism, may be associated with some clinical manifestations involving the nervous system, in particular, mental retardation.

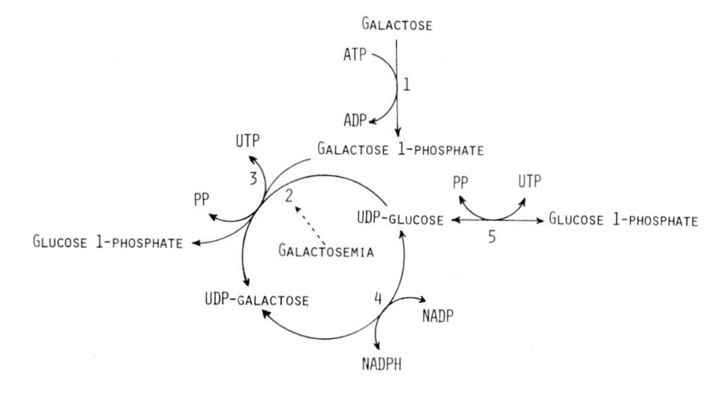

FIGURE 8. Mechanism of galactose metabolism and its relationship with galactosuria. Galactose is converted to galactose 1-phosphate catalyzed by galactokinase (1) in the presence of ATP. The conversion of galactose 1-phosphate to glucose 1-phosphate involves the participation of uridine diphosphoglucose by galactose 1-phosphate uridyl transferase (2). From UTP uridine diphosphoglucose is also formed with the loss of pyrophosphate catalyzed by pyrophosphorylase (3). Uridine diphosphoglucose and uridine diphosphogalactose are interconverted by a reductive pathway (4). Uridine glucose pyrophosphorylase (5) can synthetize uridine diphosphoglucose from glucose 1-phosphate. The enzymatic defect in galactosemia is connected with the deficiency of galactose 1-phosphate uridyl transferase (2) present in human erythrocytes. (Data from Isselbacher, K. J., Anderson, E. P., Kurahashi, I., and Kalckar, H. M., *Science*, 123, 635, 1956.)

Due to the deficient availability of uridine diphosphoglucose, galactosemic patients often develop jaundice. This compound is the precursor of the uridine diphosphoglucuronic acids required for the elimination of bilirubin. Uridine diphosphoglucuronic acid deficiency is accompanied by hyperbilirubinemia. In severe form there is a marked retention of bile in the liver — the bile canaliculi are blocked with bile plugs or thrombi. Consequently, cirrhosis and destruction of liver parenchymal cells may occur.

4. Glucose 6-Phosphate Dehydrogenase Deficiency

This is a frequently occurring condition due to the deficiency of glucose 6-phosphate dehydrogenase in red cells. The disease only becomes apparent, and clinical symptoms such as hemolysis of the erythrocytes only develop, if these patients are exposed to certain drugs. The antimalarial, primaquine, was the first example of drug action, but many other drugs and other compounds cause hemolysis in sensitive persons, including sulfonamides, sulfones, aminopyrine, acetanilid, fava bean products, and viral and bacterial toxins. Many drugs can activate oxidizing processes and produce methemoglobinemia independently from glucose-6-phosphate dehydrogenase deficiency. Kidney and liver disease increase sensitivity.

In association with glucose 6-phosphate, dehydrogenase deficiency and subsequent enzymatic reactions of the production of NADPH are reduced — which is the main consequence of this disorder. There are other abnormalities in the erythrocytes, but the disease manifests itself only when these patients are exposed to primaquine or any other drug that is known to activate the abnormal process.

5. Glycogen Storage Disease

These are connected with a specific enzyme defect associated with glycogen metabolism resulting in structural alterations and deposition of glycogen in the tissue.[16,59,61,64] Currently, there are at least eight different types of storage diseases known where glycogen is deposited

Table 9
DISEASES OF GLYCOGEN STORAGE (GLYCOGENOSIS)

Classification (Cori type)	Enzyme defect	Organ affected	Glycogen structure	Name of disease
I	Glucose 6-phosphatase	Liver, kidney, intestine	Normal	Von Gierke
II	α-1,4-Glucosidase (acid maltase)	Generalized	Normal	Pompe
III	Amylo-1,6-glucosidase (debrancher)	Liver, heart, leukocytes, erythrocytes	Abnormal; outer chains missing, or very short	Forbes
IV	Amylo (1,4 → 1,6) transglucosidase (brancher)	Liver	Abnormal; very long unbranched chains	Andersen
V	Phosphorylase	Muscle	Normal	McArdle-Schmid Pearson
VI	Phosphorylase	Liver, leukocytes	Normal	Hers
VII	Phosphofructokinase	Muscle		
	Phosphorylase b kinase	Liver	Normal	

in excess (Table 9). These are classified according to the specific enzyme defect and the resulting alterations in the structure of the accumulated tissue glycogen. The biochemical lesion in each of these diseases is due to the genetically determined absence or deficiency of a single enzyme concerned with the degradation or structural modification of glycogen. As a consequence of the deficiency, glycogen is stored in some or all organs.[88]

Glycogen is a polydisperse polysaccharide (Figure 9) present in almost all tissues of the body. Certain tissues appear to have specialized systems connected with the biotransformation of glycogen and, therefore, it is necessary to characterize the metabolism in each tissue independently. Two major enzymes, namely, phosphorylase and debranching enzyme (amylo-1,6-glucosidase) are responsible for the complete degradation of the glycogen molecule to glucose 1-phosphate. Nonphosphorolytic pathways also exist — hydrolytic enzymes, amylases, glucosidases in the liver and muscle — but their role in glycogen storage diseases has not been established.

The amount of glycogen in a cell represents the equilibrium between the rate of synthesis and breakdown (Figure 10). Several reactions are involved in the activation of phosphorylase which initiates the degradation of glycogen (Figure 11). Glycogen synthetase is also activated by a kinase enzyme. It is dependent on the presence of glucose 6-phosphate and inactivated by phosphatase.

Many factors regulate tissue glycogen. Its major place of storage is the liver. Fasting causes rapid depletion, whereas diets high in carbohydrate rapidly fill up the stores; hyperalimentation causes excessive deposition, even in normal subjects. Hormones and other regulatory factors can influence glycogen metabolism including insulin, glucagon, epinephrine, pituitary hormones, thyroid hormones, and adrenal and ovarian steroids. Insulin enhances utilization of glycogen in the muscle. In the presence of adequate amounts of insulin somatotrophin promotes glycogen storage in the liver. Adrenal steroids decrease peripheral uptake of glucose and increase glycogen deposition in the liver by affecting gluconeogenesis from protein. Epinephrine and glucagon are also powerful activators of gluconeogenesis in the liver and this action is the basis of tests applied for the diagnosis of glycogen storage diseases. Thyroid hormones accelerate hepatic glycogen depletion and increase the blood

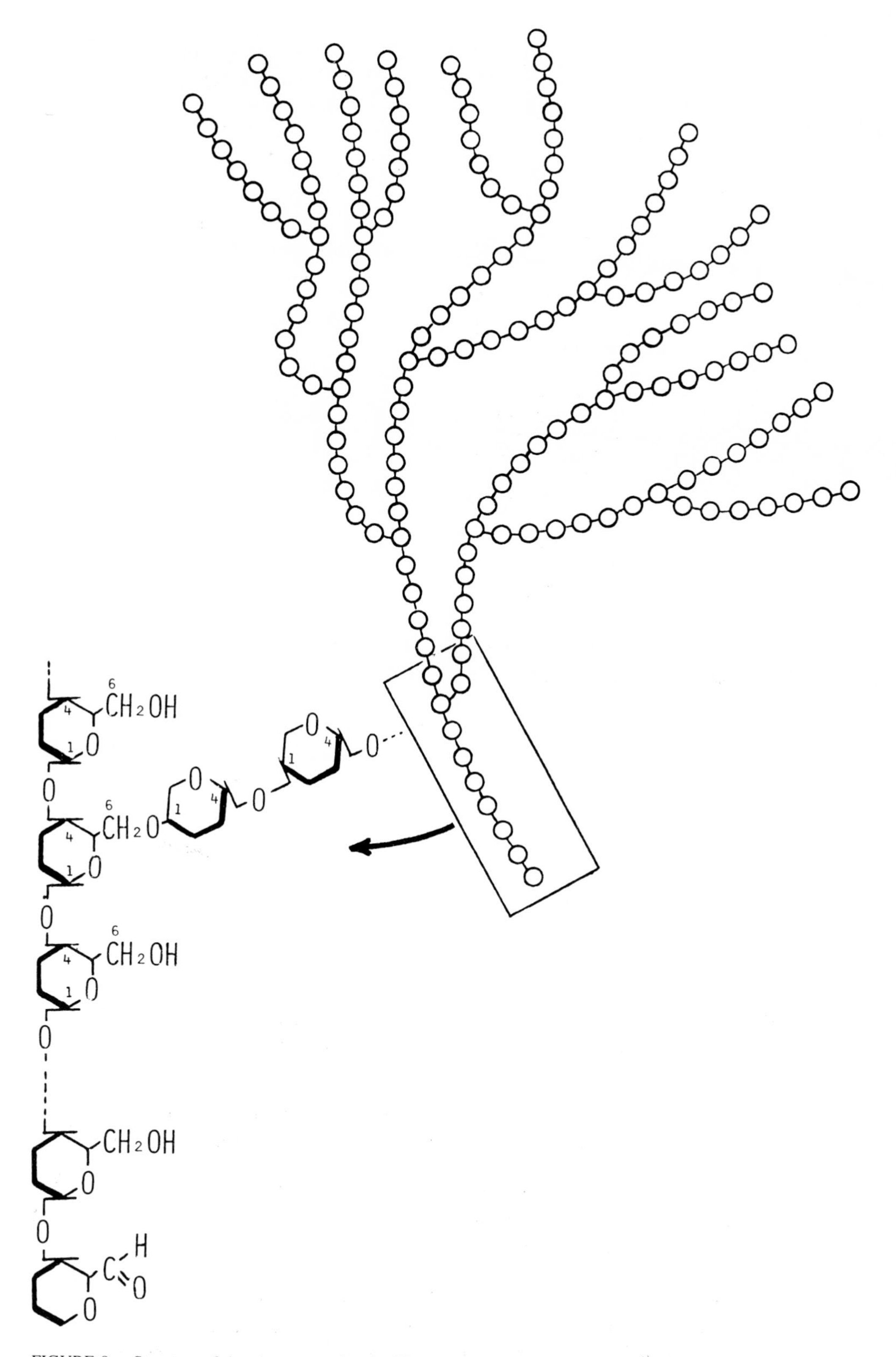

FIGURE 9. Structure of the glycogen molecule. The branching in the structure is more variable than illustrated; the ratio of 1,4 to 1,6 bonds varies between 12 and 18. Section in the box shows the enlargement of the structure at a branch point and the primary glucose residue with the free reducing aldehyde group.

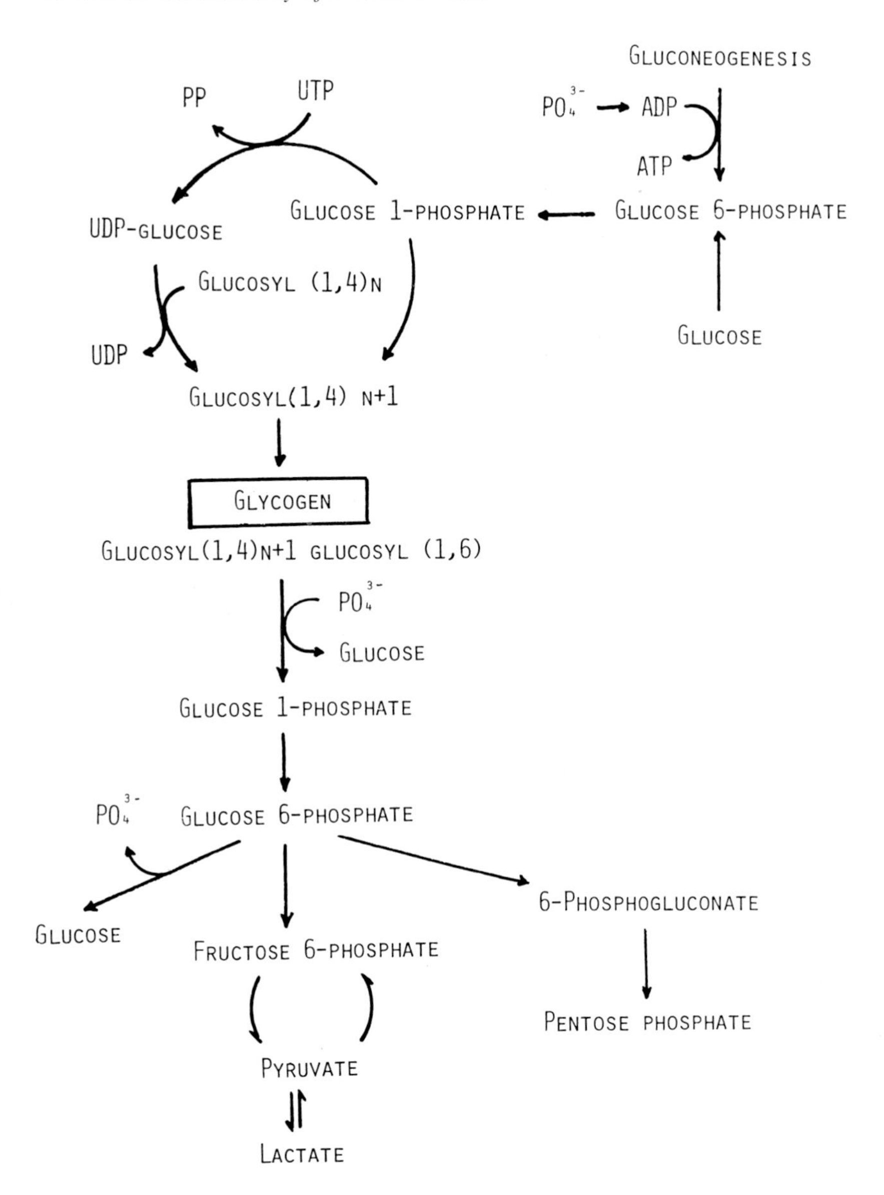

FIGURE 10. Basic mechanisms regulating the synthesis and metabolism of glycogen. Top part of the scheme represents pathways of glyconeogenesis, bottom part shows the degradation. These processes manifest in liver and muscle.

glucose level. In contrast, progesterone brings about an increased deposition of glycogen in various tissues. Some diseases may contribute to the variations in hepatic storage, such as liver cirrhosis and diabetes.

There are several forms of glycogen storage disease or glycogenesis (Table 9). Von Gierke's disease or Cori Type I glycogenesis was the first established form. This disease is due to the congenital absence of glucose 6-phosphatase (Figure 12). Glycogen is not metabolized in this disease and the liver synthesizes glycogen on the account of blood glucose, which is subsequently reduced. The excess glycogen produced is accumulated in the liver and spleen, causing enlargement of these organs. Intestinal function is impaired and renal tubules are also affected by this disease and parenchymal cells are swollen and filled with glycogen. Generally, metabolic processes dependent on glycogen are inhibited in Von Gierke's disease (Plate 8).

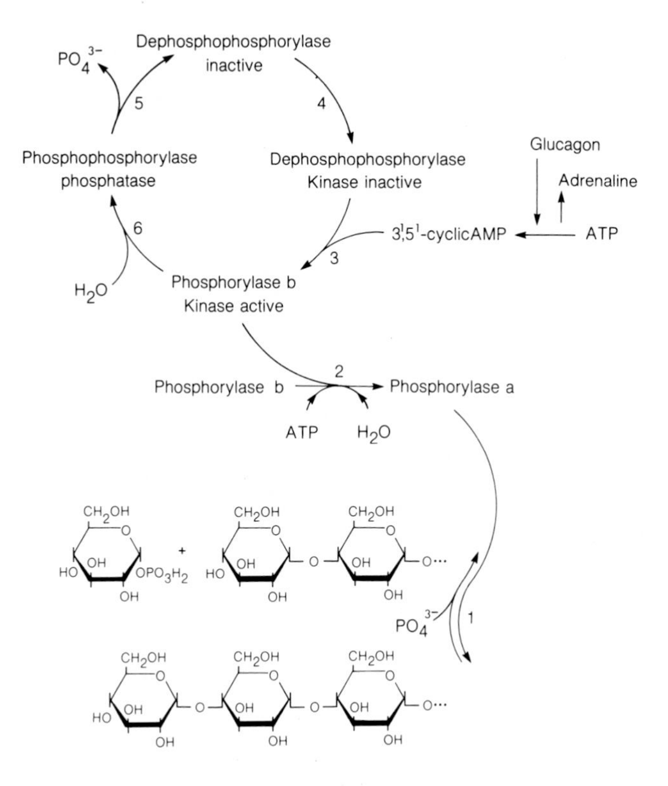

FIGURE 11. Mechanism of phosphorylase action. Phosphorylase a (1) is responsible for addition and breaking of the α-1,4 linkage in the glycogen molecule. This enzyme is a tetramer containing four molecules of pyridoxal phosphate per molecule of protein. It is produced from the precursor phosphorylase b activated by phosphorylase b kinase (2). Inactivation of phosphorylase b requires four molecules of water, which splits the tetramer into two dimers. Conversely, recondensation of the dimers into an active tetramer requires four molecules of ATP. Phosphorylase b kinase is derived from an inactive form in the presence of 3′5′ cyclic AMP by phosphorylation catalyzed by dephosphophosphorylase kinase (3). This enzyme is also present in inactive form and the activation probably represents a conformational changes (4) since both the active and inactive forms have the same molecular weight. Inactive dephosphophosphorylase is formed from phosphophosphorylase phosphatase by the cleavage of a phosphate molecule (5) which in turn is probably derived from phosphorylase b kinase by hydrolysis (6). (From Baranowsky, T., Illingworth, B., Brown, D. H., and Cori, C. F., *Biochim. Biophys. Acta*, 25, 16, 1957; Krebs, E. G., Kent, A. B., and Fisher, E. H., *J. Biol. Chem.*, 231, 73, 1958.) Liver and muscle both contain these systems, but they are biochemically and immunologically different from each other. (Data from Henion, W. F. and Sutherland, E. W., *J. Biol. Chem.*, 224, 477, 1957.)

The clinical signs of this storage disease appear very early in life and are attributed directly or indirectly to severe episodes of hypoglycemia. The blood glucose level is sometimes only 10 mg/dℓ. The mortality is high, but if the affected children survive the first 4 years some adaptation occurs through metabolic processes although the enzyme defect still remains permanent. In glucose 6-phosphatase deficiency the liver sometimes contains a glycogen content of more than 6% of its weight. The hypertrophied liver causes abdominal discomfort. All hepatic function tests are normal, with the exception of those involving carbohydrate metabolism. The kidney is also enlarged, due to the accumulated glycogen, but the enlargement has no recognizable effect on renal function. The frequent episodes of prolonged hypoglycemia bring about convulsions and mental retardation. Often the brain utilizes other available substrates for its function or glucose uptake may be increased by a raised cerebral blood flow.

There are several less frequently occurring forms of glycogen storage disease. One form of glycogen storage disease is connected with the absence of glycogen debranching enzyme,

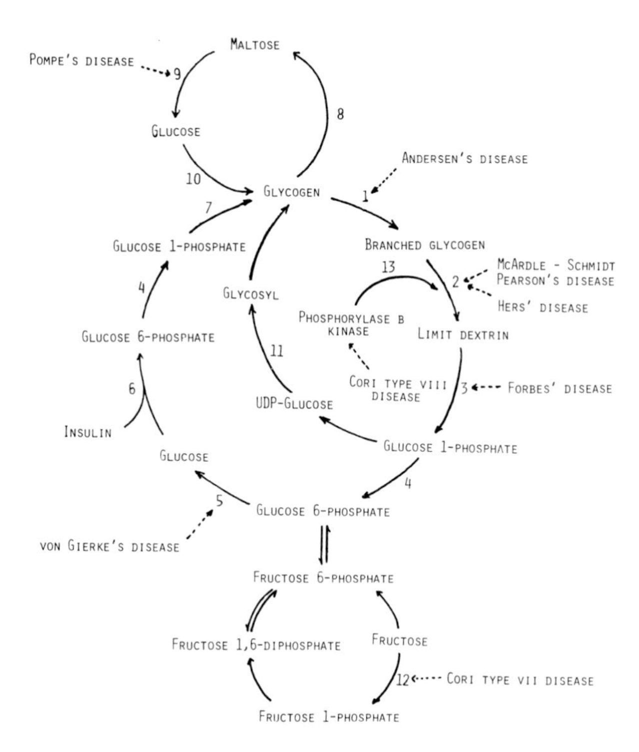

FIGURE 12. Mechanisms of glycogen storage diseases. Enzymes involved in these schematic pathways are amylo-(1,4 → 1,6)-transglucosidase (1), phosphorylase (2) amylo-1,6-glucosidase (3), phosphoglucomutase (4) glucose 6-phosphatase (5), hexokinase (6), glycogen neogenesis (7), α-glucosidase (8), acid maltase (9), amylo-1,4-glucosidase (10), uridine diphosphoglucose glycogen transglucosidase (11). Various sites of genetically determined enzyme defects causing excessive glycogen deposition are connected with glycogenosis. These are Andersen's disease or brancher deficiency amylopectinosis. The defect causes accumulation of abnormal, very long, embranched glycogen in the liver. Lesion of myophosphorylase is connected with McArdle-Schmid-Pearson's disease; lesion of hepatophosphorylase is connected with Hers' disease. These phosphorylase deficiency glycogenoses are associated with the deposition of normal glycogen in the muscle or in the liver, respectively. Forbes' disease or debrancher deficiency limit dextrinosis is connected with the accumulation of abnormal glycogen with missing outer chain in liver, muscle, and erythrocytes. In Von Gierke's disease, normal glycogen is stored in the kidney and liver. Pompe's disease is associated with generalized disease and accumulation of glycogen in several organs. Cori Type VII glycogenosis is connected with fructophosphokinase (12) deficiency and accumulation of glycogen in muscle. Phosphorylase b kinase (13) deficiency leads to the accumulation of normal glycogen in the liver. (Data from Cori, G. T., *Mod. Probl. Paediat.*, 3, 344, 1958; Hers, H. G., *Advan. Metab. Disord.*, 1, 1, 1964.)

amylo-1,6-glycosidase, and is known as Forbes' or Cori Type III disease. Patients suffering from this disorder are able to metabolize glycogen partially by the hydrolysis of the 1,4-glycoside linkage which form the main skeleton of the macromolecule, but cannot hydrolyze the 1,6-glycoside bond. Some glycogenolysis occurs in these patients and they may maintain a fairly normal blood glucose level. In Forbes' disease, liver and muscle accumulate abnormally high levels of glycogen which contain short outer chains. The clinical symptoms of the disease is manifest in the liver, heart, and skeletal muscle and leukocytes.

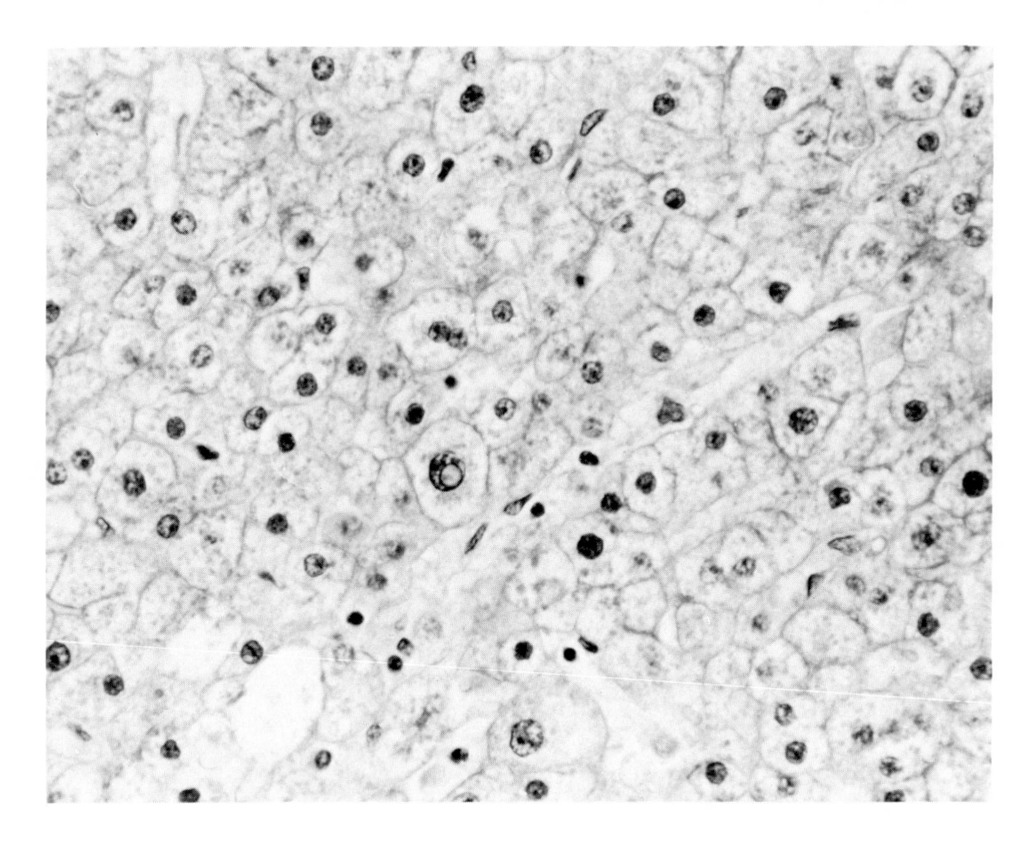

PLATE 8. Von Gierke's disease. Note large vacuolated liver cells with centrally placed nuclei storing glycogen. This accumulation is caused by lack of glucose 6-phosphatase.

Another form is related to the defect of branching enzyme which catalyzes the conversion of amylo-1,4 to the amylo-1,6 linkage; this defect is known as Andersen's or Cori Type IV disease. The glycogen formed in this condition contains abnormally long side chains with relatively little branching. Due to these long chains, the abnormal glycogen is only partially metabolized by phosphorylase. In this condition the most affected organs are the liver, spleen, intestinal mucosa, and lymph nodes.

Hepatic phosphorylase deficiency is associated with Hers' disease and muscular phosphorylase deficiency with McArdle-Schmid-Pearson's disease. In Hers' disease glycogen accumulates in the liver due to reduced phosphorylase activity. This enzyme represents the rate-limiting steps in glycogen breakdown. Low phosphorylase activity has also been found in some patients with increased hepatic glycogen storage. Hers and associates described one case of glyconeogenesis which is associated with the presence of inactive phosphorylase due to the absence of phosphorylase B kinase. Addition of this activating enzyme restores normal phosphorylase activity. In McArdle's disease, glycogen is accumulated in the muscle, resulting from the reduced activity of muscle phosphorylase. Mild myopathy is the characteristic symptom of this condition. In McArdle's disease hepatic phosphorylase is normal, indicating that different genes are responsible for the control of muscle and liver phosphorylase synthesis.

Storage disease also occurs in a generalized form, with predominant accumulation of glycogen in the heart or neuromuscular system. This is Pompe's disease or Cori Type II glycogenesis. If the symptoms manifest in the heart a crisis of cyanosis and polypnea develop soon after birth and results in death from heart failure. In the neuromuscular form of glycogenesis, muscular hypotonia and mental retardation are the main symptoms. The liver

Table 10
INBORN ERRORS OF LIPID METABOLISM

Disease	Biochemical manifestation or enzyme defect	Clinical symptoms
Triglyceridemia	Lipoprotein lipase	Increased incidence of atherosclerosis
Hypercholesterolemia		Xanthomatosis, high incidence of coronary atherosclerosis
Ganglioside storage	Ganglioside metabolism	Mental retardation, convulsions, death
Sphingomyelin storage	Sphingomyelin metabolism	Mental retardation

and kidney also contain glycogen which is structurally normal. Glycogen accumulations in the liver produce certain morphological structures which resemble the formation of secondary lysosomes. In Pompe's disease, the accumulation of glycogen, therefore, is probably the consequence of the absence of acid α-glucosidase, a specific lysosomal enzyme involved in the breakdown of glycogen.

Glycogen storage diseases are often followed by related symptoms. Prolonged hypoglycemia causes secondary disturbances of lipid metabolism and synthesis in the liver. Lipolysis is increased in the adipose tissue. In response to hypoglycemia, epinephrine, growth hormone, and cortisol are released. These substances also stimulate lipolysis. The increased amounts of fatty acids and glycerol increases cholesterol synthesis, cholesterol level in the blood, and deposition in the skin. Sometimes, more fat is stored in the liver cell than glycogen.

When the liver capacity for oxidizing fatty acids is exceeded, various metabolites accumulate and cause metabolic acidosis. The lactic acid concentration produced by the muscle and other tissue is also high because its conversion to glycogen is impaired. Acidosis and hyperlactic acidemia cause further metabolic complications; they compete with uric acid for renal excretion and subsequent deposition of uric acids in the large joints and skin — thus gout may develop as a side effect.

F. Defects of Lipid Metabolism

Inherited disorders in lipid or lipoprotein metabolism may cause abnormal increases of certain lipids or reduction and absence of essential lipoproteins in the plasma.[12,22,29,34] Other sequelae are the deposition of abnormal amounts of lipids in various tissues, especially in the brain (Table 10).

Plasma lipoprotein abnormalities known generally as familial hyperlipoproteinemia may be separated according to the lipoprotein pattern. These are associated with various enzyme deficiencies and independent metabolic and environmental factors. Only in one type of hyperlipoproteinemia has the relationship to the lack of an enzyme been firmly established. In fat-induced hypertriglyceridemia the removal of dietary fat present as chylomicrons from the blood is defective. This is connected with the deficiency of plasma lipolytic activity due to the lack of lipoprotein lipase. Xanthemas, pancreatitis, lipemia, and liver and spleen enlargements are the frequently accompanying clinical complications.

Tangier disease, a rare lipoprotein deficiency, is characterized by a very low level or the complete absence of high-density lipoproteins. Cholesterol is stored in several tissues and lipid deposition is often accompanied by enlargements of the spleen, liver and lymph nodes, and vascular abnormalities. The function of the high-density lipoprotein is associated with the mobilization of cholesterol and cholesterol esters from various tissues (see Atherosclerosis, in Volume III); the lack of this carrier, therefore, precipitates the disease, particularly in the foam cells of the reticuloendothelial tissues.

Abetalipoproteinemia is also an infrequent disease, and is characterized by the absence of low-density lipoproteins in the plasma. Fatty acid and phospholipid content of the blood is also abnormal. The altered plasma composition and the presence of unusual erythrocytes precipitate episodes of hemolysis. In abetalipoproteinemia, the cerebellum progressively degenerates causing weakness, ataxia, and downhill progress which leads to severe disability. The probable biochemical lesion of this disease is manifest in the inability to synthesize apoprotein molecules specific for low-density lipoproteins.

Lipid storage diseases are usually fatal. Death occurs by the age of 3 to 4 years. These include ganglioside lipidoses: Tay-Sachs disease, Niemann-Pick disease, Gaucher's disease, Fabry's disease, and leukodystrophies. Since the characteristic pathological changes are mostly restricted to changes in membranes in the nervous system, these diseases are presented in a later chapter.

REFERENCES

1. **Ambrecht, H. J.,** Changes in hepatic microsomal membrane fluidity with age, *Exp. Gerontol.*, 17, 41, 1982.
2. **Arias, I. M., Doyle, D., and Schimke, R. T.,** Studies on the synthesis and degradation of proteins of the endoplasmic reticulum of rat liver, *J. Biol. Chem.*, 244, 3303, 1969.
3. **Ashworth, L. A. E. and Green, C.,** Plasma membranes phospholipids and sterol content, *Science*, 151, 210, 1966.
4. **Bailey, J. M. and Dunbar, L. M.,** Essential fatty acid requirement of cells in tissue culture. A review, *Exp. Mol. Pathol.*, 18, 142, 1973.
5. **Blobel, G. and Dobberstein, B.,** Transfer of proteins across membranes. II. Reconstitution of functional rough microsomes from heterologous components, *J. Cell Biol.*, 67, 852, 1975.
6. **Bretscher, M. S.,** Asymmetrical lipid bilayer structure for biological membranes, *Nature New Biol.*, 236, 11, 1972.
7. **Bruckdorfer, K. R. and Graham, J. M.,** The exchange of cholesterol and phospholipids between cell membranes and lipoproteins, in *Biological Membranes*, Vol. 3, Chapman, D. and Wallach, D. F. H., Eds., Academic Press, New York, 1976, 103.
8. **Capaldi, R. A.,** The structure of mitochondrial membranes, in *Mammalian Cell Membranes*, Vol. 2, Jamieson, G. A. and Robinson, D. M., Eds., Butterworths, Reading, Mass., 1977, 141.
9. **Carlson, S. A. and Gelehrter, T. D.,** Hormonal regulation of membrane phenotype, *J. Supramol. Struct.*, 6, 325, 1977.
10. **Caspar, D. L. D. and Kirschner, D. A.,** Myelin membrane structure at 10 Å resolution, *Nature New Biol.*, 231, 46, 1971.
11. **Chakrabarti, P. and Khorana, H. G.,** A new approach to the study of phospholipid-protein interactions in biological membranes. Synthesis of fatty acids and phospholipids containing photosensitive groups, *Biochemistry*, 14, 5021, 1975.
12. **Chapman, D.,** Some recent studies of lipids, lipid-cholesterol and membrane systems, in *Biological Membranes*, Vol. 2, Chapman, D. and Wallach, D. F. H., Eds., Academic Press, New York, 1973, 152.
13. **Chauhan, V. P., Shikka, S. C., and Kalra, V. K.,** Phospholipid methylation of kidney cortex bush border membranes, *Biochim. Biophys. Acta*, 688, 357, 1982.
14. **Cherlow, B. S.,** The role of lysosomes and proteases in hormone secretion and degradation, *Endocrinol. Rev.*, 2, 137, 1981.
15. **Clarke, J. T.,** The glycosphingolipids of human plasma lipoproteins, *Can. J. Biochem.*, 59, 412, 1981.
16. **Collipp, P. J., Chen, S. Y., Maddaiah, V. T., Thomas, J., and Huijing, F. J.,** Diagnosis of glycogen storage disease type I by serum immunodiffusion, *Pediatrics*, 53, 71, 1974.
17. **Colombo, B. and Martinez, G.,** Hemoglobinopathies including thalassemia. II, *Clin. Hematol.*, 10, 730, 1981.
18. **Danks, D. M.,** Inborn errors of metabolism. A review of some general concepts, *Aust. N.Z. J. Med.*, 11, 309, 1981.
19. **De Pierre, J. W. and Ernster, L.,** Enzyme topology of intracellular membranes, *Ann. Rev. Biochem.*, 46, 201, 1977.

20. **Dhami, M. S. I., de la Iglesia, F. A., and Feuer, G.,** Fatty acid content and composition of phospholipids from endoplasmic reticulum in developing rat liver, *Res. Commun. Chem. Pathol. Pharmacol.*, 32, 99, 1981.
21. **Dhami, M. S. I., Feuer, C. F., and Feuer, G.,** Fatty acid changes in the hepatic endopasmic reticulum during pregnancy in the rat, *Res. Commun. Chem. Pathol. Pharmacol.*, 23, 383, 1979.
22. **DiMauro, S. and Eastwood, A. B.,** Disorders of glycogen and lipid metabolism, *Adv. Neurol.*, 17, 123, 1977.
23. **Emmelot, P.,** The organization of the plasma membrane of mammalian cells structure in relation to function, in *Mammalian Cell Membranes,* Vol. 2, Jamieson, G. A. and Robinson, D. M., Eds., Butterworths, Reading, Mass., 1977, 1.
24. **Farber, J. L.,** Biology of disease: membrane injury and calcium homeostasis in the pathogenesis of coagulative necrosis, *Lab. Invest.*, 47, 114, 1982.
25. **Farias, R. N., Bloj, B., Morero, R. D., Sineriz, F., and Trucco, R. E.,** Regulation of allosteric membrane-bound enzymes through changes in membrane lipid composition, *Biochim. Biophys. Acta,* 415, 231, 1975.
26. **Fernandes, J., Koster, J. F., Grose, W. F. A., and Sorgedrager, N.,** Hepatic phosphorylase deficiency. Its differentiation from other hepatic glycogenoses, *Arch. Dis. Child.*, 49, 186, 1974.
27. **Ferrero, M. E., Orsi, R., and Bernelli-Zazzera, A.,** Cell repair after liver injury. Membranes of the endoplasmic reticulum ribosomes in post-ischemic livers, *Exp. Mol. Pathol.*, 32, 32, 1980.
28. **Feuer, G.,** Role of phospholipids in the development of the hepatic endoplasmic reticulum associated with drug metabolism, *Res. Commun. Chem. Pathol. Pharmacol.*, 22, 549, 1978.
29. **Fisher, K. A.,** Analysis of membrane halves cholesterol, *Proc. Natl. Acad. Sci. U.S.A.*, 73, 173, 1976.
30. **Froesch, E. R.,** Disorders of fructose metabolism, *Clin. Endocrinol. Metabol.*, 5, 599, 1976.
31. **Froesch, E. R.,** Essential fructosuria and hereditary fructose intolerance, in *The Metabolic Basis of Inherited Disease,* 3rd ed., Stanbury, J. B., Wyngaarden, J. B., and Fredrickson, D. S., Eds., McGraw-Hill, New York, 1972.
32. **Gahmberg, C. G.,** Cell surface proteins changes during cell growth and malignant transformation, in *Dynamic Aspects of Cell Surface Organization,* Poste, G. and Micolson, G. S., Eds., Elsevier/North-Holland, Amsterdam, 1977, 371.
33. **Gelbart, D. R., Brewer, L. L., Fajardo, L. F., and Weinstein, A. B.,** Oxalosis and chronic renal failure after intestinal bypass, *Arch. Intern. Med.*, 137, 239, 1977.
34. **Glomset, J. A. and Verdery, R. B.,** Role of LCAT, in *Cholesterol Metabolism and Lipolytic Enzymes,* Polonovski, J., Ed., Masson Publ., New York, 1977, 136.
35. **Goldberg, D. M.,** Hepatic protein synthesis in health and disease with special reference to microsomal enzyme induction, *Clin. Biochem.*, 13, 216, 1980.
36. **Gray, M. W. and Doolittle, W. F.,** Has the endosymbiont hypothesis been proven? *Microbiol. Rev.*, 46, 1, 1982.
37. **Greequest, A. C. and Shohet, S. B.,** Phosphorylation in erythrocyte membranes from abnormally shaped cells, *Blood,* 48, 877, 1976.
38. **Grunfeld, J. P., Ganeval, D., Chanard, J., Fardeau, M., and Dreyfus, J. C.,** Acute renal failure in McArdle's disease. Report of two cases, *N. Engl. J. Med.*, 286, 1237, 1972.
39. **Haest, C. W. M., Fischer, T. M., Plasa, G., and Deuticke, B.,** Stabilization of erythrocyte shape by a chemical increase in membrane shear stiffness, *Blood Cells,* 6, 539, 1980.
40. **Hatefi, Y.,** The enzyme and the enzyme complexes of the mitochondrial oxidative phosphorylation system, in *The Enzymes of Biological Membranes,* Vol. 4, Martonosi, A., Ed., Plenum Press, New York, 1976, 3.
41. **Hanover, J. A. and Lennarz, W. J.,** Transmembrane assembly of membrane and secretory glycoproteins, *Arch. Biochem. Biophys.*, 211, 1, 1981.
42. **Harry, D. S.,** Lipids, lipoproteins and cell membranes, *Prog. Liver Dis.*, 7, 319, 1982.
43. **Hayashi, H. and Capaldi, R. A.,** The proteins of the outer membrane of beef heart mitochondria, *Biochim. Biophys. Acta,* 282, 166, 1972.
44. **Hinton, R. H. and Reid, H.,** Enzyme distribution in mammalian membranes, in *Mammalian Cell Membranes,* Vol. 1, Jamieson, G. A., and Robinson, D. M., Eds., Butterworths, Reading, Mass., 1976, 161.
45. **Holloway, C. T. and Garfield, S. A.,** Effect of diabetes and insulin replacement on the lipid properties of the hepatic smooth endoplasmic reticulum, *Lipids,* 16, 525, 1981.
46. **Honey, N. K., Miller, A. L., and Shows, T. B.,** The mucolipidoses identification by abnormal electrophoretic patterns of lysosomal hydrolases, *Am. J. Med. Genet.*, 9, 239, 1981.
47. **Hosey, M. M. and Tao, M.,** Altered erythrocyte membrane phosphorylation in sickle cell disease, *Nature (London),* 263, 424, 1976.
48. **Hynes, R. O.,** Cell surface proteins and malignant transformation, *Biochim. Biophys. Acta,* 458, 73, 1976.

49. **Hynes, R. O., Destree, A. T., and Mautner, V.,** Spatial organization at the cell surface, *Progr. Clin. Biol. Res.*, 9, 189, 1976.
50. **Jahnig, G.,** Critical effects from lipid-protein interaction in membranes, *Biophys. J.*, 36, 347, 1981.
51. **Kenyon, K. R.,** Conjuctival biopsy for diagnosis of lysosomal disorder, *Prog. Clin. Biol. Res.*, 82, 103, 1982.
52. **Kleinig, H.,** Nuclear membranes from mammalian liver. II. Lipid composition, *J. Cell. Biol.*, 46, 396, 1970.
53. **Koenig, R. J.,** Glycohemoglobins in the adult erythrocyte, *Curr. Top. Hematol.*, 2, 59, 1979.
54. **Korn, E.,** Cell membranes structure and synthesis, *Ann. Rev. Biochem.*, 38, 263, 1969.
55. **Krebs, J. J.,** The topology of phospholipids in artificial and biological membranes, *J. Bioenerg. Biomembr.*, 14, 141, 1982.
56. **Lee, T. C., Stephens, N., Moehl, A., and Snyder, F.,** Turnover of rat liver plasma membrane phospholipids. Comparison with microsomal membranes, *Biochim. Biophys. Acta*, 291, 86, 1973.
57. **Lee, T. C. and Snyder, F.,** Phospholipid metabolism in rat liver endoplasmic reticulum. Structural analyses turnover studies and enzyme activities, *Biochim. Biophys. Acta*, 291, 71, 1973.
58. **LeFevre, P. A.,** The degranulation test. Six tests for carcinogenicity, *Br. J. Cancer*, 37, 937, 1978.
59. **Lubran, M. M.,** McArdle's disease. A review, *Ann. Clin. Lab. Sci.*, 5, 115, 1975.
60. **MacLennan, D. H. and Holland, P. C.,** The calcium transport ATPase of sarcoplasmic reticulum, in *The Enzymes of Biological Membranes*, Vol. 3, Martonosi, A., Ed., Plenum Press, New York, 1976, 221.
61. **Mahler, R. F.,** Disorder of glycogen metabolism, *Clin. Endocrinol. Metab.*, 5, 579, 1976.
62. **Malhotra, S. K.,** Molecular structure of biological membranes functional characterization, *Subcell. Biochem.*, 5, 221, 1978.
63. **Mancinella, A.,** Enkephalins and endorphins, peptide neurotransmitters regulating cerebral homeostasis: biological actions, biochemical properties, therapeutic prospects, *Clin. Teratol.*, 100, 401, 1982.
64. **McAdams, A. J., Hug, G., and Bove, K. C.,** Glycogen storage disease Types I to X, *Hum. Pathol.*, 5, 463, 1974.
65. **Mehler, M. and DiMauro, S.,** Late onset acid maltase deficiency. Detection of patients and heterozygotes by urinary enzyme assay, *Arch. Neurol.*, 33, 692, 1976.
66. **Melnick, R. L., Tinberg, H. M., Maguire, J., and Packer, L.,** Studies on mitochondrial proteins. I. Separation and characterization by polyacrylamide gel electrophoresis, *Biochim. Biophys. Acta*, 311, 230, 1973.
67. **Morre, D. J., Kartenbeck, J., and Franke, W. W.,** Membrane flow and interconversions among endomembranes, *Biochim. Biophys. Acta*, 559, 71, 1979.
68. **Nicolson, G. L.,** Transmembrane control of the receptors on normal and tumor cells. I. Cytoplasmic influence over cell surface components, *Biochim. Biophys. Acta*, 457, 57, 1976.
69. **Paine, A. J.,** Hepatic cytochrome P-450, *Essays Biochem.*, 17, 85, 1981.
70. **Palek, J., Church, A., and Faibanks, G.,** Transmembrane movements and distribution of calcium in normal and hemoglobin S erythrocytes, in *Membranes and Disease*, Bolis, L., Hoffman, J. F., and Leaf, A., Eds., Raven Press, New York, 1976, 41.
71. **Parry, G.,** Membrane assembly and turnover, *Subcell. Biochem.*, 5, 261, 1978.
72. **Perheentupa, J., Raivio, K. O., and Nikkita, E. A.,** Hereditary fructose intolerance, *Acta Med. Scand.*, Suppl. 542, 65, 1972.
73. **Pontremoli, S., Melloni, E., Michetti, M., Sparatore, B., and Horecker, B. L.,** Localization of two lysosomal proteinases on the external surface of the lysosomal membrane, *Bichem. Biophys. Res. Commun.*, 106, 903, 1982.
74. **Racker, E. and Eytan, E.,** A coupling factor from sarcoplasmic reticulum required for the translocation of Ca^{2+} ions in a reconstituted Ca^{2+} ATPase pump, *J. Biol. Chem.*, 250, 7533, 1975.
75. **Rex, A. P. and Preus, M. A.,** A diagnostic index for Down syndrome, *J. Pediatr.*, 100, 903, 1982.
76. **Richardson, J. C. and Agutter, P. S.,** The relationship between the nuclear membrane and the endoplasmic reticulum in interphase cells, *Biochem. Soc. Trans.*, 8, 459, 1980.
77. **Rose, I. A. and Warms, J. V. B.,** Control of glycolysis in the human red blood cell, *J. Biol. Chem.*, 241, 4848, 1966.
78. **Roseman, M., Litman, B. J., and Thompson, T. E.,** Transbilayer exchange of phosphatidylethanolamine for phosphatidylcholine and A-acetimidoylphosphatidylethanolamine in single-walled bilayer vesicles, *Biochemistry*, 14, 4826, 1975.
79. **Rosenberg, R. N.,** Biochemical genetics of neurologic disease, *N. Engl. J. Med.*, 305, 1181, 1981.
80. **Rothman, J. E. and Kennedy, E. P.,** Rapid transmembrane movement of newly synthesized phospholipids during membrane assembly, *Proc. Natl. Acad. Sci. U.S.A.*, 74, 1821, 1977.
81. **Rouser, G., Nelson, G. J., Fleischer, S., and Simon, G.,** Lipid composition of animal cell membrane organelles and organ, in *Biological Membranes Physical Fact and Function*, Chapman, D., Ed., Academic Press, New York, 1968, 5.

82. **Rousselet, A., Guthmann, C., Matricon, J., Bienvenue, A., and Devaux, P. F.,** Study of the transverse diffusion of spin labeled phospholipids in biological membranes. I. Human red blood cells, *Biochim. Biophys. Acta,* 426, 357, 1976.
83. **Shih, V. E., Levy, H. L., Karolkewicz, V., Houghton, S., Efron, M. L., Isselbacher, K. J., Beutler, E., and MacCready, R. A.,** Galactosemia screening of newborns in Massachusetts, *N. Engl. J. Med.,* 284, 753, 1971.
84. **Siekevitz, P.,** Biological membranes, the dynamics of their organization, *Ann. Rev. Physiol.,* 34, 117, 1972.
85. **Singer, S. J.,** The fluid mosaic model of membrane structure, in *Structure of Biological Membranes,* Abrahamsson, S. and Pascher, I., Eds., Plenum Press, New York, 1977, 443.
86. **Singer, S. J., Ash, J. F., Bourguignon, L. Y. W., Heggeness, M. H., and Louvard, D.,** Transmembrane interactions and the mechanism of transport of proteins across membranes, *J. Supramol. Struct.,* 9, 373, 1978.
87. **Spector, A.,** Fatty acid glyceride and phospholipid metabolism, in *Growth, Nutrition and Metabolism of Cells in Culture,* Vol. 1, Rothblat, G. and Cristafalo, V., Eds., Academic Press, New York, 1972, 257.
88. **Starzl, T. E., Putnam, C. W., Porter, K. A., Halgrimson, C. G., Corman, J., Brown, B. I., Gotlin, R. W., Rodgerson, D. O., and Green, H. L.,** Portal diversion for the treatment of glycogen storage disease in humans, *Ann. Surg.,* 178, 525, 1973.
89. **Steck, T. L. and Fox, C. F.,** Membrane proteins, in *Molecular Biology,* Fox, C. F. and Keith, A. D., Eds., Sinauer Associates, Stamford, Conn., 1972, 27.
90. **Steck, T. L.,** Cross-linking the major proteins of the isolated erythrocyte membrane, *J. Mol. Biol.,* 66, 295, 1972.
91. **Tanner, M. J. A.,** Erythrocyte glycoproteins, *Curr. Top. Membranes Transp.,* 11, 279, 1978.
92. **Van Hoeven, R. P. and Emmelot, P.,** Studies on plasma membranes, *J. Membrane Biol.,* 9, 105, 1972.
93. **Varsanyi, M. and Heilmeyer, L. M. G.,** The protein kinase properties of calsequestrin, *FEBS Lett.,* 103, 85, 1979.
94. **Vincenzi, F. F. and Hinds, T. R.,** Plasma membrane calcium transport and membrane-bound enzymes, in *The Enzymes of Biological Membranes,* Vol. 3, Martonosi, A., Ed., Plenum Press, New York, 1976, 261.
95. **Wasi, P.,** Hemoglobinopathies including thalassemia. I, *Clin. Hematol.,* 10, 707, 1981.
96. **Watson, M. L.,** The nuclear envelope. Its structure and relation to cytoplasmic membranes, *J. Biophys. Biochem. Cytol.,* 1, 257, 1955.
97. **Weber, A. L. and Miller, S. L.,** Reasons for the occurrence of the twenty coded protein amino acids, *J. Mol. Evol.,* 17, 273, 1981.
98. **Weed, R. I., LaCelle, P. L., and Merrill, E. W.,** Metabolic dependence of red cell deformability, *J. Clin. Invest.,* 48, 795, 1969.
99. **Williams, R. M. and Chapman, D.,** Phospholipids, liquid crystals and cell membranes, *Prog. Chem. Fats,* 11, 1, 1970.
100. **Wolf, B.,** Propionic acidemia: a clinical update, *J. Pediatr.,* 99, 835, 1981.

FURTHER READINGS

Andreoli, T. E., Hoffman, J. F., and Fanestil, D. D., Eds., *Physiology of Membrane Disorders,* Plenum Press, New York, 1978.

Bittar, E. E., Ed., *Cell Biology in Medicine,* John Wiley & Sons, New York, 1973.

Capaldi, R. A., Ed., *Membrane Proteins and Their Interactions With Lipids,* Marcel Dekker, New York, 1977.

Cohen, G. N., *The Regulation of Cell Metabolism,* Holt, Rinehart & Winston, New York, 1968.

Hommes, F. A. and Van den Berg, C. J., Eds., *Inborn Errors of Metabolism, Academic Press, New York, 1973.***Stanbury, J. B., Wyngaarden, J. B., and Fredrickson, D. S., Eds.,** *The Metabolic Basis of Inherited Disease, 4th ed., McGraw-Hill, New York, 1978.***Wallach, D. F. H.,** *Proteins of Animal Cell Plasma Membranes, Eden Press, Westmount, Quebec, 1979.*

Chapter 2

ABNORMALITIES OF PROTEIN SYNTHESIS AND METABOLISM

I. INTRODUCTION

All living organisms contain proteins, which are essential components of life. These are fundamental constituents of every cell and are involved in their structure and function. Proteins are complex substances, of molecular weights ranging from several thousand to several million daltons, composed of simple amino acid units. The hydrolysis of proteins isolated from human or animal tissue yields 23 different amino acids (Figure 1). Whereas some proteins are made up from all 23, in some proteins 1 or 2 amino acids are missing, while others are constructed from the combination of only a few amino acids. Various other amino acids fulfill important roles in metabolic processes in free or combined states or as constituents of proteins (Figure 2). Smaller units composed of a limited number of amino acids are also found in various tissues: these are the peptides. Many peptides are synthesized as end products, others are intermediates in protein synthesis. Several peptides play essential biologic roles and possess an important pharmacologic activity. In proteins and peptides, amino acids are linked through the specific peptide bond to form chains of varying lengths. There are great variations in the frequency of each amino acid in the protein molecule as well as in the combination of proteins with lipids, carbohydrates, and nucleic acids, producing lipoproteins, glycoproteins, or nucleoproteins. In most proteins, variations in the amino acid sequence of the peptide chain do not bring about any essential changes in the biological properties and functions. However, there are many examples where the replacement of a single amino acid can drastically alter the characteristics and role of the given molecule and thus may lead to disease.[89,172,183]

Proteins are involved in many ways in our biochemical processes.[119,133,137] Complex lipoproteins serve as structural elements and take part in the construction of the cells and cellular organelles as separate units.[106,159,161] Glycoproteins are involved in the formation of cell surfaces in mammalian tissues and are important constituents of many tissues. Proteins, in association with nucleic acids, are involved in the basic formation of body constituents, in particular in the synthesis of *molecules* responsible for the activity of tissues and organs in the form of enzymes. Enzyme systems are the key molecules in the maintenance of homeostatic control over the normal distribution of cell constituents throughout life. Impairment of homeostasis is associated with abnormal changes in protein synthesis and abnormalities in enzyme formation.[60,80,123] Proteins regulate the storage of many crucial materials such as oxygen, lipids, and many electrolytes, and direct their transport to the site of utilization.[166] They are involved in transport processes necessary for the excretion from the cell of breakdown products which are formed during metabolism as well as in the catalysis of the biotransformation of foreign compounds, and carry out the elimination of these foreign substances by detoxification mechanisms.[37,58]

Proteins form antibodies essential for the defense against infection.[82,148] Many hormones which regulate various processes in our body are simple amino acid derivatives, small peptides, or proteins.[65] Basic hereditary characteristics of each individual are dependent on the activity and structure of complex nucleoproteins. Proteins have many diverse functions and disease may follow an impairment of any of these functions.[88,103] There are, further, interrelationships between the various types of proteins. For example, a structural unit may possess catalytic properties, or may participate in transport processes. In some cases some enzyme action can be substituted by another enzyme — various pathways of glucose 6-phosphate degradation can transform glucose 6-phosphate to other products.

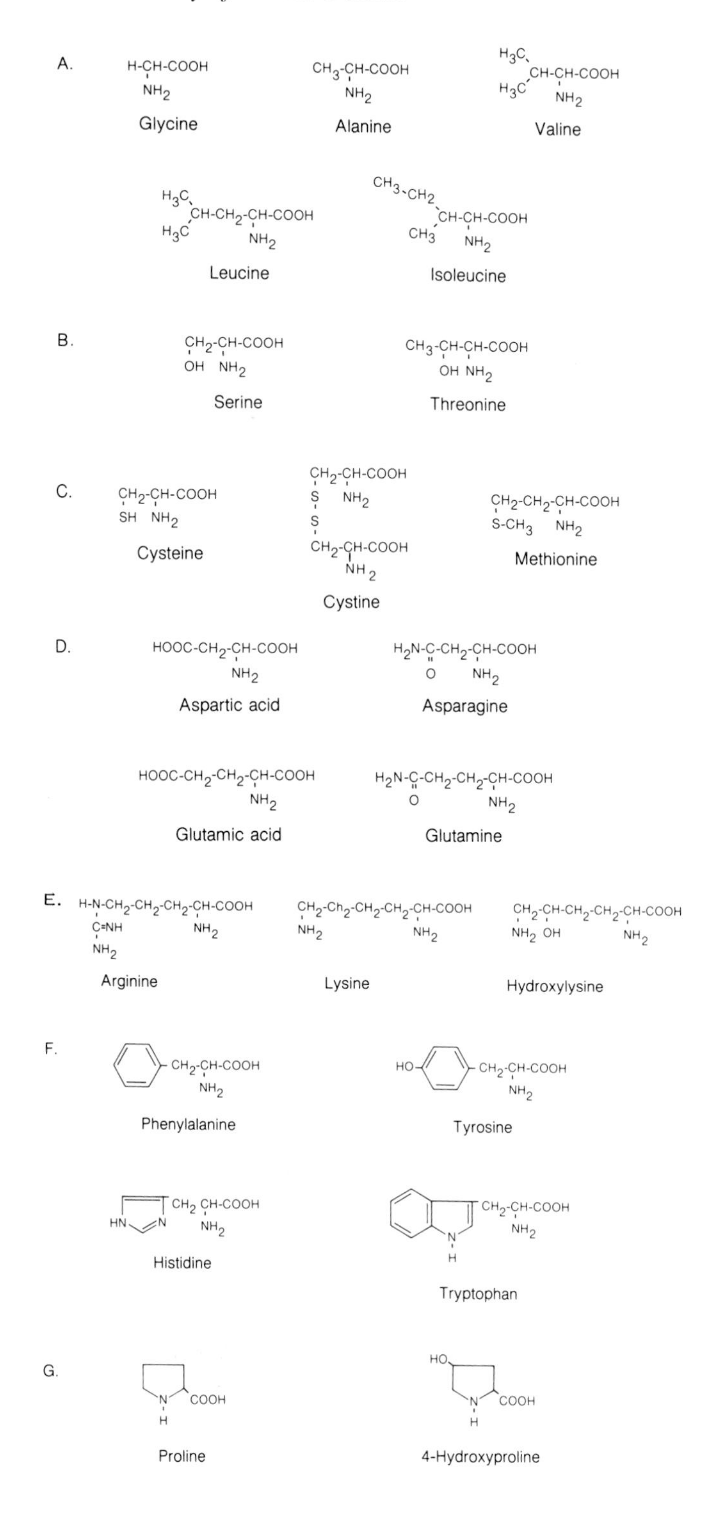

FIGURE 1. Major L-α-amino acids found in proteins. Various groups include amino acids A: with aliphatic side chains; B: with side chains containing hydroxyl groups; C: with side chains containing sulfur atom; D: with side chains containing acid groups or acidamides; E: with side chains containing basic groups; F: with aromatic rings, and G: with pyrrolidine ring.

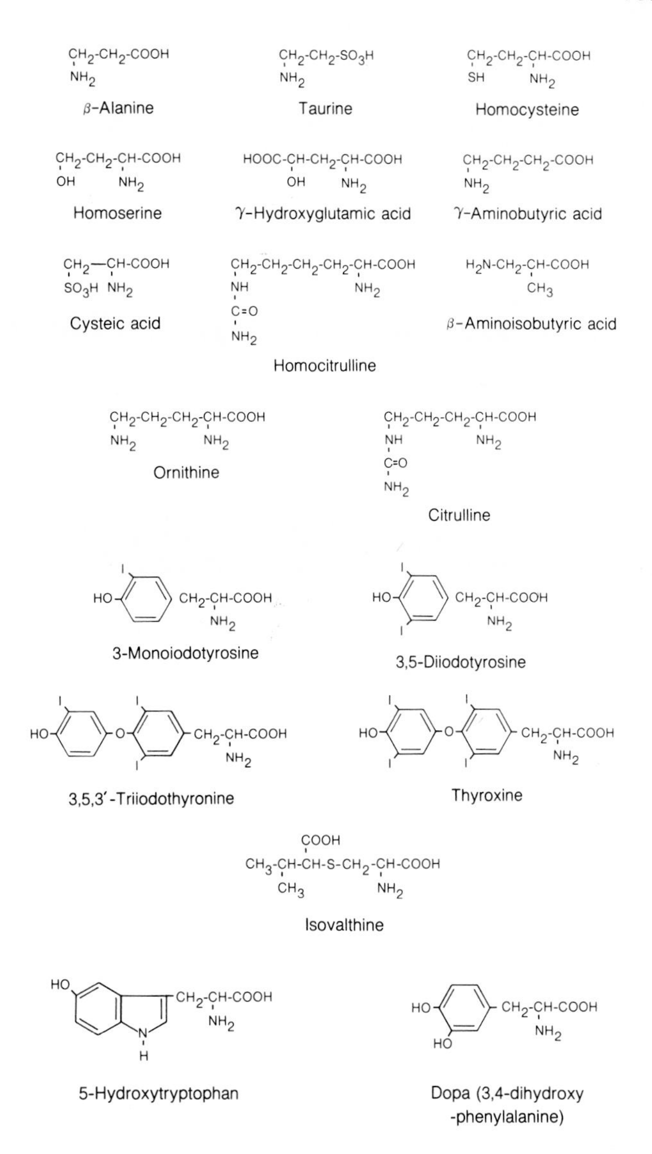

FIGURE 2. Naturally occurring acids which are not present in proteins. Some occurs as part of coenzymes (β-alanine: coenzyme A and pantetheine); or combined with bile acids (taurine: taurocholic acid); intermediate in the biosynthesis (homocysteine: methionine or metabolism of amino acids (homoserine: threonine, methionine, and aspartate); γ-hydroxyglutamic acid: 4-hydroxyproline; γ-aminobutyric acid: glutamine in brain tissue; cysteic acid: cysteine in hair; homocitrulline: in the urine of normal children); or pyrimidine (β-aminoisobutyric acid: in the urine of patients with metabolic disorder); intermediate in the urea cycle (ornithine, citrulline); present in thyroid tissue and serum from the thyroid gland (monoiodotyrosine, diiodotyrosine, triiodothyronine, thyroxine); occurs in certain hypothyroid patients (isovalthine); or precursors of pharmacologically active derivatives (5-hydroxytryptophan: serotonin; dihydroxyphenylalanine: melanin).

Proteins and peptides are synthesized from amino acids. Although we can produce amino acids from intermediates of carbohydrate and lipid metabolism, the major source of amino acids for protein synthesis comes from natural foods containing proteins. These can be absorbed directly from the intestinal tract in infants. In normal adults the major source of body proteins are di- and tripeptides and amino acids released by hydrolysis from dietary proteins which enter the circulation. Lack of protein in the diet causes many serious symptoms

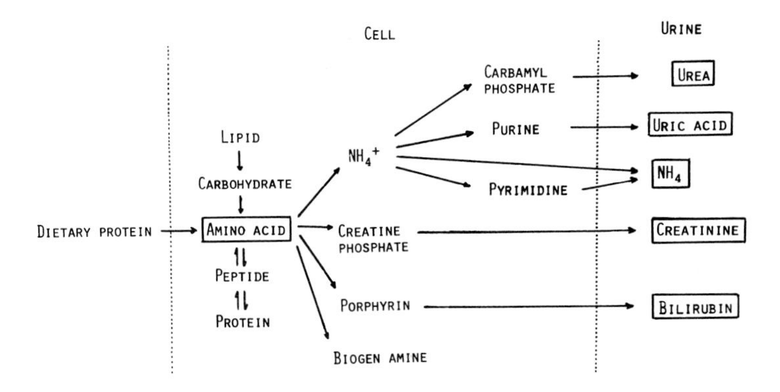

FIGURE 3. Scheme of protein and amino acid metabolism. This figure represents only the major routes of metabolism. Metabolites are eliminated in substantial quantities in the urine.

of protein deficiency, particularly in newborn babies, pregnant and lactating women, and in older people.[108] Abnormally low intake of dietary protein results in inadequate amounts of essential amino acids and, subsequently, a limitation of the production of body proteins which leads to serious metabolic derangements.

Amino acids are not only necessary for the synthesis of proteins, but specific ones are also important in the production of a variety of nitrogen-containing metabolites: creatine, porphyrines, purines, pyrimidines, and several biologically active amines (Figure 3). Catecholamines, histamine, serotonin, thyroxine, melatonin, and urea-cycle intermediates are formed from specific amino acids. Cysteine and methionine provide sulfur in taurine, coenzyme A, and other sulfur-containing compounds, organic sulfate esters (which are formed in the liver from various steroids), indoles, phenols, and many drugs. Methionine is also important as a methylating agent in the synthesis of lecithin, acetylcholine, epinephrine, and creatine.

II. AMINO ACIDS

A. Body Pool and Nitrogen Balance

Amino acids are indispensable for the synthesis of proteins.[4] Some of these are essential building stones and have to be supplied by our diet, some of them are not found in dietary proteins in appreciable amounts yet play important roles in metabolism (Table 1). Proteins of our body are continually hydrolyzed to amino acids and resynthesized. The rate of endogenous protein synthesis is 80 to 100 g/day. Turnover is highest in the intestinal mucosa and lowest in collagenous proteins. The breakdown of exogenous and endogenous proteins to amino acids follows the same pattern. Together, both sources comprise the body's amino acid pool, supplying our total need. This amino acid pool is not a storage depot of substantial quantity, but is composed of the entire circulating amino acids in the blood and small amounts present in the tissues. If the dietary protein intake and metabolism are constant, the amino acid pool involved in protein production does not change much since, in the kidney, almost the total amount of amino acids is reabsorbed.[50] During growth more proteins are synthesized and the equilibrium between the amino acids and proteins shifts toward the latter. A small amount of protein is lost as hair, unreabsorbed amino acids from protein breakdown products in the feces, and protein hormones in the urine. There are some diseases where aminoaciduria is present, e.g., Fanconi syndrome, due to congenital defects in the tubular reabsorption capability of the kidney, or where, in association with an inborn error due to the lack of special enzymes, certain amino acids are not metabolized.[43,53,81]

Table 1
ESSENTIAL AND NONESSENTIAL AMINO ACIDS

Essential	Nonessential	Derivative
Valine	Glycine	
Leucine	Alanine	
Isoleucine	Serine	
Threonine	Cysteine	
Methionine	Aspartic acid	Asparagine
Phenylalanine	Glutamic acid	Glutamine
Tryptophan	Tyrosine	
Lysine	Hydoxylysine	
Arginine	Proline	
Histidine	Hydroxyproline	
	Citrulline	

Note: These data are mainly derived from growth-rate experiments in animals[185,186] since it is difficult to exclude any essential amino acids from the human diet without endangering health. However, this classification covers well man's requirement. In the human organism essential amino acids are not synthesized in adequate amounts and must be obtained from dietary sources. The first eight amino acids in the table are required during the entire life. Arginine and histidine are essential in childhood. If the diet is deficient in phenylalanine, tyrosine is necessary. In premature babies some enzyme systems which synthesize special amino acids may not be present at birth; these infants, therefore, need cysteine and tyrosine.[187]

Normally, the intake of a small amount of dietary protein is required to replace the loss of amino acids and proteins. This need can be met by feeding pure amino acids. The loss of protein and derivatives in the stool is very small, less than 2 g daily. Nitrogen excreted in the urine appears mainly in the form of urea. When renal function is normal this can be used as a reliable index to establish the balance of protein breakdown. Other nitrogen-containing compounds in the urine are derived from the metabolism of creatine phosphate, purines, pyrimidines, and porphyrins. Nitrogen balance is in equilibrium when the eliminated amino acid nitrogen in the urine is equal to the nitrogen content of the protein in the diet. In a normal individual, in order to maintain the normal balance if increased amounts of protein are digested, the extra amino acids are eliminated and urea production is increased. However, in some circumstances nitrogen losses may exceed intake, representing a negative nitrogen balance.[56,169,170] These conditions occur during starvation, at terminal stages of many diseases, in cachexia, or when the secretion of insulin is decreased or that of catabolic hormones of the adrenal cortex is increased.[74] In contrast, during growth or recovery from severe disease or as a consequence of anabolic steroid administration, the excretion of nitrogen is reduced and intake is greater, thus resulting in a positive nitrogen balance.

B. Essential Amino Acids

Since the amino acid composition of the various human proteins differs from that of proteins present in food, it is necessary that they be rearranged in the body by many metabolic routes. Moreover, in the human body, nitrogen balance is not maintained without the adequate intake of essential amino acids (Table 1). Some essential amino acids, such as histidine and arginine, are only required for growth during infancy. Other amino acids can be synthesized from carbohydrate and lipid metabolites by amine transfer reactions. All these sources and the rate of synthesis are sufficient to provide the proteins necessary in a normal individual.

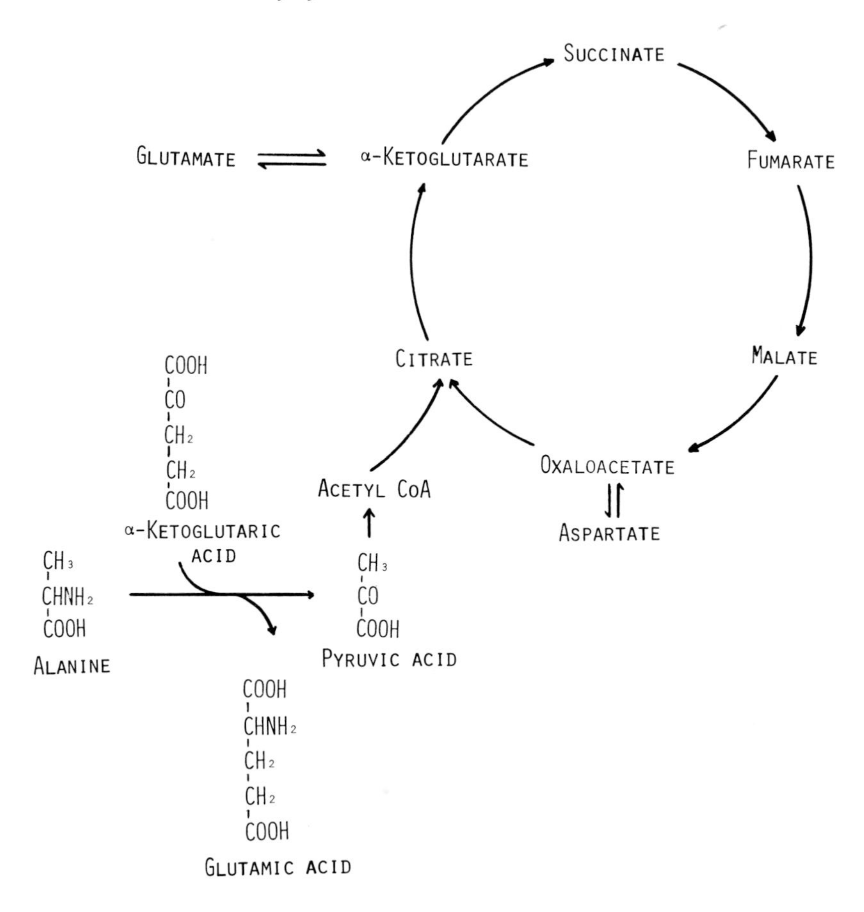

FIGURE 4. Relationship between amino acid synthesis and the citric acid cycle.

Lack of essential amino acids causes sickness and eventually death.[81,151] When an essential amino acid necessary for the synthesis of a particular protein is absent from the amino acid pool, that protein is not formed. The other amino acid constituents of this protein, therefore, represent a surplus; they are metabolized and excreted as urea and a negative nitrogen balance develops. Some diseases are associated with this condition. Protein-calorie deficiency results in either marasmus or Kwashiorkor, depending on the composition of the diet. Marasmus is the consequence of total calorie deficiency.[56,102] In Kwashiorkor, calorie intake may be adequate, but the diet is deficient of protein. The common occurrence of Kwashiorkor in tropical countries is the consequence of diets deficient in animal protein. This disorder is especially severe in children and characterized by growth failure, enlarged fatty liver, anemia, edema, and dermatosis. Inadequate protein intake also brings about anemia and other disturbances related to the absence of important proteins.

During starvation, proteins are catabolyzed for energy and the excretion of urea nitrogen is greatly reduced. Most protein depletion occurs in the liver, spleen, and skeletal muscle — relatively little in the brain and heart. Liver glycogen is reduced preceding or accompanying the depletion of proteins and the blood glucose level falls; however, no hypoglycemic symptoms are produced indicating that the glucose is still maintained above a certain level. Neutral fat is rapidly metabolized and the onset of ketosis is apparent. When fat stores are exhausted, the catabolism of proteins is even further enhanced and soon results in death.

C. Disorders of Amino Acid Metabolism

Several amino acids are synthesized from the intermediates of the Krebs cycle by transamination reactions (Figure 4). The transaminases function with pyridoxal phosphate as

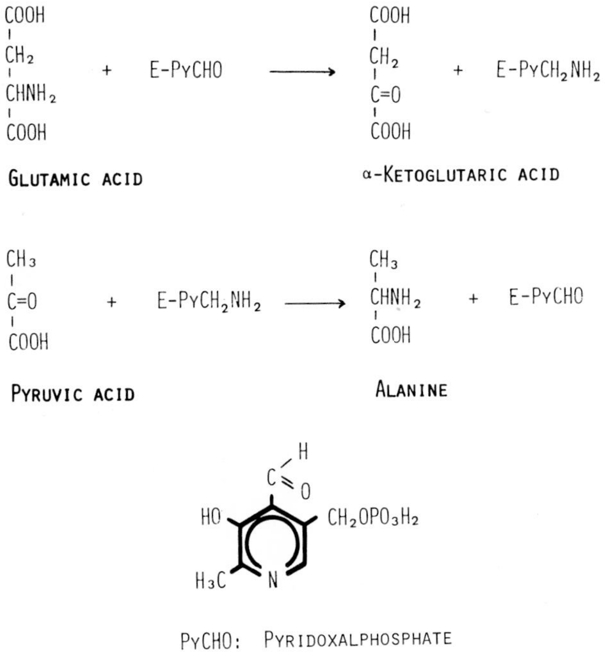

$$\underset{\text{GLUTAMIC ACID}}{\begin{matrix}COOH\\|\\CH_2\\|\\CHNH_2\\|\\COOH\end{matrix}} + \text{E-PyCHO} \longrightarrow \underset{\alpha\text{-KETOGLUTARIC ACID}}{\begin{matrix}COOH\\|\\CH_2\\|\\C{=}O\\|\\COOH\end{matrix}} + \text{E-PyCH}_2\text{NH}_2$$

$$\underset{\text{PYRUVIC ACID}}{\begin{matrix}CH_3\\|\\C{=}O\\|\\COOH\end{matrix}} + \text{E-PyCH}_2\text{NH}_2 \longrightarrow \underset{\text{ALANINE}}{\begin{matrix}CH_3\\|\\CHNH_2\\|\\COOH\end{matrix}} + \text{E-PyCHO}$$

H
C=O
HO
$CH_2OPO_3H_2$
H_3C
N

PyCHO: PYRIDOXALPHOSPHATE

FIGURE 5. Mechanism of alanine aminotransferase enzyme system. Pyridoxal phosphate coenzyme participates in the transfer of the alanine amino group to pyruvic acid.

coenzyme and the enzymes derived from various tissues may have different substrate specificities.[86,134] Some of these enzymes are important in clinical diagnosis; elevated levels indicate release from damaged tissues. Inadequate pyridoxal intake in the diet or abnormal synthesis of pyridoxal phosphate can bring about impairment of aspartate or glutamate synthesis catalyzed by transaminases. Some drugs, most notably isoniazid, phenothiazine derivatives, and cefazolin occasionally produce low serum transaminase levels. Their action may be related to the trapping of tissue pyridoxal by their hydrazine groups and, hence, the decreased availability of the coenzyme (Figure 5 and Figure 6). In contrast there are drugs which cause an elevated serum transaminase activity associated with overt jaundice. In this instance, particular attention has been paid to the hepatic action of contraceptives. These adverse reactions on tissue transaminase levels and amino acid metabolism are related to changes in the metabolic control of mitochondria where several intracellular compartments may be affected. Due to the impairment, cytoplasmic components may not be able to penetrate mitochondrial membranes, consequently the synthesis of amino acids which can be derived from Krebs cycle intermediates by simple transamination becomes inefficient (Figure 7).

Methionine has a major role in the transfer of methyl groups to produce many important compounds. In the hepatic endoplasmic reticulum, methionine adenosyl transferase catalyzes the formation of *S*-adenosylmethionine, which is the principal methyl donor in our body. The methyl group of the activated methionine is transferred to a variety of acceptor molecules (Figure 8). The formation of membrane-bound phosphatidylcholine is essential for the function of membrane-bound enzymes. Drugs and hepatotoxic compounds can adversely influence this process. Lack of methionine or other methyl donor is associated with fatty degeneration of the liver.

Both arginine and histidine are considered as nonessential amino acids for the adult, but both are required for the rapid growth of infants. It is probable that the synthesis of arginine from glutamate via ornithine and citrullin is not fully developed in infancy. Dietary histidine

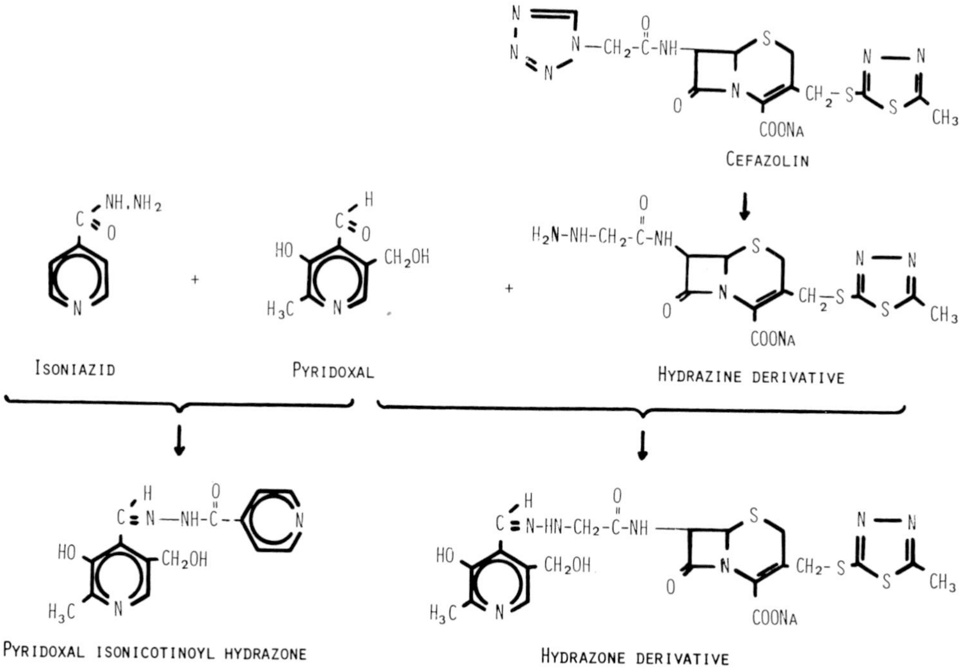

CEFAZOLIN

ISONIAZID + PYRIDOXAL + HYDRAZINE DERIVATIVE

PYRIDOXAL ISONICOTINOYL HYDRAZONE

HYDRAZONE DERIVATIVE

FIGURE 6. Coupling reaction between pyridoxal and isoniazid or cefazolin.

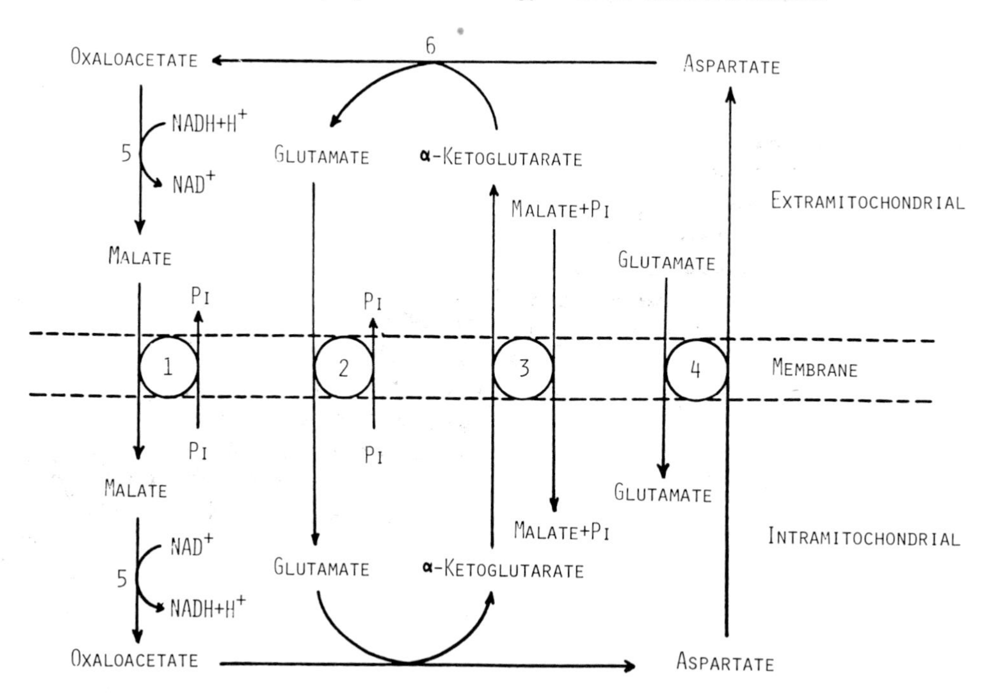

FIGURE 7. Interrelationship between translocase enzyme systems and the regulation of transaminase activity. Translocases are present in the inner membrane of most mitochondria. These enzymes organize the movement of many substances into or out of these cellular organelles; they include malate (1), glutamate (2) translocases which function with inorganic phosphate counterions, α-ketoglutarate (3), and another glutamate (4) translocase which function with malate and inorganic phosphate or aspartate counterions, respectively. The action of the four translocases is synchronized. The concerted effect is dependent on the activity of malate dehydrogenase (5) and glutamate oxaloacetate transaminase (6). Defect of transaminase activity brings about an impairment of amino acid synthesis and transport through the membrane as well as a disturbance of many mitochondrial functions.

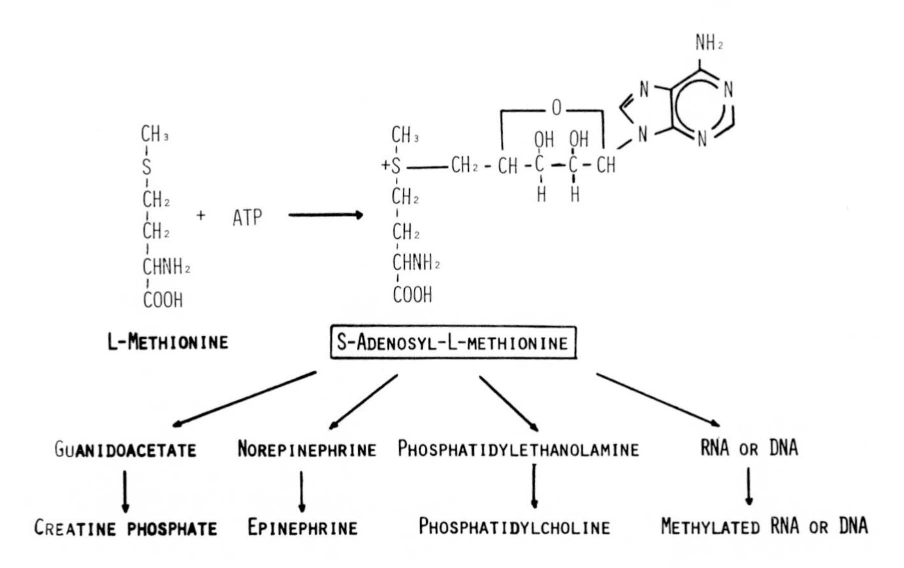

FIGURE 8. Role of methionine in methyl transfer reactions.

is stored in great amounts in muscle in the form of carnosine and anserine. The infant is not able to mobilize enough derivatives from these stores to satisfy the demand of growth. The metabolism of tryptophan is important in the synthesis of alanine, acetyl-CoA, nicotinic acid (niacin), and the production of nicotinamide adenosine dinucleotide (Figure 9). This indicates the importance of the growth-supporting role of tryptophan. Pellagra is caused by the deficiency of niacin and can be treated by tryptophan which is converted to niacin; however, the rate is inadequate to maintain good health. During pregnancy there is an increased transformation of tryptophan, but the synthesis requires pyridoxal phosphate as coenzyme of kynureninase which catalyzes an intermediate reaction. Lack of these important vitamins creates an impairment in the metabolism of this amino acid and complications of pellagra.

The final breakdown products of amino acids are the ammonium ions.[84] These are formed by catabolism, but they are also released from amino acids in the intestines by bacterial degradation and they are removed through the liver. In the portal blood, ammonium ion concentration is high, while its level is low in the peripheral blood. It is detoxified in the liver, otherwise the very toxic ammonium ions exert their adverse action on the central nervous system. In severe liver disease the elimination of ammonium ions is decreased and the higher blood concentration may cause brain disorder. The central nervous system also has the capacity to detoxify the ammonium ion by converting it to glutamine via glutamine synthetase. Other tissues also contain this enzyme, and the elevated glutamine levels found in the blood after the ingestion of protein-rich food represent the stores of circulating ammonia. Glutamine is hydrolyzed to glutamic acid and ammonium ions in the kidney; this is an important reaction in acid-base balance. Glutamine is also broken down by hepatic glutaminase to yield ammonium ions which are further converted to urea. Urea is produced in a cycle (Figure 10); this compound is much less toxic than ammonia. Portal obstruction or liver disease can cause a failure in urea synthesis, leading to ammonia intoxication. The symptoms of this complex condition — hepatic coma — are tremors, blurred vision, confusion, and coma leading to death.

Elevated blood ammonia levels are connected with inherited diseases. In these abnormalities, defects occur in various enzymes of the urea cycle. In congenital hyperammonemia type I, carbamyl phosphate synthetase is missing; in type II, ornithine transcarbamylase is missing. Citrullinuria and argininosuccinic acidemia are associated with the lack of enzymes

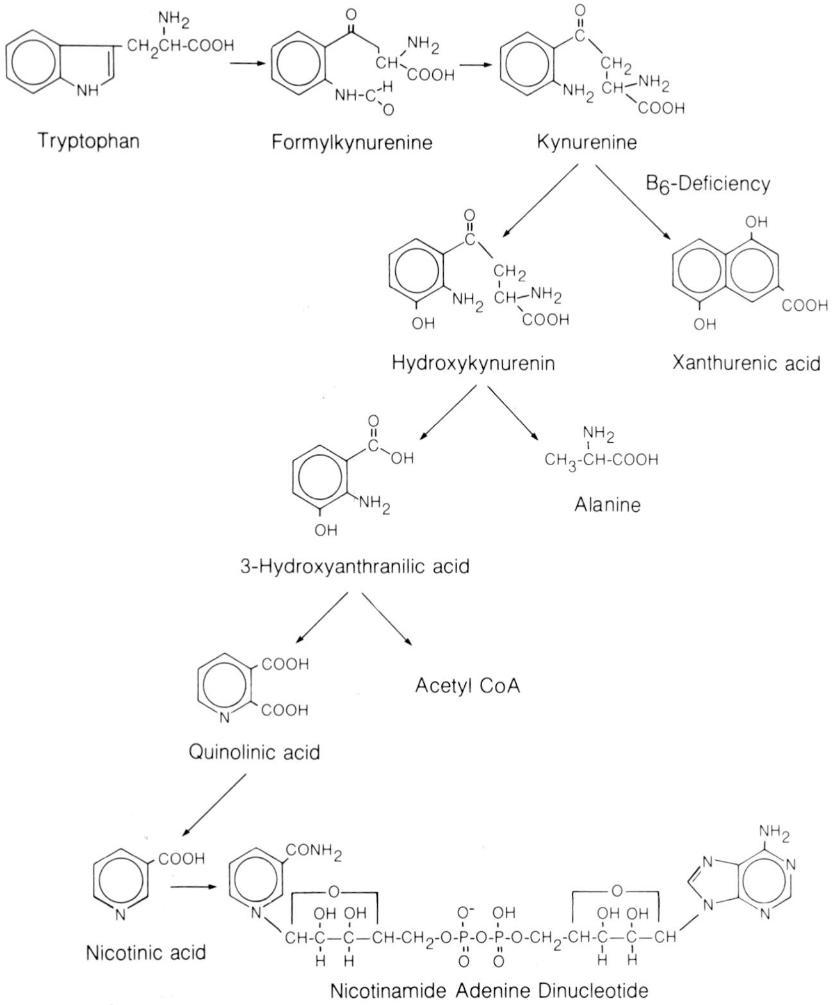

Tryptophan
Formylkynurenine
Kynurenine
B6-Deficiency
Hydroxykynurenin
Xanthurenic acid
Alanine
3-Hydroxyanthranilic acid
Acetyl CoA
Quinolinic acid
Nicotinic acid
Nicotinamide Adenine Dinucleotide

FIGURE 9. Synthesis of nicotinamide adenine dinucleotide from tryptophan.

L-ARGININO SUCCINATE
ATP
L-ASPARTATE
AMP + PPi
L-CITRULLINE
FUMARATE
ATP
CO2 + NH4
CARBAMYL PHOSPHATE
L-ARGININE
ADP + Pi
L-ORNITHIN
UREA

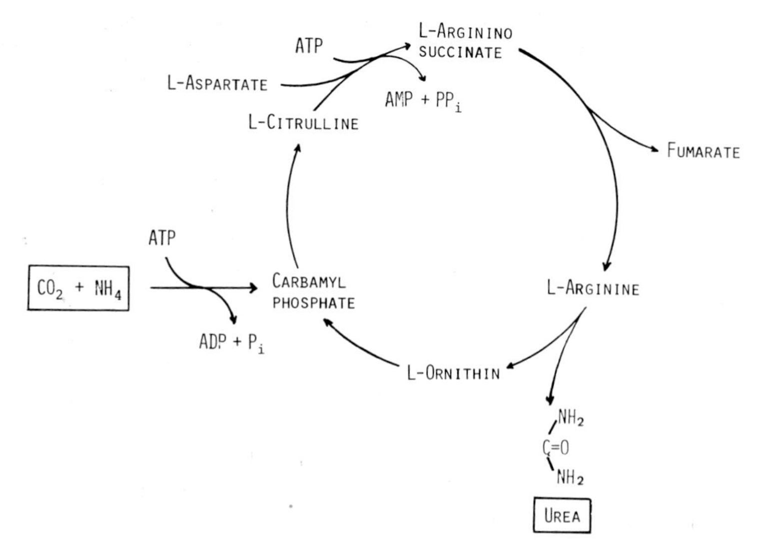

FIGURE 10. Synthesis of urea from amino acid precursors.

which metabolize these substances. Most of these diseases are fatal in the first few years, some can be partially controlled by restricted protein diet and the symptoms — episodes of vomiting and psychomotor retardation — can be reduced by this treatment.

III. PEPTIDES

A. Peptide Hormones

Several biologically active compounds are formed from amino acids.[147] Among these are neurotransmitters: acetylcholine, dopamine, norepinephrine, γ-aminobutyric acid, serotonin; important mediators such as histamine and hormones of the thyroid gland: thyroxine, triiodothyronine; adrenal medulla: epinephrine, melatonin (a hormone derived from the pineal body and peripheral nerves). These compounds are produced by simple reactions, some requiring only a single enzyme action (Figure 11). Neurotransmitters are essential substances in the function of the central and peripheral nervous systems, histamine plays a role in the inflammatory process and allergy, and hormones are necessary for the normal metabolism and development. Melatonin blocks the action of melanocyte-stimulating hormone and adrenocorticotropin and lightens the color of the skin. Failure in the synthesis of these compounds causes overproduction or impairment in the regulatory mechanisms and leads to disorder and disease.

A variety of pepides and polypeptides possess hormone activity. Their molecular weight varies from a few hundred to 41,000 daltons and they are produced in various glands, such as the posterior and anterior pituitary, thyroid, parathyroid, and adrenal cortex and in specific cells of many organs such as the kidney, pancreas, mucosa of stomach and duodenum, hypothalamus, and in various nerve cells of the brain. They are also synthesized from serum proteins. There are many similarities in the chemical structure of these hormones and some overlap in their biological actions. Their roles include control of many physiological processes.

1. Pituitary Peptides

Adrenocorticotropic hormone is produced in the anterior pituitary.[77] Its secretion is partially controlled by a feedback mechanism as it is discharged into the circulation when the blood corticosteroid level is decreased. Adrenocorticotropic hormone is also released as a response to stress associated with the function of the central nervous system. Both chemical and nervous stimulations act directly on the hypothalamus, initiating the release of a corticotropin releasing factor which is a peptide hormone, not yet completely identified. The corticotropin-releasing hormone causes the secretion of adrenocorticotropic hormone from the pituitary.

Human adrenocorticotropic hormone has been isolated and prepared by chemical synthesis. It contains 39 amino acid residues (Figure 12). The amino terminal serine is needed for hormone activity. If this amino acid is acetylated or replaced by D-serine, the activity is increased due to greater resistance of these derivatives to hydrolytic degradation. From the carboxy terminal 19 amino acids can be cleaved off without any change of activity. The peptide remaining after the removal of residue 20 still has 30% of the original hormone activity. The activity is reduced to nil after residue 16 is separated. The presence of basic amino acids, lysine and arginine, plays an important role in hormone activity.

The formation of some steroid hormones, such as aldosterone and cortisol from cholesterol is stimulated by the adrenocorticotropic hormone in the adrenal cortex. The synthesis of these steroids involves the incorporation of oxygen into the nucleus. This process is probably mediated by ascorbic acid which is present in high concentration, since during the synthesis adrenal ascorbic acid content is reduced. Adrenocorticotropic hormone has a darkening effect on skin and initiates the release of fatty acids from the adipose tissue. Probably the latter action is a response to the energy requirement brought about by stress conditions. Considering the interrelationship between the adenohypophysis and adrenal cortex, any impairment of

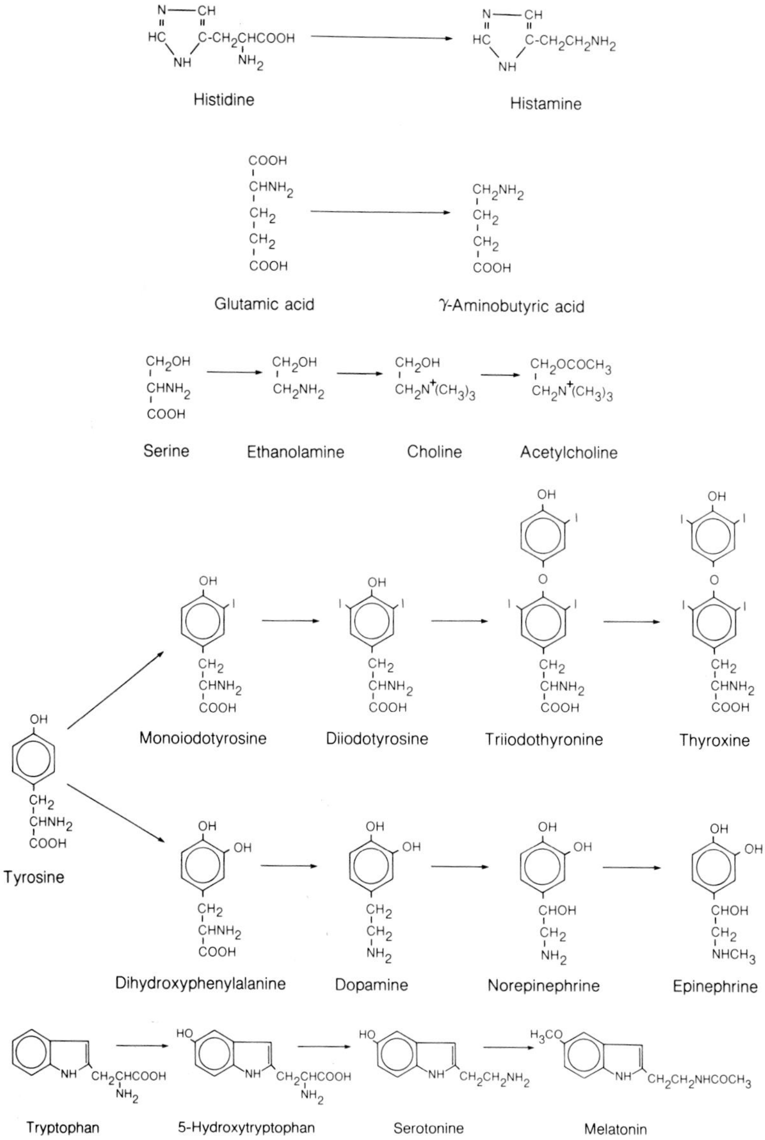

Histidine → Histamine

Glutamic acid → γ-Aminobutyric acid

Serine → Ethanolamine → Choline → Acetylcholine

Tyrosine → Monoiodotyrosine → Diiodotyrosine → Triiodothyronine → Thyroxine

Tyrosine → Dihydroxyphenylalanine → Dopamine → Norepinephrine → Epinephrine

Tryptophan → 5-Hydroxytryptophan → Serotonine → Melatonin

FIGURE 11. Synthesis of hormones from single amino acids.

the feedback mechanism is associated with malfunction of these organs. If the function of the anterior pituitary is inadequate, the administration of adrenocorticotropic hormone can correct the conditions of ill health, but if the activity of the adrenal cortex is inadequate, the release of corticotropin may be accompanied by a reduction in steroid excretion.

Growth hormone, or somatotropin, is also formed in the anterior pituitary in response to various stimuli such as exercise, fasting, hypoglycemia, or sleep.[96,120] The production of this hormone is regulated in the hypothalamus by the secretion of growth hormone-releasing factor.[39] The latter substance is a low-molecular-weight peptide; its structure, however, has not yet been established. There is no apparent feedback mechanism; over- or underproduction may occur in abnormal conditions. Overproduction is associated with a tumor of the anterior pituitary. If the tumor develops at an early age, it causes gigantism, and in older patients, acromegaly. In the latter condition, overgrowth occurs in bones of the face, hands, and feet. Deficient production of growth hormone brings about an abnormally short stature, dwarfism.

SER-TYR-SER-MET-GLU-HIS-PHE-ARG-TRYP-GLY-LYS-PRO-VAL-GLY-LYS-LYS-ARG-ARG-PRO-VAL-

LYS-VAL-TYR-PRO-ASP-ALA-GLY-GLU-ASP-$GLUNH_2$-SER-ALA-GLU-ALA-PHE-PRO-LEU-GLY-PHE

ADRENOCORTICOTROPIC HORMONE

PHE-PRO-THR-ILEU-PRO-LEU-SER-ARG-LEU-PHE-ASP-$ASPNH_2$-ALA-MET-LEU-ARG-ALA-HIS-ARG-LEU-

HIS-$GLUNH_2$-LEU-ALA-PHE-ASP-THR-TYR-$GLUNH_2$-GLU-PHE-GLU-GLU-ALA-TYR-ILEU-PRO-LYS-GLU-$GLUNH_2$-

LYS-TYR-SER-PHE-LEU-$GLUNH_2$-$ASPNH_2$-PRO-$GLUNH_2$-THR-SER-LEU-CYS-PHE-SER-GLU-SER-ILEU-PRO-THR-

PRO-SER-$ASPNH_2$-ARG-GLU-GLU-THR-$GLUNH_2$-LYS-SER-$ASPNH_2$-LEU-$GLUNH_2$-LEU-LEU-ARG-ILEU-SER-LEU-LEU-

LEU-ILEU-$GLUNH_2$-SER-TRYP-LEU-GLU-PRO-VAL-$GLUNH_2$-PHE-LEU-ARG-SER-VAL-PHE-ALA-$ASPNH_2$-SER-LEU-

VAL-TYR-GLY-ALA-SER-$ASPNH_2$-SER-ASP-VAL-TYR-ASP-LEU-LEU-LYS-ASP-LEU-GLU-GLU-GLY-ILEU-

GIN-THR-LEU-MET-GLY-ARG-LEU-GLU-ASP-GLY-SER-PRO-ARG-THR-GLY-$GLUNH_2$-ILEU-PHE-LYS-$GLUNH_2$-

THR-TYR-SER-LYS-PHE-ASP-THR-$ASPNH_2$-SER-HIS-$ASPNH_2$-ASP-ASP-ALA-LEU-LEU-LYS-$ASPNH_2$-TYR-GLY-

LEU-LEU-TYR-CYS-PHE-ARG-LYS-ASP-MET-ASP-LYS-VAL-GLU-THR-PHE-LEU-ARG-ILEU-VAL-$GLUNH_2$-

CYS-ARG-SER-VAL-GLU-GLY-SER-CYS-GLY-PHE

GROWTH HORMONE

FIGURE 12. Structure of peptide hormones synthesized in the hypothalamus.

AC-SER-TYR-SER-MET-GLU-HIS-PHE-ARG-TRYP-GLY-LYS-PRO-VAL-NH_2

α-MELANOCYTE-STIMULATING HORMONE

ASP-GLU-GLY-PRO-TYR-LYS-MET-GLU-HIS-PHE-ARG-TRYP-GLY-SER-PRO-PRO-LYS-ASP

β-MELANOCYTE-STIMULATING HORMONE

FIGURE 13. Structure of peptide hormones synthesized in the anterior pituitary.

Structural studies have established that human growth hormone contains 190 amino acids (Figure 12). The total amino acid sequence is not essential for hormone activity. Partial enzymatic degradation products of the hormone and shorter synthetic peptides show activity. These fragments may be effective alone or the side chain may be further elongated by the addition of more amino acid residues. The structure of human chorionic somatomamotropin is similar to that of growth hormone — 160 amino acids are in identical positions in both hormones. Chorionic somatomamotropin is produced in the placenta and elicits action similar to growth hormone.

The effects of growth hormone are complex. Most tissues are influenced directly by this peptide and, at the same time, it potentiates the action of other hormones. It increases protein synthesis and stimulates the mobilization of fats and carbohydrates from storage depots.

The middle lobe of the pituitary secretes hormones which increase melanin deposition by melanocytes in the human skin. Two peptides, α- and β-melanocyte-stimulating hormone, have been identified.[76] α-Melanocyte-stimulating hormone consists of 18 amino acids. Part of the amino acid sequence of this peptide — between 7 and 13 amino acids — shows a similarity with adrenocorticotropin (Figure 13). A synthetic hexapeptide comprising the amino acid chain from 8 to 13 has been produced and this peptide elicits melanocyte-stimulating activity. Removal of the glutamic acid moiety brings about complete inactivation. It can be replaced by glycine, indicating that the length of the peptide chain is essential for hormone action.

(Pyro) Glu-His-Pro-NH_2

Thyrotropin-releasing hormone

(Pyro) Glu-His-Tryp-Ser-Tyr-Gly-Leu-Arg-Pro-Gly-NH_2

Gonadotropin-releasing hormone

H_2N-Cys-Tyr-Phe-GluNH_2-AspNH_2-Cys-Pro-Arg-Gly-NH_2

Antidiuretic Hormone (Vasopressin)

H_2N-Cys-Tyr-Ileu-GluNH_2-AspNH_2-Cys-Pro-Leu-Gly-NH_2

Oxytocin

FIGURE 14. Structure of peptide hormones synthesized in the middle lobe of the pituitary.

2. Hypothalamic Peptides

The formation and release of thyroid hormones are controlled by the thyroid-stimulating hormone or thyrotropin, which in turn is regulated by two control mechanisms: a feedback connected with increased levels of thyroxine and triiodothyronine and a response to an external stimulus mediated through the hypothalamus.[8,121] This organ contains the thyrotropin-releasing hormone which is secreted into blood vessels going directly to the anterior pituitary, where it triggers off the discharge of the thyrotropin into the blood stream. Thyrotropin-releasing hormone was first isolated from porcine tissue and identified as pyroglutamyl-histidylprolinamide (Figure 14). The formation of this tripeptide takes place in the hypothalamus.[8] It has also been chemically synthesized. Both the natural and synthetic hormones elicit the same biological activity in man. It appears that women are more sensitive to the action of this hormone, which may be associated with the more frequent occurrence of thyroid hypertrophy in females.

Gonadotropin-releasing hormone is also formed in the hypothalamus.[49] This peptide activates the secretion of both luteinizing and follicle-stimulating hormones from the pituitary. Gonadotropin-releasing hormone is a decapeptide, its structure has been established by synthesis. The pyroglutamylhistidine amino terminal of this molecule is identical with the thyrotropin-releasing hormone and, similarly, it has an amide group at the carboxyl terminal (Figure 14). Animal experiments indicate that sex hormones regulate the serum level of the gonadotropin-releasing factor released from the hypothalamus.

In the hypothalamus two other important peptide hormones are formed: antidiuretic hormone or vasopressin, and oxytocin.[34,35,52] These substances migrate along to the posterior pituitary gland where they are stored. The chemical structures of these two hormones are very similar, both made up from eight amino acids which form a cyclic pentapeptide closed up by the cystine disulfide bond with a tripeptide side chain (Figure 14). Reduction of the disulfide bond yields a linear nonapeptide which has no hormone activity. Many similar peptides have been synthesized and structure-activity studies revealed that most alterations cause a reduction or produce qualitatively different activities from those of the parent compounds.

In spite of the close structural similarity, these two peptides have different actions. The antidiuretic hormone increases water permeability of the distal tubules and collecting ducts

ASP-ARG-VAL-TYR-ILEU-HIS-PRO-PHEN

ANGIOTENSIN II

LYS-ARG-PRO-PRO-GLY-PHE-SER-PRO-PHE-ARG

KALLIDIN

ARG-PRO-PRO-GLY-PHE-SER-PRO-PHE-ARG

BRADYKININ

FIGURE 15. Structure of pharmacologically active peptide hormones.

of the kidney and thus participates in the regulation of body water balance. This hormone stimulates the action of the adenylate cyclase system which converts adenosine triphosphate to cyclic 3′,5′-adenylate. This action is closely related to antidiuretic potency. Oxytocin stimulates the contractions of the uterus during parturition and it initiates the secretion of milk from the mammary gland. Due to the similar structure there is some overlap in their biological activity. The release of both hormones from the neurohypophysis is induced by reflex action of the central nervous system. Increase in the blood osmotic pressure triggers off the secretion of antidiuretic hormone. Oxytocin release starts by the dilatation of the birth canal at term and suckling.

3. Serum Peptides

Enzymatic hydrolysis of α-globulin in the blood results in pharmacologically active peptide hormones. One of these peptides is angiotensin II, an octapeptide (Figure 15). This is a very potent pressor substance; it constricts blood vessels and causes increased blood pressure. It exerts an indirect action on blood pressure and enhances the release of aldosterone from the adrenal gland, which in turn causes sodium and water retention leading to hypertension.

Angiotensin II is inactivated fairly rapidly by peptidases in the blood which destroy the intact peptide. However, the activity is not strictly related to the entire amino acid sequence. Replacement of isoleucine with valine results in change in hormone activity. The removal of the asparagyl-arginine amino terminal also causes no significant loss, but if asparagyl-arginyl-valine is cleaved off the pentapeptide has no activity. The presence of proline, phenylalanine, or tyrosine is essential to the biological action.

A similar enzymatic process produces bradykinin, a nonapeptide, and kallidin, a decapeptide from glycoproteins (Figure 15).[18,22,54] These peptide hormones cause dilatation and increased permeability of blood vessels, enhance leukocyte migration. Bradykinin represents a factor in the onset of pain. Its involvement in the processes of inflammation is similar to that of histamine, but the effect is not inhibited by antihistamines. The analgesic action of aspirin is associated with its ability to block the pain induced by bradykinin. The effect of wasp venom is probably related to the presence of a nonapeptide identical to bradykinin, in concert with histamine and serotonin.

(Pyro) Glu-Gly-Pro-Trp-Met-Glu-Glu-Glu-Glu-Glu-Ala-Tyr-Gly-Tryp-Met-Asp-Phe-NH_2
|
SO_3H

Gastrin

Lys-Ala-Pro-Ser-Gly-Arg-Val-Ser-Met-Ileu-Lys-AspNH_2-Leu-GluNH_2-Ser-Leu-Asp-Pro-Ser-His-

Arg-Ileu-Ser-Asp-Arg-Asp-Tyr-Met-Gly-Tryp-Met-Asp-Phe-NH_2
|
SO_3H

Cholecystokinin

His-Ser-Asp-Gly-Thr-Phe-Thr-Ser-Gly-Leu-Ser-Arg-Leu-Arg-Asp-Ser-Ala-Arg-Leu-GluNH_2

Arg-Leu-Leu-GluNH_2-Gly-Leu-Val-NH_2

Secretin

FIGURE 16. Structure of peptide hormones synthesized in gastric and duodenal mucosa.

Tests with various synthetic peptides have revealed that replacement of proline at position 3, or of serine with alanine causes no alteration; the conversion of the carboxyl terminal at position 9 to amide, however, reduces activity. The presence of both arginine, phenylalanine and proline in the molecule is essential; when any of these amino acids is replaced by alanine, inactivation takes place.

4. Gastric and Duodenal Peptides

In the gastric mucosa, gastrin, a hormone peptide, is produced near the junction of stomach and small intestine.[64,66,113,114] The presence of food in the stomach stimulates cholinergic nerves which, in turn, activate the release of gastrin into the blood. This hormone then stimulates an increase of acid and pepsinogen secretion in the gastric mucosa. The fluid flow from the pancreas is also enhanced by this hormone.

Gastrin is a linear peptide constituting 17 amino acids (Figure 16). It occurs with or without ether bond with the sulfate group of the tyrosine moiety. Both peptides elicit equal hormone action. The carboxyl terminal tetrapeptide, tryptophanyl-methionyl-aspartyl-phenylalanylamide, is mainly responsible for the activity. The tetrapeptide alone possesses about 10% of the original activity. It appears that the entire molecule is not essential for the hormone action; full activity can be achieved from the tetrapeptide by the addition of 12 other amino acids of the natural hormone. In the tetrapeptide amide the presence of the aspartyl residue is the most important. Exchange with glutamic acid causes inactivation. The absence of the amide group leads to complete loss of activity, but its methylation has no effect at all. Similarly, tryptophan can be replaced with 5-hydroxytryptophan, methionine with norleucine, and phenylalanine with methylphenylalanine without loss of activity. Many synthetic derivatives have been prepared, some of which have even greater activity than the tetrapeptide amide. One of these compounds, tert-butyloxycarbonyl-β-Ala-Tryp-Met-Asp-Phen-NH_2, or pentagastrin, is frequently applied in therapy.

The duodenal mucosa produces two important peptide hormones: cholecystokinin and secretin.[19] Secretin stimulates the secretion of pancreatic juice, cholecystokinin causes gall-bladder contraction and triggers the release of pancreatic enzymes. The latter activity has been associated with the action of pancreozymin, but current investigations have shown that both activities are found in the same peptide unit. The release of secretin is induced by the presence of acid in the duodenum and the stimulation of cholinergic nerves by partially digested food in the duodenum causes the release of cholecystokinin into the blood. Secretin enhances pancreatic secretion and the activity of digestive enzymes. The acid coming from the stomach is neutralized by secretin and this provides a shift of the pH in the alkaline direction, which is essential for the function of pancreatic enzymes. In addition to the action of secretin, the release of cholecystokinin greatly elevates the enzyme content of the pancreatic juice.

Secretin contains 27 amino acids joined together in a linear chain (Figure 16). This peptide is apparently very sensitive to any treatment. Removal of the amino terminal histidine causes great loss of activity. Hydrolysis of any peptide bond results in inactive fragments. On the action of thrombin, secretin is separated into two peptides between arginine and asparagine at position 14; these peptides show no secretin activity.

Cholecystokinin also forms a single linear peptide chain containing 33 amino acids (Figure 16). The carboxyl terminal pentapeptide amide is identical with that of gastrin. In contrast to secretin, despite successive degradation of this peptide its activity is still maintained. The smallest unit which shows hormone activity is the carboxyl terminal heptapeptide amide, although it is less potent than the intact molecule. Addition of one amino acid to this peptide produces full activity. Further addition of two amino acids to the octapeptide amide to form a decapeptide raises the activity tenfold. The position of tyrosine sulfate is important in hormone activity; if its neighboring amino acids are replaced on both sides, complete loss of activity occurs. Moreover, the presence of a sulfate moiety on the tyrosine is essential; if it is absent from the octapeptide most activity disappears. This characteristic of cholecystokinin is different from that of gastrin, where the removal of the sulfate group leads to no loss of function.

5. Pancreatic Peptides

In the pancreas three very important peptides are synthesized: insulin, glucagon, and somatostatin.[70,181] These hormones play an important role in the regulation of the blood glucose level. Glucagon is formed in α-cells and insulin is the product of β-cells of the islets of Langerhans. Glucagon is released into the circulation in response to low glucose; in contrast, insulin is released when glucose concentration is elevated. Moreover, increased concentrations of fatty acids or amino acids also stimulate the discharge of insulin into the blood. Insulin enhances the uptake and storage of these substances by various cells. Stimulation of the vagus nerve which innervates the pancreas also initiates insulin release. The discharge of insulin is mainly under the regulatory action of changes in blood glucose level (see Chapter 4) and is also related to the action of norepinephrine. This compound inhibits the release of insulin and enhances the metabolism of glucose, fatty acids, and amino acids for utilization as energy sources.

Insulin is synthesized in the pancreas as proinsulin which contains 81 to 83 amino acids (Figure 17).[17] This precursor possesses no hormonal activity. Due to the action of proteolytic enzymes, proinsulin is broken up between residues 30 and 31 and between 60 and 61 of the peptide chain into smaller fragments, providing the insulin A chains, the peptide between units 61 and 81 (C peptide), and the B chain which is the fragment between units 1 and 30. These two chains are joined together by two disulfide bridges, and there is another disulfide bond within the A chain. In addition to these interconnecting bridges, hydrogen and hydrophobic bonds and linkages to bivalent ions confer stability to the two chain system. The disulfide bridges can be reduced by the action of mercaptoethanol and urea, and thus the A and B chains can be separated. After this treatment, recombination of the two chains or individual chains alone do not restore insulin activity; reoxidation results in only 2% of the original activity of the hormone molecule. In contrast, if the disulfide bridges are eliminated by mercaptoethanol treatment of proinsulin, by reversing the action 70% of the original hormone precursor is recovered. It appears that the connecting peptide in proinsulin prevents the conformational changes associated with the sulfhydryl group and, thus, on oxidation there is less possibility of inactive isomer formation. Further studies have shown that in the

A CHAIN → GLY-ILEU-VAL-GLU-GLUNH$_2$-CYS-CYS-ALA-SER-VAL-CYS-SER-LEU-TYR-GLUNH$_2$-LEU-GLY-ASPNH$_2$-TYR-CYS-ASPNH$_2$

B CHAIN → PHE-VAL-ASPNH$_2$-GLUNH$_2$-HIS-LEU-CYS-GLY-SER-HIS-LEU-VAL-GLU-ALA-LEU-TYR-LEU-VAL-CYS-GLY-

GLU-ARG-GLY-PHE-PHE-TYR-THR-PRO-LYS-ALA-ARG-ARG-GLU-VAL-GLU-GLY-PRO-GLUNH$_2$-VAL-GLY-

ALA-LEU-GLU-LEU-ALA-GLY-GLY-PRO-GLY-ALA-GLY-GLY-LEU-GLU-GLY-PRO-PRO-GLUNH$_2$-LYS-ARG

BOVINE PROINSULIN

HIS-SER-GLUNH$_2$-GLY-THR-PHE-THR-SER-ASP-TYR-SER-LYS-TYR-LEU-ASP-SER-ARG-ARG-ALA-GLUNH$_2$-

ASP-PHE-VAL-GLUNH$_2$-TRYP-LEU-MET-ASPNH$_2$-THR

GLUCAGON

FIGURE 17. Structure of peptide hormones synthesized in the pancreas.

B chain the carboxyl terminal alanine is unnecessary to the hormone action. Even if an octapeptide is removed from this end the remaining residue still retains 5 to 15% of the hormone activity of the whole insulin molecule.

The major role of insulin is regulation of the blood glucose level. In juvenile diabetes mellitus, insulin secretion is greatly reduced or totally absent, producing symptoms of hyperglycemia, lipemia, and ketonuria. The development of these symptoms in maturity onset diabetes, and the mechanism of action of insulin will be dealt with in another chapter.

Antagonistic to insulin, glucagon raises blood glucose. Its action is mediated by the enzyme adenyl cyclase; the production of cyclic 3′,5′-adenylic acid stimulates phosphorylase activity which is responsible for the synthesis of glucose 1-phosphate from glycogen and, in turn, for the production of glucose. Besides giving glucose, this mechanism has been applied clinically for increasing blood glucose levels in order to reverse the effect of insulin-induced hypoglycemia. It is also considered for use in cardiac disorders, because in abnormal pathological conditions it enhances the strength and rate of myocardial contractions. Glucagon contains 29 amino acids and its structure shows a close resemblance to that of secretin (Figure 17). For both hormones, the entire peptide chain is required for activity. If the amino terminal of histidine is removed, glucagon activity is lost completely. None of the many fragments produced by hydrolysis exhibit any hormone action. In contrast to the chemical similarity of glucagon and secretin, their physiological roles and target organs are very different, and there is no overlap in their action.

6. *Thyroid and Parathyroid Peptides*

Apart from the production of thyroxine and triiodothyronine,[139] some specialized cells in the thyroid gland (parafollicular A or C) synthesize calcitonin which decreases calcium concentration in the blood. Calcitonin thus influences the availability of calcium for membrane function and muscular contraction. The hormone is secreted from the thyroid if the blood calcium is high and then it inhibits the amount of calcium released from the bones into the circulation. Parathyroid hormone has the opposite action; if the blood calcium level is low it is released from the parathyroid gland, subsequently enhancing calcium release from bones. There is some association between calcitonin and gastrin; synthetic pentagastrin causes calcitonin release.

The structure of calcitonin has been established. It is formed by a single peptide chain consisting of 32 amino acid residues with an interchain sulfide bridge forming a ring. Reduction of this bridge destroys the activity essential for hormone action. The ring is at the amino terminal and the terminal pentapeptides are commonly occurring constituents in calcitonins of various origins (Figure 18). The carboxy terminal prolinamide is also identical. There are many great species variations in the amino acid composition in the middle section of this molecule which has no apparent effect on hormone activity in man. However,

Cys-Gly-AspNH$_2$-Leu-Ser-Thr-Cys-Met-Leu-Gly-Thr-Tyr-Thr-GluNH$_2$-Asp-Phe-AspNH$_2$-Lys-Phe-His-

Thr-Phe-Pro-GluNH$_2$-Thr-Ala-Ileu-Gly-Val-Gly-Ala-Pro-NH$_2$

Calcitonin

Ala-Val-Ser-Glu-Ileu-GluNH$_2$-Phe-Met-His-AspNH$_2$-Leu-Gly-Lys-His-Leu-Ser-Ser-Met-Gly-Arg-

Val-Glu-Tryp-Leu-Arg-Lys-Lys-Leu-GluNH$_2$-Asp-Val-His-AspNH$_2$-Phe-Val-Ala-Leu-Gly-Ala-Ser-

Ileu-Ala-Tyr-Arg-Asp-Gly-Ser-Ser-GluNH$_2$-Arg-Pro-Arg-Lys-Lys-Glu-Asp-AspNH$_2$-Val-Leu-Val-

Glu-Ser-His-GluNH$_2$-Lys-Ser-Leu-Gly-Glu-Ala-Asp-Lys-Ala-Asp-Val-Asp-Val-Leu-Ileu-Lys-

Ala-Lys-Pro-GluNH$_2$

Bovine parathyroid hormone

FIGURE 18. Structure of peptide hormones synthesized in the thyroid and parathyroid glands.

conformational changes and differences in breakdown rate may influence the duration of hormone action.

Parathyroid hormone[132] reduces elimination of calcium by the kidney by increased tubular reabsorption, increases the reabsorption from the bones, and enhances the absorption from the small intestine indirectly. The structure of the bovine parathyroid hormone shows a straight-line peptide with 84 amino acids (Figure 18). The active part contains probably the first 34 amino acids from the amino terminal. Fragments smaller than 13 amino acids are totally inactive. It is possible that the extra 50-amino-acid residue in the native hormone protects the active nucleus from metabolic destruction. The human hormone has a similar structure to bovine parathyroid hormone.

Diseases are associated with the reduced production of parathyroid hormone such as hypoparathyroidism or with the reduced production of calcitonin such as osteoporosis, characterized by increased porosity of the bones.[35] Paget's disease is associated with soft and deformed bone formation.

A tetrapeptide has been shown to be essential for the phagocytosis exercised by leukocytes. These cells are coated with a protein, leukokinin, migrating with the γ-globulin fraction. Leukocyte cell membranes contain a proteolytic enzyme which breaks off a tetrapeptide from leukokinin. This tetrapeptide threonyl-lysyl-prolyl-arginine is named tuftsin and is involved in phagocytosis. It has been isolated from blood, but the synthetic peptide has the same activity as the natural hormone.

7. Brain Peptides

Nerve cells use electrical signals for communication between cells and, in addition, they communicate chemically with neurotransmitters; most of them are peptides or simple amino acid derivatives.[64,93,131] Acetylcholine was first discovered as a neurotransmitter. At present, about 20 peptides are firmly established as neurotransmitters and there are about another 20 which are under investigation as transmitters of nerve signals. The most widely studied peptide is substance P, which is considered the transmitter of the brain signals carried by sensory nerves into the spinal cord and then relayed to the brain. Endorphins, the brain's built-in opiate-like components, may exert their analgesic action by suppressing the release of substance P in the spinal cord. Substance P contains a sequence of 11 amino acids (Figure 19).

Substance P is present in nerve fibers that carry pain signals. The small-diameter nerve fibers of the dorsal horn and the tooth pulp, which carry many pain fibers, contain the peptide. In addition to the dorsal horn of the spinal cord, substance P is concentrated in several areas of the brain. One brain region, the substantia nigra, which helps to control movement, contains particularly high concentrations in neurons of the autonomic neuron

ARG-PRO-LYS-PRO-GLN-GLN-PHE-PHE-GLY-LEU-MET

FIGURE 19. Structure of substance P.

system which controls more or less involuntary activities such as breathing and the beating of the heart. It is in the myenteric plexus, a nerve network innervating the intestinal tract that may regulate the undulating movements that directs food through the intestines. Substance P has been found in some endocrine-like cells of the intestinal lining, where its function is not yet fully established.

Many neurotransmitter brain peptides are also found in other regions of the body. These peptides include endorphins and enkephalins which are also found in many peripheral areas of the body such as the myenteric plexus of the gut. These endogenous opiates possibly act as neurotransmitters in the spinal cord and the brain, suppress pain perception under some conditions, and may have other effects on behavior.

Neurotensin is a peptide consisting of 13 amino acid residues and exerts several actions on the periphery. Neurotensin elicits an effect on a number of hormones including insulin and glucagon, thus it helps to regulate blood glucose concentrations.[64] There is evidence that the central nervous system contains pain-suppressing pathways which are independent from endorphins. In these cells neurotensin may mediate some of the endorphin-independent analgesia.

Angiotensin, a powerful peptide in increasing blood pressure, is found originally in the blood stream. It is produced from α_2 globulin by an enzyme, renin, formed in the kidney. All components of the renin angiotensin system have been identified in the brain and there are indications that this system is partly responsible for the mechanism in the brain that regulates blood pressure.

Cholecystokinin is a peptide hormone formed primarily in the small intestines in response to the movement of food from the stomach into the intestines. This peptide is responsible for the contraction of the gallbladder and the release of bile into the small intestine, where enzymes, bile acids, and other components of the bile promote digestion. Cholescystokinin has been found in the brain; its role in this organ is not clear. There are indications that it may be involved in the sensation of hunger and may help to regulate feeding.

Vasoactive intestinal peptide is also present in both the gut and the brain. Its peripheral actions include the reduction of blood pressure by causing vasodilation, suppression of the secretion of stomach acid, and stimulation of the secretion in the small intestine and colon. The vasoactive intestinal peptide is a possible neurotransmitter that may play a role in affecting arousal. This peptide also stimulates the release of a number of pituitary hormones, including prolactin and growth hormone, thus may exert a regulatory action on the endocrine system.

Many pituitary and hypothalamic hormones are present in several areas of the brain. These include adrenocorticotropin and melanocyte-stimulating hormones which are formed in the pituitary gland, and vasopressin formed in the hypothalamus but transported to and released by the pituitary. These hormones facilitate some facets of learning and memory. Releasing factors which are synthesized in the hypothalamus and transported to the pituitary where they induce the release of appropriate hormones are also found in some other brain regions. Among these, the thyroid hormone-releasing factor and luteinizing hormone-releasing hormone are connected with behavioral changes. In experimental animals the administration of luteinizing hormone-releasing hormone enhances mating behavior.

There are many peptides in the brain which participate in neural transmission; some exert well-defined functions as hormones or other regulators of physiological processes. Some of these peptides are not neurotransmitters of nerve impulses between specific neurons but influence neural activity as neuromodulators or neurohormones. These peptides modify the response of the nerve cell to a neurotransmitter. Some peptides or other agents (precursors or analogues of peptides) could affect the activity of nerve cells by increasing or decreasing the production, release, or metabolism of neurotransmitters, neuromodulators, or neurohormones.

Some of the neurotransmitter substances are involved in transferring impulses though the neuron system into the brain, some of these show opposite actions. Substance P appears to excite neurons, whereas serotonin usually inhibits them. The existence of two transmitters in the same nerve cell has been increasingly documented. Abnormalities of sending out these chemical signals may be connected with neurological and degenerative disorders.

Huntington's chorea is a degenerative disease characterized by uncontrollable movements and progressive mental deterioration. In patients suffering from this condition, loss of neurons containing γ-aminobutyric acid and loss of substance P-containing neurons of the substantia nigra have been observed.

In the development of schizophrenia, another often intractable disorder, the two neurotransmitters involved are dopamine and substance P. Overproduction of dopamine has been implicated as the cause of this disorder. Substance P stimulates the release of dopamine in certain nerve cells.

B. Functional Peptides

Normally, the digestion of exogenous proteins or the degradation of endogenous proteins proceeds to the liberation of amino acid constituents. The endogenous molecules are in a state of flux and their metabolites are constantly mixing with the exogenous materials drawn from catabolic reactions. The resulting amino acid pool is utilized for the synthesis of tissue proteins and intermediate peptides. There are, however, some peptides that not only serve as precursors of larger units but, as peptides, participate in cellular metabolism. A few, such as glutathione, carnosine, and anserine are widely distributed in various cells and elicit important actions in some cellular processes.[5,57,72,94,105]

1. Deficiency of Glutathione Synthesis

Glutathione and several related glutamyl compounds are found predominantly within individual cells in relatively high concentrations.[61,140] Glutathione is a tripeptide, γ-glutamyl-cysteinyl-glycine, and has two characteristic structural features: a sulfhydryl radical and a γ-glutamyl linkage. It is the most abundant sulfur-containing compound and important within the cell as a donor of γ-glutamyl residue. Glutathione is converted to glutathione disulfide catalyzed by glutathione peroxidase. This reaction takes place extracellularly. The intracellular concentration of the reduced compound is much greater than that of glutathione disulfide, consistent with the presence of a highly active enzyme, glutathione reductase.[12,13,99] The intracellular localization of the reduced substance is essential in glutathione function. It protects cell membranes and proteins by maintaining their sulfhydryl groups and by destroying hydrogen peroxide, other peroxides, and free radicals.[72,117] It acts as a coenzyme for certain enzymes such as glyoxylase, and as a catalyst of disulfide exchange reactions. It participates in the transfer of amino acids, some small peptides and amines across cell membranes, and in the detoxication of foreign compounds by the mercapturic acid pathway. The synthesis and metabolism of glutathione are linked with the γ-glutamyl cycle (Figure 20). The kidney contains a fairly high concentration of two enzymes required for the synthesis of glutathione; glutamyl-cysteine synthetase[90,112] and glutathione synthetase[21,78] which are responsible for the maintenance of a substantial steady-state concentration of glutathione in this organ. These enzymes are present in many other tissues. The initial step in glutathione breakdown is catalyzed by γ-glutamyl transpeptidase. This reaction is reversible and a wide variety of amino acids serve as acceptors of the γ-glutamyl group. Although early studies associated this enzyme with cellular particles, it is actually bound to cell membranes of certain epithelial structures.

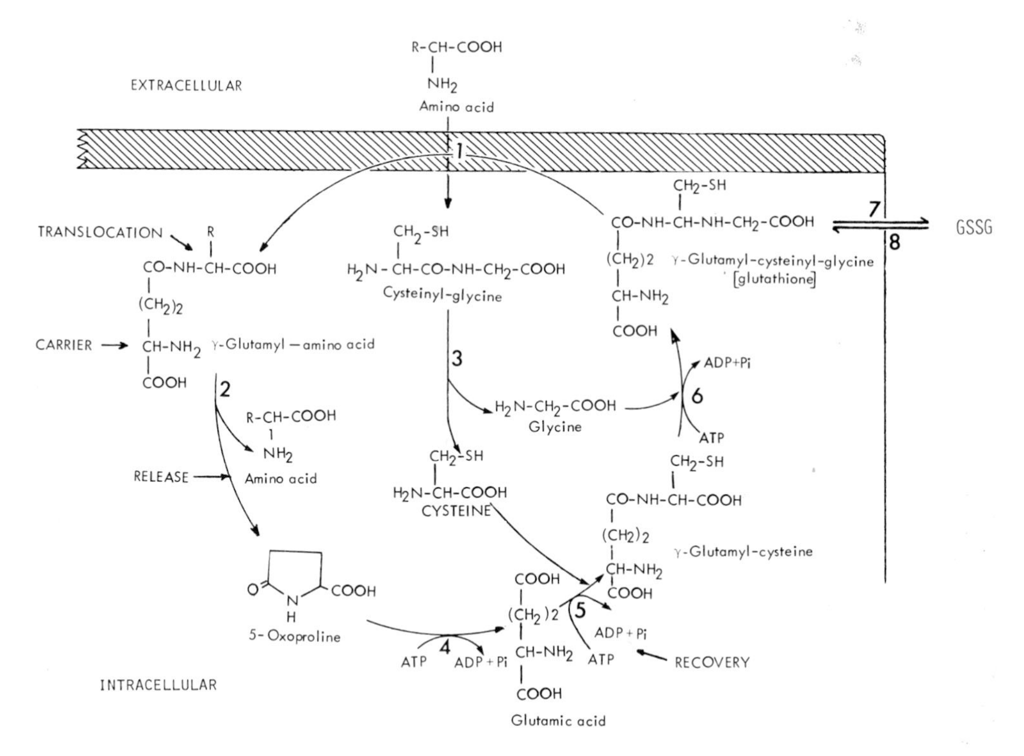

FIGURE 20. Metabolism of glutathione via the γ-glutamyl cycle. The synthesis and utilization of this compound is linked with a series of enzyme-catalyzed reactions.[188] Throughout the cycle, glutathione is continually synthesized, metabolized, and resynthesized. At equilibrium considerable amounts of glutathione are present. The operation of the γ-glutamyl cycle requires amino acids and ATP. The essential feature of this mechanism is the active transport into the cell by the membrane-bound γ-glutamyl transpeptidase enzyme (1). The amino acid becomes bound to γ-glutamyl cyclotransferase (2). This process is specially important in the kidney where very large amounts of amino acids are transported across the tubules — about 50 g daily in man. The high energy requirement is essential for the high efficiency of amino acid transport. Cysteinyl-glycine is hydrolyzed by peptidase (3) to the constituting amino acids which are then available for the resynthesis of glutathione. Each turn of the cycle produces one molecule of 5-oxoproline which is converted to glutamate and finally to glutathione. Defects of this mechanism are associated with pronounced 5-oxoprolinuria where 5-oxoprolinase is inhibited (4). Some other diseases are linked with the deficiency of γ-glutamylcysteine synthetase (5) and glutathione synthetase (6) activities. Generalized disorders are associated with decreased glutathione synthesis in many tissues.[90] Secondary disorders are related to deficiency in glutathione reductase (7).[99,189] Low levels of glutathione peroxidase (8) in the newborn is linked with hemolytic disorders.[117]

γ-Glutamyl transpeptidase is localized in the brush border of the proximal tubules of the kidney. It represents about 1.5% of total membrane protein.[14,15] Additional locations include the epithelia of the bile ductules, bronchioles, choroid plexus, uterus, fallopian tubes, epididymis, prostate, and seminal vesicles. Glutamyl transpeptidase is present in high concentrations in epithelial structures, such as the proximal renal tubule, the blood-cerebrospinal fluid barrier, jejunal mucosa, and the blood-aqueous humor barrier. These structures also contain high concentrations of glutathione. γ-Glutamyl transpeptidase is also localized in the capillaries of brain, spleen, lung, ovaries, testes, and mucosa.[71]

The enzymes of the γ-glutamyl cycle are widely distributed in various parts of the brain, but they are present in much higher concentration in the choroid plexus which is involved in the formation of cerebrospinal fluid. The very low glutamyl transpeptidase activity of the kidney in the fetus increases rapidly, coinciding with the development of the adult amino acid transport systems. The action of transpeptidase is probably connected with the transport of amino acids. Transpeptidase has been purified from the kidney, liver, and seminal vesicles, showing very similar properties. The amino acid specificity is broad, favoring neutral compounds. The γ-glutamyl transpeptidase present in normal human urine is probably liberated from renal tubule cells, indicating its location on the outer surface of the cell membrane. Transpeptidase of human serum probably arises mainly from the epithelial cells of the hepatic bile ductules. The measurement of release of this enzyme into the blood is used as an important test in the laboratory for the diagnosis of many liver and biliary disease conditions.[151]

The function of the glutamyl cycle is very important in translocating amino acids within the cells. Increased amino acid concentration generally brings about an enhanced glutathione utilization, although some glutamyl amino acids may undergo hydrolysis and are used in additional transpeptidation reactions. The intracellular concentration of glutathione is much higher than that of cysteine, cystine, or glutamyl-cysteine; it seems that cysteine is stored in the form of glutathione and protected in a metabolically less active form. However, the rate of glutathione production is influenced significantly by the L-cysteine level of the cell. Probably a physiologically important end-product feedback mechanism also controls glutathione synthesis. Several amino acids and amino acid derivatives inhibit this process such as D-glutamate, α-aminomethyl-glutarate (a β-amino acid analog of glutamate), glycylglycine, and some methionine derivatives. These compounds inhibit the activity of glutamyl-cysteine synthetase. The administration of methionine, glutamine, and other amino acids elicits no depression of glutathione levels.

Methionine sulfoximine blocks glutathione production by inhibiting glutamine synthetase and glutamyl-cysteine synthetase actvities.[21] This inhibition is related to a competition with the transient formation of enzyme-bound tetrahedral intermediates. The impairment of glutathione synthesis in the nervous system leads to convulsions. This observation provided some help in the elucidation of the mechanism by which convulsive states are produced.

The presence of glutathione in the cell is essential to normal cell function. Inhibition of any enzyme participating in the glutamyl cycle causes an aberration. Determination of the intermediate 5-oxo-L-proline level in the plasma and daily urinary excretion are applied as tests in the assessment of normal glutathione metabolism. The normal plasma 5-oxoproline content is 0.7 to 0.25 mg/dℓ and the normal daily excretion is only a few milligrams. Patients with renal failure and anephrotic patients have been shown to have 3 to 4 times higher 5-oxo-L-proline and 25 times higher 5-oxo-D-proline levels than normals. The elevated amount of L-isomer indicates a reduced kidney 5-oxo-L-prolinase activity.

5-Oxo-D-proline is probably derived from dietary and bacterial D-glutamate which is cyclized by the D-glutamate cyclotransferase enzyme present in kidney and liver. The enhanced excretion of 5-oxo-D-proline, therefore, represents a sign of renal or hepatic lesion. Enzyme studies of the glutamyl cycle have revealed several cases suffering from marked deficiency of glutathione synthesis. Some patients exhibiting mental retardation, psychosis, spinocerebellar degeneration, and spastic tetraparesis accompanied with metabolic acidosis have often been found to have high blood and cerebrospinal fluid concentration of 5-oxo-L-proline and excreted 25 to 30 g daily. These findings are consistent with a block at the 5-oxoprolinase step of the γ-glutamyl cycle, particularly when showing an extremely low activity of glutathione synthetase — only about 2% of that of the control. The acidosis of 5-oxoprolinuria is related to the accumulation of 5-oxoproline in body fluids and, presumably, in various tissues (Figure 21). The enzyme deficiency probably occurs in the kidney, but enzyme studies on placenta, erythrocytes and cultured skin fibroblasts also show low activities. The glutathione content of red and white cells and muscles is markedly reduced, suggesting a generalized deficiency of glutathione synthesis. Some of these patients also have decreased erythrocyte concentration of glutamyl-cysteine synthetase and glutathione, and increased rate of hemolysis linked with anemia.

In some cases symptoms of the central nervous system involvement are not observed and the only clinical problem is hemolytic anemia associated with the significant reduction of erythrocyte glutathione synthetase. This lesion is apparently restricted to the red blood cells,

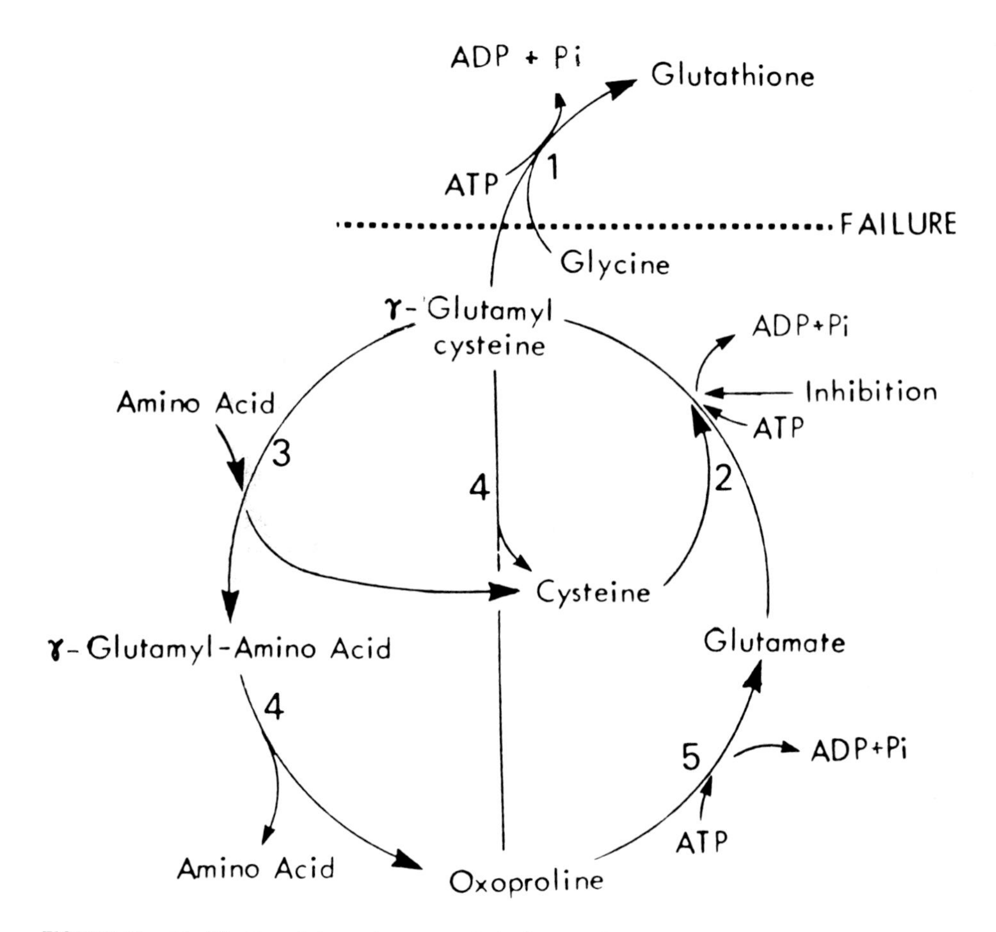

FIGURE 21. Modification of the γ-glutamyl cycle in 5-oxoprolinuria. A generalized defect of glutathione synthetase activity (1) leads to glutathione deficiency. Overproduction of γ-glutamyl cystein by γ-glutamyl-cysteine synthetase (2) results in increased formation of 5-oxoproline through the action of γ-glutamyl transpeptidase (3), γ-glutamyl cyclotransferase (4), and 5-oxoprolinase (5). Glutathione normally exerts a fedback inhibition of this cycle by inhibiting γ-glutamylcysteine synthetase. Due to reduced tissue concentration there is a failure in the regulatory mechanism.[178,190]

but it may be that a rapid turnover of the defective protein synthesis in most tissues can compensate the deficiency.[61]

Some patients have been observed with generalized glutamyl-cysteine deficiency. They have hemolytic anemia, central nervous system disorder and aminoaciduria.[112,151] These findings are consistent with a block of glutamyl-cysteine synthetase as it is found reduced in the erythrocytes and cultured fibroblasts. Some mentally retarded patients exhibit glutathionemia and glutathionuria associated with marked reduction of glutamyl transpeptidase activity. The presence of glutathione in the urine probably reflects its enhanced level in the blood and leakage from the kidney. These patients do not have a significant aminoaciduria, but the glutamine and asparagine excretion is much higher than in normals. Thus, the deficiency of kidney glutamyl transpeptidase activity may represent an adaptation. The kidney is likely to have several amino acid transport systems and the loss of one system can be compensated for by an elevated activity of other systems.

A marked deficiency in the activity of glutathione reductase of erythrocytes is often observed in patients with a variety of clinical disorders.[13,99,122,154,158] Due to the variety of symptoms, it is probable that the enzyme defect is the secondary rather than the primary etiologic cause. Glutathione reductase is a flavoprotein with flavine adenine dinucleotide as prosthetic group. Lack of riboflavin associated with the lower reductase activity in many patients can be restored to normal or nearly normal by the addition of flavine adenine dinucleotide. Riboflavin nutrition may play a role in some hemolytic disorders through the activity of red blood cell glutathione reductase. No change in hematologic status has been observed in patients with glutathione reductase deficiency associated with primaquine therapy although riboflavin administration restored the enzyme activity of erythrocytes to normal. Moreover, in several hemolytic disorders red cell glutathione is not altered; the determination of this component, therefore, may not serve as a reliable parameter for diagnosis. Glutathione peroxidase is important in the protection of erythrocytes against the deleterious action of peroxides.[117] This enzyme forms a link between the hexose monophosphate shunt and detoxication of hydrogen peroxide. Defects of the glutathione-linked pathway bring about an enhanced susceptibility to oxidizing agents usually characterized by an increased number of Heinz bodies and hemolytic episodes. In some newborn infants a moderate decrease of peroxidase activity has been found.[171] Particularly low glutathione peroxidase can lead to hemolysis and secondary hyperbilirubinemia. The disease may become symptomatic in the presence of oxidizing agents. Erythrocyte glutathione peroxidase deficiency in older children and adults is usually symptomless.

2. Glutathione in Detoxication

The involvement of glutathione in the protection of the organism against foreign substances is one of the most important biological functions of this compound. It is used by the cell in a variety of detoxication processes which lead to the production of mercapturic acids.[36,58] The first step in this reaction is a binding of the foreign compound with the sulfhydryl group of glutathione. This may be a spontaneous reaction or catalyzed by the enzyme glutathione S-transferase.[37,97] These enzymes appear to function as cellular protective units. Glutathione S-transferase activity is associated with the intracellular carrier protein, ligandin. Ligandin is a dimeric cytoplasmic protein with a basic character. It binds various large anions by noncovalent bonds and some carcinogens by covalent binding. The carrier role of ligandin protects the cell from the damaging action of endogenous and exogenous toxic substances such as bilirubin, drugs, and other foreign compounds. The ligandin-glutathione transferase catalyzes the binding and transport of these compounds coupled to glutathione. The S-substituted glutathione derivatives are metabolized and release the glutamyl group by the action of glutamyl transpeptidase.[48] Next, the glycine moiety is removed and the cysteine residue is acetylated. These enzyme reactions are principally carried out by hepatic cytosol proteins. However, in the detoxication mechanism a link with microsomal hydroxylase has been suggested (Figure 22).

Double-bond-containing compounds produce an anionic intermediate with glutathione (Figure 23). The presence of the negative charge will accelerate further reaction, therefore these derivatives are usually reactive compounds. Several natural products possessing antitumor activity, such as quinones, contain methylene lactone groups which are considered to be responsible for the biological activity. The reactivity of these compounds with glutathione, and their activation, increases the importance of glutathione in cellular anticancer processes. The use of glutathione in conjugation reactions with foreign compounds catalyzed by glutathione S-transferase leads to loss of hepatic glutathione. Since this tripeptide can be considered as an endogenous protective agent through its capacity to bind and eliminate drugs, pesticides, and other potentially toxic foreign compounds, the measurement of hepatic or extrahepatic depletion of glutathione may provide a useful test to establish drug-induced disorders.

Recently γ-glutamyl transpeptidase has been found in lymphoid cells of mesenchymal origin. It has been suggested that the secretory component of immunoglobulin A may also be identical with glutamyl transpeptidase. Proteins and large polypeptides may serve as

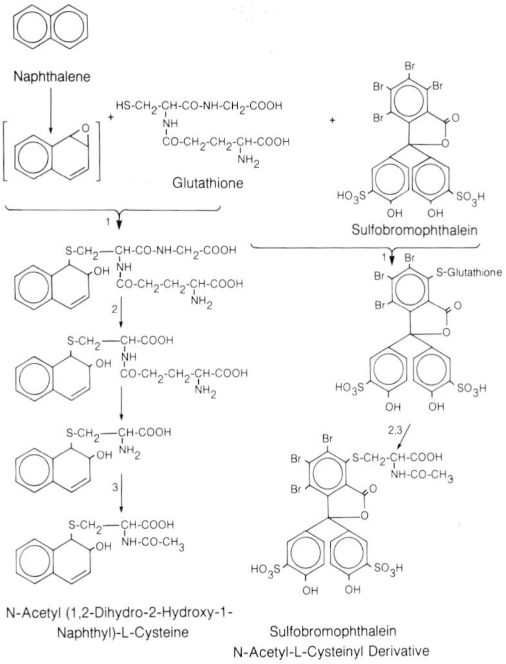

Naphthalene

Glutathione

Sulfobromophthalein

N-Acetyl (1,2-Dihydro-2-Hydroxy-1-Naphthyl)-L-Cysteine

Sulfobromophthalein N-Acetyl-L-Cysteinyl Derivative

FIGURE 22. Mechanism of detoxication through the biosynthesis of mercapturic acid. The initial stage involves the catalytic action of glutathione S-transferase (1). The transfer of the γ-glutamyl group (2) results in a cysteinyl glycine derivative, further hydrolysis produces the cystein conjugate which then undergoes N-acetylation (3) to provide the mercapturic acid.[191,192] The glutathione conjugate and its metabolites are usually excreted in the bile, whereas mercapturic acid is eliminated in the urine. The formation of epoxide may be essential in the detoxication of foreign compounds associated with microsomal hydroxylation.[193] In some additional elimination reactions, the sulfhydryl group of the glutathione is replaced by aromatic substituents such as in the conjugation of bromosulfophthalein.

$$GSH + R^1\text{-}CH{=}CHR^2 \longrightarrow GS\text{-}CH{=}CH\text{-}R^2 + R^1H$$

FIGURE 23. Formation of an anionic intermediate from glutathione and unsaturated compounds.

acceptors of the glutamyl moiety and they are secreted as glutamyl derivatives in the N-terminal position by epithelial cells. This concept, however, has been recently disproved. In both the epithelial and lymphoid cell, the enzyme is membrane bound. The transpeptidase activity of human lymphocytes is markedly increased by stimulation with mitogenic compounds. Studies on neoplastic lymphoid cells indicate that there may exist a significant relationship between the activity of this enzyme and the development of neoplasia.[61]

3. Glutathione in Membrane Protection

Hemolysis caused by defects in glutathione metabolism shows the importance of this compound for the normal viability of red blood cells. The biochemical lesion in the glutathione content or metabolism alone do not cause significant destruction. Shortening of the life span of erythrocytes is induced by additional actions such as some foreign compounds, fava beans, and severe infections. In these circumstances, the administration of a special drug can cause rapid oxidation of glutathione, leading to cell injury. The membrane damage is usually associated with alterations of membrane lipids and the appearances of lipid peroxidation products, such as those resulting from the peroxidation of unsaturated fatty acids, and with the loss of phosphatidylethanolamine and an increase of phosphatidic acid fractions. Membrane damage of glutathione-poor cells shows increased severity in diminution of surface charges. The oxidation of membrane thiol groups causes changes in the rate of ion transport.

4. Glutathione in Function and Structure

Glutathione exists in two forms in the intracellular space; a substantial amount in sulfhydryl form, a much smaller amount as disulfide, the ratio being 20 to 25:1. About one third of the reduced glutathione is bound to proteins with a disulfide bridge. Reversible or irreversible changes may cause perturbation in the status of reduced/oxidized glutathione in the cell. Reversible changes are induced by enzymes — increased by glutathione peroxidase, decreased by glutathione reductase, directly or indirectly through riboflavin deficiency, or lack of glucose 6-phosphate dehydrogenase activity due to inborn error. Irreversible perturbations are associated with the failure of glutathione synthesis or reactions which bind glutathione into stable form.

One of the assumed roles of glutathione is its possible involvement in mitosis and cell growth. This is dependent on the presence of a high level of the reduced form. Oxidation of the sulfhydryl radical has an inhibitory effect on growth of microorganisms. In reticulocytes at low glutathione concentrations, the two major stages of protein synthesis, namely the initiation of polypeptide chain formation and chain elongation, are diminished, mainly linked with the inhibitory action of glutathione disulfide. With partial restoration of the reduced glutathione, only polypeptide chain elongation starts operating. The initiation process is more sensitive, it recovers only after most cellular glutathione is regenerated. When reduced glutathione is decreased, deoxygenation of hemoglobin is accompanied by conformational changes associated with a decrease in reactivity of the protein sulfhydryl groups, loss of haptoglobin binding capacity, and increased affinity to protons, resulting in changes in the pK values.

Perturbation of the reduced/oxidized glutathione ratio results in many changes in physiological functions leading to pathological effects. High concentration of reduced glutathione is related to the operation of muscular contraction; the contractility of actomyosin fibers is optimal at maximum glutathione concentrations. The release of neurotransmitters and radiation protection is influenced by the glutathione status of the cell. If this system is altered, pathological effects develop such as hemolysis and anemia, cataract and lens opacity, and abnormalities in ion transport.

5. Carnosine and Anserine

These peptides are not derived from protein as is glutathione. They are produced by synthesis from the component amino acids. Carnosine is a dipeptide of histidine and β-alanine, anserine is 1-methylcarnosine (Figure 24). These compounds occur in muscle in fairly large amounts, but their functions are not known. Histidine can be replaced by carnosine in the diet. It has a circulatory depressant activity similar to histamine, but not as potent. Large doses of carnosine may cause vascular collapse.

Anserine is metabolized to 1-methylhistidine, which is excreted in the urine. Rabbit muscle is particularly rich in anserine; presence in the diet increases urinary 1-methylhistidine content. 3-Methylhistidine also appears in the urine; it is probably derived from a peptide similar to anserine. However, no evidence has so far been presented to identify this peptide. The urinary excretion of 3-methylhistidine is very low in patients with Wilson's disease.

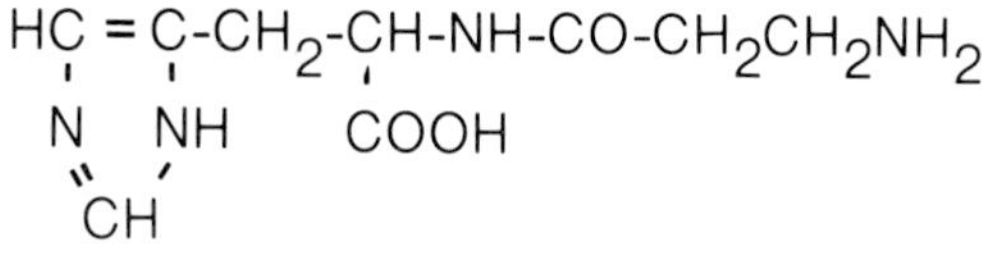

HC = C-CH_2-CH-NH-CO-$CH_2CH_2NH_2$
N NH COOH
CH

Carnosine

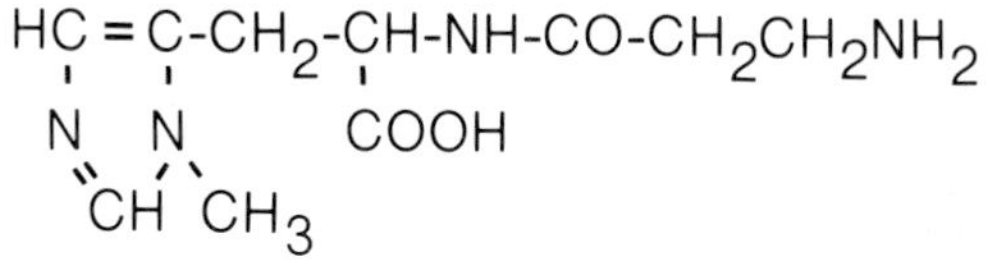

HC = C-CH_2-CH-NH-CO-$CH_2CH_2NH_2$
N N COOH
CH CH_3

Anserine

FIGURE 24. Structure of carnosine and anserine.

IV. PROTEINS

A. Protein Degradation

Proteins taken in the food undergo digestion in the stomach and small intestine. This process is essentially a hydrolysis resulting in the formation of polypeptides, then of progressively smaller peptides, and finally yielding the constituent amino acids. Digestion largely stops at di- and tripeptides; further metabolism results in amino acids. These are reutilized for the synthesis of peptides and endogenous proteins.[138,184] Endogenous proteins are also continually degraded and replaced by new synthesis. All these processes are well organized and interrelated.

The living system is in a continuous state of turnover at several levels from the whole organism to cells and to small molecules within the cell. The various constituents of the body are in a permanent state of flux, that is, all proteins are continually metabolized and reproduced. The turnover rate of protein breakdown and synthesis shows great variations between organs; it may be slow as in the muscle and red blood cells, taking several months, or fast as in liver and intestines where more than half of the total protein is reconstituted within 10 days. Protein turnover even occurs when an apparent steady balance exits, or in deficiency states. Active degradation of proteins persists during periods of negative nitrogen equilibrium and active resynthesis occurs in prolonged starvation. In starvation, serum proteins and hemoglobin may supply the requirements of other organs. Exogenous proteins are generally digested completely under normal circumstances, only very little (3 to 5%) is excreted in the feces.[156,160,163] Some insoluble fibrous proteins like keratin, which are not hydrolyzed by proteolytic enzymes of the human digestive tract, are eliminated from our body unchanged. The end products of protein degradation, mainly amino acids and small peptides, are absorbed from the intestines. There is some evidence, however, that intact protein molecules may also escape digestion in an unaltered state. If an intact foreign protein of other than human origin enters the circulation it acts as an antigen, initiating the production of specific antibodies and sensitizing the host.

The sensitization reaction to a protein with a species specificity different from that of human proteins constitutes the basis for allergic reactions in certain cases. Subsequent exposure to the same foreign protein then may results in a severe, sometimes fatal anaphylactic reaction. Therefore, exogenous proteins must be altered to smaller units possessing no such specificity and without any ability to induce a sensitization reaction. The proper metabolism of exogenous proteins is essential to produce building stones of body proteins, but at the same time it also represents a defense reaction. Patients allergic to food proteins and protein products may show some deficiency in the protein-digesting activity of pancreatic or intestinal secretions.

The complete continuous hydrolysis of intracellular proteins down to their component amino acids seems wasteful, but this process represents some selective advantage to the organism. The catabolism of endogenous proteins is important to remove abnormal proteins which may be produced by denaturation, chemical modification, and as a consequence of deficient synthesis due to mutations or errors in gene expression.[88] The cells have an ability to selectively hydrolyze abnormal constituents and eliminate potentially harmful substances. This degradation is especially important in slow-growing mammalian cells, which do not have the ability to reduce their quantity by dilution or fast elimination.

The degradation of enzymes, biologically active polypeptides, and polypeptide hormones plays a significant physiological and pathological role. The rates differ over a wide range. The average half-life of liver proteins is 3.5 days, but that of specific enzymes varies from 11 min for ornithin decarboxylase to 19 days for lactic dehydrogenase isoenzyme type 5. The life span of various lactic dehydrogenase isoenzymes depends on the distinct half-lives of the subunits. Different protein constituents of cellular organelles such as endoplasmic reticulum, ribosomes, mitochondria, plasma membranes, and chromosomes are metabolized and replaced simultaneously. In the heart, the turnover rate of myofibrillar proteins is variable; in mitochondria, δ-aminolevulinic acid synthetase and ornithin transcarbamylase are degraded and resynthesized much faster than cytochromes and other protein constituents of the inner membrane.

The size of proteins is an essential factor in their catabolism. Larger protein molecules of certain organelles are degraded more rapidly than the smaller components. This may be related to more protease-sensitive sites on the larger units acting as targets for proteolytic action. The supply of substrates and coenzymes also influences the rate of degradation. These effects are often related to ligand-induced conformational changes leading to stabilization. Sometimes substrate binding increases the rate of catabolism. For example, oxygen binding of hemoglobin increases its susceptibility to carboxypeptidase, and glucose binding of glycogen phosphorylase raises the rate of its cleavage by trypsin. Pyridoxal phosphate reduces the sensitivity of some aminotransferases to proteolytic action, similarly to the effect of NAD in NAD-requiring enzymes. The sensitivity of several such enzymes is increased when their cofactor is removed. Hepatic enzymes inducible by glucocorticoids have dissociable coenzymes and undergo rapid turnover, while poorly or noninducible enzymes have long half-lives and tightly bound cofactors. This may indicate that the formation of apoenzymes is the rate-limiting step in the degradation process influenced by the rate of coenzyme dissociation. Thus, the strength of ligand binding may have a general effect by prolonging protein catabolism. The greater susceptibility of cellular enzymes to proteolysis and the low levels of vitamin-requiring enzymes in vitamin B_6 and niacin deficiency diseases, as well as the drug-induced disorders involving cofactor competition, may support the concept that cellular proteins show an increased resistance to degradation when protected by their coenzymes.

Differences in the rate of protein metabolism are dependent upon physiological and pathological circumstances.[88,89,109,111] The degradation of proteins in the muscle is influenced by changes in nutrient supply and contractile activity due to denervation and variation of hormones. An adaptive mechanism allows the organism to withstand changes associated with abnormal circumstances. The muscle and the liver are the major protein reservoirs of our body supply of protein precursors, mobilizing these proteins as energy sources in times of starvation when the caloric intake is reduced. The supply of these amino acids is regulated through protein degradation and by hormones. Insulin can retard protein catabolism in cardiac and skeletal muscle, liver, adipose tissue, and fibroblast. During fasting, the liver and muscle enzymes needed for catabolism are synthesized at a greater rate (i.e., those enzymes related to the processes of gluconeogeneis), or the turnover is reduced as with arginase, or they are degraded more rapidly such as in the case of acetyl coenzyme A carboxylase. An interrelationship exists in the reutilization of amino acids between organs in the intact organism. Amino acids released from one tissue can be transported and incorporated into the proteins of another tissue. Injected lysine is found to be accumulated primarily in hepatic proteins. Subsequently, it is slowly released from the liver into the blood and incorporated into peripheral tissues, especially into muscle proteins. During fasting the opposite phenomenon occurs, peripheral tissues lose amino acids which are utilized by other organs more advantageously. These adaptive regulatory processes primarily serve physiological functions, but they may play important roles in acute and chronic heart and liver conditions and other disorders of protein catabolism.

The importance of intracellular proteolysis in the elimination of unwanted proteins indicates further implication of protein degradations.[98,137] This process is relevant in reducing the incidence of mutational events. Mammalian cells can selectively decompose aberrant proteins resulting from abnormal cell divisions. For example, as a consequence of faster catabolism some human hemoglobin variants, such as sickle cell hemoglobin HbS, are more rapidly degraded by the cell than the normal hemoglobin. The increased degradation of the abnormal proteins may involve initial denaturation or chemical modification.

B. Protein Deficiency

Changes in plasma protein levels, in particular reduction and deficiency, are frequent signs of disease.[23,123] Many clinical laboratory tests are based almost entirely on the study of plasma proteins, although abnormalities often occur in the formation of many proteins found only in the cell. By and large, proteins are involved in all diseases by undergoing either structural or functional changes. This chapter, however, deals only with the relationship between serum proteins and disease. The association between lesions of a specific intracellular protein and the accompanying disease is discussed in specific sections. A great variety of proteins present in blood originate from several cell types, but the liver cell is the site of synthesis of most plasma proteins except for immunoglobulins. The site of the individual immunoglobulin production has not yet been completely established. Plasma cells and lymphocytes are the principal locations of synthesis, but many other cells of the reticuloendothelial system, such as lymphoblasts and reticular cells, also produce immunoglobulins. Albumin is produced entirely in the liver in ribosomes bound to the rough endoplasmic reticulum. In this cellular event, the initial step is a very rapid synthesis in microsomes and a subsequent slow accumulation in the Golgi apparatus. This organ is also the only demonstrated area for the protein synthesis which appears following an inflammatory response brought about by microbial infection or any stimulation which is not normally present such as acute-phase α_1 and α_2 proteins and C-reactive protein. The liver cell produces proteins in high amounts, for example, haptoglobulin, ceruloplasmin, and α_1-acid glycoprotein. Some of these proteins have been localized in the site of inflammation. Many clotting factors of protein nature are synthesized in the liver as well, such as fibrinogen and prothrombin.

Abnormal production of plasma proteins is related to pathological conditions.[60,61] Diseases are most often associated with a decrease, invariably due to a decreased albumin synthesis. The total globulin fraction may also be diminished, but more often it is increased simultaneously with the reduction in albumin concentration, particularly if protein deficiency is associated with certain chronic infections. In general, decreased plasma proteins are replenished simultaneously. In certain cases, such as nephrotic syndrome, the loss of albumin is not only compensated by the increased synthesis of this protein, but an increased production of several globulins in the liver is also stimulated.

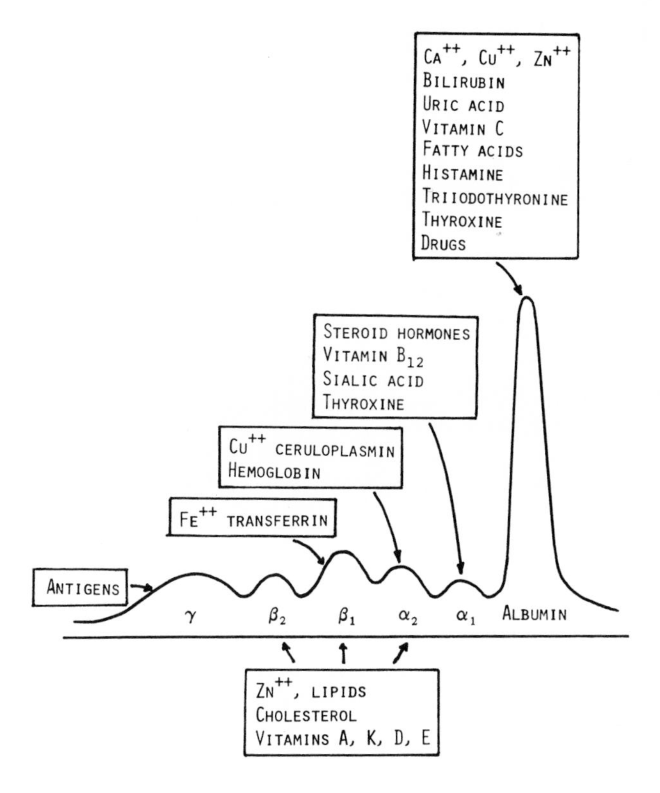

FIGURE 25. Binding of various substances by plasma proteins. Several proteins can bind normal physiological constituents as well as some foreign compounds. In this interaction certain specificities exist. Albumin can bind various nonsteroid hormones, smaller molecules representing the end point of metabolism, vitamin C and drugs, irrespective in the latter group, whether they are acidic compounds such as phenylbutazone, neutral such as digitoxin, or basic such as imipramine. Globulin fractions bind various lipids and these lipoproteins are very important in the transport of lipid-soluble compounds such as cholesterol, steroid hormones, and lipid-soluble vitamins. γ-Globulin fraction specifically interacts with antigens. Plasma proteins transport bound ions. Most are found in the albumin fraction but there are specific interactions like the binding of Fe^{++} to transferrins in β-globulin fraction, binding of Cu^{++} to ceruloplasmin in α_2-globulin fraction, and binding of Zn^{++} to lipoproteins.[194,195]

Plasma proteins have many major functions and any alteration in their level is associated with a reduced capacity to fulfill these activities. They form part of the body protein pool and thus any loss represents a reduction in the source of protein precursors as nutrients for all tissues. Some plasma proteins are antibodies and others possess enzyme activity essential in our protective system against changes in the internal environment. They carry many important constituents to the site of utilization, such as hormones, vitamins, lipids, trace metals, metabolites, and drugs (Figure 25).[125,128] They constitute most of the blood coagulation factors. Proteins are important in the plasma buffering system and in regulating the distribution of the extracellular fluid. Total plasma osmolality is mainly associated with the level of sodium and associated anions; however, in the regulation of the colloid osmotic pressure proteins play a major role. The intravascular protein concentration is much greater than the extravascular one, causing higher hydrostatic pressure on the inside of the capillary wall. The difference is responsible for attracting extracellular fluid at the venous end of capillaries and maintaining the intracellular fluid compartment. Among the various plasma

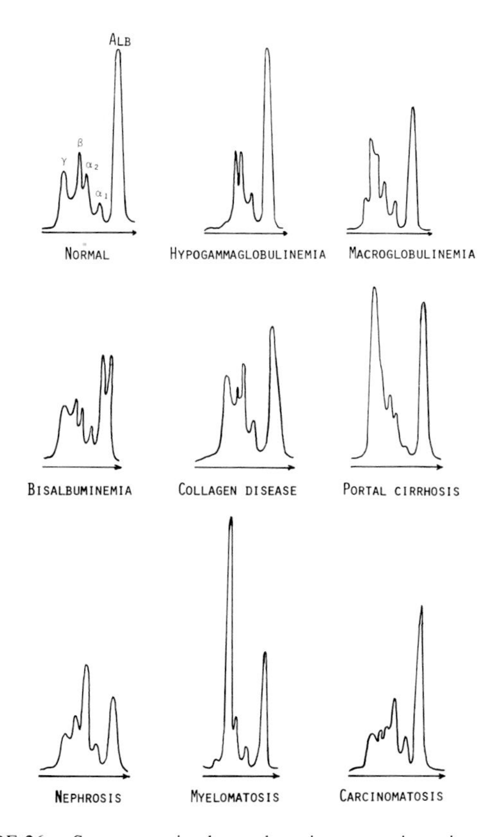

FIGURE 26. Serum protein electrophoresis patterns in various diseases. Compared to the normal pattern, lack or increase of various serum protein fractions characterize many diseases. Hypogammaglobulinemia: reduced γ-globulin represent the lack of immune response by the lymphoreticular system to the antigen stimulus. In macroglobulinemia, the production of β_2-globulin is excessive; in bisalbuminemia the serum albumin level is elevated. In collagen disease the antibody-antigen complex formation is abnormal. In nephrosis the kidney is degenerating without inflammation, serum albumin level is low, but α_2-globulin is increased. In multiple myeloma, a primary malignant tumor of the bone marrow, the specific distribution of the globulin fraction is characteristic — β increased, γ reduced; carcinomatosis, a condition associated with the widespread dissemination of cancer throughout the body, alters the appearance of the globulin fractions.

proteins, albumins are more important contributors to osmolarity than globulins because the latter are smaller molecules.

The concentration of plasma proteins is constant in the normal individual.[108,141] Plasma proteins are in equilibrium with the proteins of various tissues, mainly with hepatic proteins which are the major source of replacement following losses. There is minor albumin production in other organs. Normally the reduction of plasma albumin is replaced by tissue proteins at a very rapid rate, while excess albumin leaves the circulating blood and is metabolized in the tissues. In disease, this equilibrium is impaired leading usually to a greater deficiency in the plasma than in the organs. Separation of serum proteins by electrophoresis or any other method reveals specific patterns which can be used for the diagnosis of various diseases and for the identification of the organ involved (Figure 26). In protein deficiency atrophy develops in all tissues, but in some conditions such as burns, protein concentration can be related to inadequate dietary intake due to food restriction or decreased absorption from the intestinal tract caused by disease, repeated vomiting resulting from gastrointestinal obstruction, or any other causes — the diminished production being brought about by liver diseases. Significant protein loss also leads to reduced plasma levels; this occurs in extensive burns, kidney diseases, and intestinal protein excretion in enteritis. Enhanced breakdown is caused by fever, necrosis, leukemia, and cancer. Prolonged administration of adrenal steroids, particularly 11-oxygenated adrenocortical hormones, and excessive amounts of thyroid hormones also cause elevated protein catabolism and thus decreased plasma protein concentration. Protein anabolism is stimulated by pituitary growth hormone, testosterone, and insulin. There are certain physiologic and pathologic conditions which are associated with increased need of protein: growth, pregnancy, and hyperthyroidism.

C. Plasma Proteins

The plasma contains a complex mixture of simple proteins, glycoproteins, lipoproteins, and metalloproteins; normal concentration is about 70 g/ℓ. Some protein exists in combination with other proteins. The molecular size of these proteins varies from 45×10^3 to 1.5×10^6 daltons. Many plasma proteins have special roles; some proteins possess enzyme activity and are secreted into the blood stream from various tissues. Others function as carriers or vehicles of substances[125,128] and exhibit the physicochemical characteristics of colloid molecules. It is probable that every plasma protein will come to have a definite function and not be merely a large colloid particle.

Factors controlling normal plasma protein levels have not yet been completely established. When a particular fraction such as albumin is reduced, it is quickly followed by an enhanced synthesis, suggesting a feedback regulation. The nature of this regulatory mechanism is not understood at the present time. It seems to be nonspecific; in particular in nephritis, the loss of proteins from the plasma usually develops faster than the replacement by albumin secretion from the tissues. Consequently, an equilibrium may never become established. Further complications are associated with the reduced protein production of the site of synthesis, especially in the liver, due to the inadequate levels of protein precursors. During recovery the conversion of these precursors to albumin is delayed and the restoration of normal plasma content is slow because the site of protein synthesis must first be sufficiently reestablished.

1. Albumin

Albumin constitutes the largest proportion of normal plasma proteins, about 65% of the total. It is essential as nutritional protein for the production of tissue proteins.[78,127] Plasma albumin is mainly responsible for the stabilization and maintenance of blood volume by regulating the exchange of fluid between vascular and extravascular compartments. Albumin is an essential carrier for a number of compounds: various normal metabolites, free fatty acids, bilirubin, drugs and their metabolites, copper, and zinc. Some of these substances are also strongly bound to specific globulins, but albumin takes part in the control of their excretion and exchange between plasma and tissues. Albumin is synthesized in the liver; reductions in plasma level, therefore, are certainly signs of hepatic disease representing an impairment of protein synthesis.

There are normal conditions associated with a reduced albumin level. Newborn infants have low plasma protein concentration; in this, both albumin and globulin are low. The reduction is even greater in premature babies. During growth the plasma protein concentration slowly rises, reaching adult levels in the third year. In pregnancy, plasma albumin is decreased. This decrease is greatest during the last trimester and mainly related to maternal protein deficiency caused by the fetal demand. The depression of the albumin level is continued for a few days after parturition due to hydremia. When the water concentration returns to normal the plasma albumin concentration follows suit.

In most diseases, plasma albumin concentration is low.[163,168,175,177] These conditions include many acute or chronic illnesses, particularly acute and chronic liver disease, malnutrition, malabsorption, gastrointestinal loss, nephrotic syndrome, and extensive burns. The major causes of hypoalbuminemia are decreased synthesis and loss due to disease. Daily catabolism of albumin involves about 4% of the total body pool. Liver disease brings about a reduced synthesis. Amino acid deficiency also impairs its production, as do conditions associated with the depletion of proteins such as exudation in extensive burns, protein-losing enteropathy, or nephrotic syndrome. There are reports that in many acute conditions, and including minor diseases such as the common cold or boils, serum albumin is slightly decreased. In these cases the production of other proteins is elevated, but upon recovery the balance of synthesis quickly normalizes. The extent of the difference from normal concentration usually reflects the severity of the clinical condition. The albumin level is decreased and α-globulin is raised in acute bacterial infections, malarial attacks, and rheumatic fever. In chronic infections plasma albumin levels are slightly lowered. In analbuminemia, a congenital inherited disease, plasma albumin is completely absent; however, a compensation by increased globulin release from the tissues prevents the formation of severe edema. There are no clinical conditions associated with increased plasma albumin levels.

Abnormally high albumin only occurs with dehydration. In bisalbuminemia two types of albumin are present in the blood. This is a congenital anomaly without any clinical consequences whatsoever.

Low serum albumin is often associated with some secondary changes. Since albumins are essential proteins in the maintenance of fluid distribution between the plasma and tissues, a reduced level involves edema formation. Hypoalbuminemia is also accompanied by decreased calcium, since albumin carries almost half of the plasma calcium. This is, however, of no biological importance.

2. Globulin

Classically, globulins are divided into α-, β-, and γ-globulins: α-globulins are further subdivided into α_1- and α_2-globulins and the β-globulins into β_1- and β_2-globulins. The α- and β-globulin fractions contain prothrombin and other fractions involved in blood coagulation synthesized in the liver and protein hormones produced by the pancreas, pituitary, or parathyroid glands; γ-globulins represent antibodies or immunoglobulins formed in plasma cells, lymphocytes, and probably in other cells of the reticuloendothelial system such as lymphoblasts, plasmoblasts, and reticular cells.[57,115,116,174] Lipoproteins are localized in these fractions; these particles contain most serum cholesterol, phospholipids, and triglycerides. Glycoproteins and metalloproteins are also present in α- and β-globulin. Transferrin associated with iron transport is bound to the β_1 fraction and ceruloplasmin, responsible for copper transport, is separated with an α_2-globulin component. Thyroxine-binding globulin, and haptoglobulin which binds hemoglobin, are also present in the α-globulin region.

Diseases where α_1- and α_2-globulins are increased include inflammatory processes, trauma, and collagen diseases; in general, conditions where tissue destruction or proliferation takes place. Low levels are found in malnutrition, malabsorption, and hepatic parenchymal disorders due to lack of precursors. In these cases the proper synthesis is diminished. β-Globulins are elevated in obstructive jaundice, nephrotic syndrome, and in certain types of atherosclerosis where the β-lipoprotein level is enhanced.

γ-Globulins contain several protein fractions including antibodies. Proteins possessing antibody properties also migrate with β-, and occasionally with the α-globulin fractions. Irrespective of their distribution pattern, these immunologically active proteins are called immunoglobulins. Human immunoglobulins can be divided into three major classes: IgG (gamma), IgM (macro), and IgA. Two minor classes have been described recently, IgD and IgE.

The basic chemical structure of all these immunoglobulins is very similar (Figure 27). Each unit contains four polypeptide chains consisting of two large polypeptides of molecular

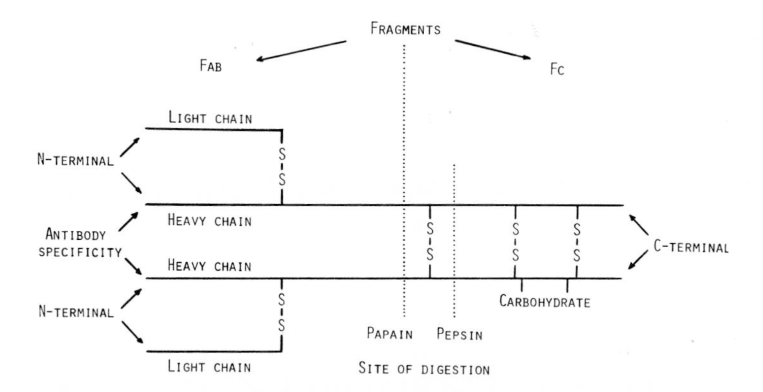

FIGURE 27. Structure of immunoglobulin peptide chains. Papain hydrolysis splits the IgG molecule into three approximately equal fragments.[196] Two fragments are identical and carry the antibody-combining site, Fab; the third fragment crystallizes easily, Fc. After papain hydrolysis almost the entire original biological activity is preserved. Peptic hydrolysis leads to different fragments, resulting in $(Fab^1)_2$ which is slightly larger than the papain Fab part. Fc splits to smaller fragments.[197]

weight about 55×10^3 called the heavy chains and two smaller polypeptides of molecular weight about 22×10^3 called light chains.[6] The heavy chains are different in all the five classes of immunoglobulins (γ in IgG, μ in IgM, α in IgA, δ in IgD, and ϵ in IgE). In each monomer unit the heavy and light chains are linked through disulfide bridges. There are only two types of light chains (κ and λ) and, therefore, two combinations of immunoglobulin structures exist in each class.

The class of immunoglobulins is determined by the heavy chain, while the type is classed according to the light chain. Immunoglobulins are produced from balanced synthesis of the heavy chains, which is complete within $2^1/_2$ min, and of the light chain, complete within 1 min. The synthesis takes place on ribosomes of two different sizes and it is most rapid during the S phase and slowest uring mitosis and in the early GI phase of the cell cycle.[6,7,103,118] After the immunoglobulin molecules have been assembled, they move through the cisternae of the endoplasmic reticulum where carbohydrate moieties are added.[4] Eventually the protein carbohydrate complex reaches the Golgi apparatus, where more carbohydrate is attached. The completed immunoglobulin molecules are then secreted into the circulation. There is some delay between synthesis and secretion due to the fact that the synthesis of the solubilizing light chains occurs in the membrane of the endoplasmic reticulum and in the Golgi region, whereas the heavy chains are produced on polyribosomes which are distributed throughout the cell.

The antibody specificity of the immunoglobulin molecule resides mainly in the heavy chain, while the light chain alone possesses minimal activity. The metabolism of these proteins is dependent on nonspecific proteolytic processes partly regulated by feedback controls. The rate of catabolism is proportional with the serum level of IgG and IgA; the breakdown of IgM is not related to its serum concentration. In disease states there are variations in the turnover of these proteins; however, there are conditions when the rate of synthesis of IgG and IgA is reduced, such as in congenital or acquired hypogammaglobulinemia. Through feedback control, the catabolism is significantly decreased in order to maintain an effective serum concentration.

IgM is the largest complex, its molecular weight being 9×10^5 daltons, given the association of five units of size similar to IgG and IgA; the relative amount in the human serum is about 7% of total protein. Sulfhydryl compounds may depolymerize the bigger unit to monomers. IgM contains about 12% carbohydrate and is usually found as a narrow

MET-HIS-GLU-ALA-LEU-HIS-$AspNH_2$-HIS-TYR-THR-

$GluNH_2$-LYS-SER-LEU-SER-LEU-SER-PRO-GLY

HUMAN IGG (γ_2B, WE)

MET-HIS-GLU-ALA-LEU-HIS-$AspNH_2$-ARG-PHE-THR-

$GluNH_2$-LYS-SER-LEU-SER-LEU-SER-PRO-GLY

HUMAN IGG (γ_2C, VI)

FIGURE 28. C-Terminal sequence of human immunoglobulin IgG heavy chains.[198] A similar peptide can be obtained from many animal immunoglobulins. The C-terminal sequences of normal and pathological human heavy chain are identical, but the N-terminal peptides are entirely different.[199]

band between the γ and β peaks, by various separation methods; its normal serum concentration ranges between 0.04 and 0.12 g/dℓ. IgD occurs only in extremely small amounts, 0.0003 g/dℓ; the amount of IgE is even less.[30]

Among these classes, IgG consists of about 70% of the total immunoglobulins ranging between 1.0 to 1.4 g/dℓ serum and is located mainly in the γ-globulin fraction. This fraction contains the majority of bacterial and viral antibodies. IgG freely crosses through the placental barrier and it is almost the only component of immunoglobulins in the blood of the newborn. It is a glycoprotein, molecular weight 16×10^4 daltons, containing about 2.5% carbohydrate including mannose, galactose, fucose, glucosamine, and sialic acid. The terminal sequence of the heavy chain of several immunoglobulins has been identified and only small differences have been noted between various IgG immunoglobulins of human and animal origin. The C-terminal sequences of normal and pathological human heavy chains are identical, but the N-terminal peptides are distinct (Figure 28). However, very different primary structures are found in the three types of heavy chains (γ, μ, and α); there are only a few common peptide patterns.[150] The polysaccharide is attached to an aspartic acid residue. IgA is also found in the γ region, its relative amount being about 22%. The concentration varies between 0.07 to 0.4 g/dℓ. This fraction includes antibodies to tetanus, diphtheria, typhoid O, paratyphoid B, thyroglobulin, and insulin. IgA is a mixture of heterogeneous glycoproteins. The average molecular weight is the same as IgG, but contains some di-, tri-, and tetramer molecules and about 11% carbohydrate.

In most diseases, in contrast to the reduction of the plasma albumin content, total serum globulin concentration shows a rise. The increase may be diffuse due to an enhanced production of antibodies but can be localized in a discrete band of paraproteins. In acute infections the α-globulins are elevated, whereas in chronic ones the increase is mainly due to γ-globulins.

The increase in γ-globulin content is great and diffuse in several diseases such as cirrhosis, systemic lupus erythematosus, sarcoidosis, rheumatoid arthritis, and myelogenous leukemia (Table 2). Discrete bands characterize myelomatosis, macroglobulinemia, and various types of paraproteinemia. In several diseases the degree of γ-globulin increase provides a fairly reliable measure for the assessment of the progress of the condition.[78]

Table 2
SERUM PROTEIN FRACTIONS IN VARIOUS DISEASES

Disease	Protein						
			Globulin				
	Total	Albumin	Total	α_1	α_2	β	γ
Normal	7.18	3.65	3.53	0.42	0.67	0.91	1.53
Newborn	5.50	3.80	1.70	—	—	—	—
Protein deficiency	4.70	2.50	2.20	—	—	—	—
Diabetes mellitus	6.50	2.82	3.68	0.33	0.87	0.97	1.51
Nephrosis	4.38	1.34	3.04	0.24	1.56	0.78	0.46
Glomerulonephritis	6.11	2.42	3.69	0.42	0.97	0.77	1.53
Cirrhosis	6.52	2.31	4.21	0.38	0.60	0.90	2.33
Viral hepatitis	7.56	3.65	3.91	0.28	0.55	0.83	2.25
Acute infections	6.62	2.73	3.89	0.48	0.88	0.86	1.67
Chronic infections	7.01	2.93	4.08	0.43	0.91	0.97	1.77
Tuberculosis	6.70	3.50	3.20	—	—	—	—
Rheumatoid arthritis	6.83	2.77	4.06	0.45	0.93	0.85	1.83
Acute rheumatic fever	7.30	3.30	4.00	0.52	0.90	0.80	1.78
Scleroderma	6.26	3.26	3.00	0.36	0.58	0.78	1.28
Lupus erythematosus	6.86	2.49	4.37	0.42	0.76	0.83	2.36
Sarcoidosis	7.84	3.29	4.55	0.39	0.84	1.10	2.22
Lymphomas	5.66	3.08	2.58	0.45	0.75	0.83	0.55
Myelogenous leukemia	7.16	2.86	4.30	0.42	0.78	0.72	2.38
Multiple myeloma	9.80	3.80	6.00	—	—	—	—
Hypogammaglobulinemia	5.51	3.60	1.91	0.38	0.50	0.73	0.30

Note: Values are expressed in g/dℓ serum.

From Sunderman, F. W., Jr. and Sunderman, F. W., *Am. J. Clin. Pathol.*, 27, 125, 1957. With permission.

In contrast to the increased γ-globulin, low serum levels are usually associated with increased sensitivity to bacterial infection. There are conditions where the serum contains very little or no γ-globulin. These are the antibody deficiency syndromes called agamma- or hypogammaglobulinemia. People suffering from this group of diseases develop low resistance against infection.

There are several forms of hypogammaglobulinemia. It is seen in infancy as a transient condition, can be a congenital disease, or may occur at a late age, or can be secondary to some other condition and the deficiency may not be complete. In the newborn the amounts of serum proteins are usually low, which is normally corrected shortly. However, some 3- to 4-month-old infants show reduced γ-globulin levels. This phenomenon, called physiologic or infancy hypogammaglobulinemia, occurs because their lympho-reticular system does not synthesize adequate quantities of γ-globulin necessary to replace the γ-globulin supply provided by the mother during *in utero* life. Most maternal immunoglobulin is freely transported through the placental barrier and, at birth, full-term babies have almost the same immunoglobulin concentration as their mother. The rate of metabolism of this protein is slow, the half-life is about 1 month, so at the end of a 3- to 4-month period only 6 to 12% of the original γ-globulin is left. In normal infants the development of the lymphocytes and plasma cells provide adequate stimulus for protein synthesis and the total production at 3 months is adequate to compensate for the gradual breakdown of maternal proteins. However, if the infant produces its own immunoglobulins slowly in response to its own bacterial and viral environment, temporary hypogammaglobulinemia develops. The consequence of the low immunoglobulin level is inadequate protection against infection.

The physiologic hypogammaglobulinemia usually does not last longer than 6 to 8 months, but in rare cases γ-globulin reaches the normal level even later.

Congenital agammaglobulinemia is characterized by the complete inability of the patient to produce immunoglobulins. This disorder is a sex-linked recessive genetic defect transmitted by mothers to their male offspring. It is usually discovered at about 6 months from birth but is manifest only after the second year. The lymph nodes of the diseased children contain no plasma cells and show an abnormal follicular organization. Many bacteria such as streptococci, pneumococci, and staphylococci can cause the onset of disease, but these organisms can be controlled by antibiotics. The incidence of rheumatoid arthritis is greater in these people. Besides antibiotic therapy, the administration of γ-globulin is usually necessary to prevent the occurrence of chronic infections.

In another type of congenital hypogammaglobulinemia that manifests in both sexes, serum γ-globulin level varies between 0 and 0.3 g/dℓ, whereas in congenital agammaglobulinemia by which only male infants are affected, γ-globulin ranges between 0 and 0.05 g/dℓ. The late-onset hypogammaglobulinemia occurs between 10 to 70 years and is associated with a disturbed or absent capability of the patient to produce immunoglobulin. The disorder affects both males and females and γ-globulin level varies between 0.005 and 0.08 g/dℓ. Secondary hypogammaglobulinemia is due to a diffuse disease of the lympho-reticular system interfering with the normal immunoglobulin synthesis. This occurs as a complication of leukemia, multiple myeloma, lymphoma, total body irradiation, and following the administration of some antitumor agents. Selective loss in the synthesis of immunoglobulins may cause a deficiency in some components of this group, while the others are produced at the normal rate. This disease is called dysgammaglobulinemia and may represent an intermediary stage in both the primary acquired and congenital sex-linked hypogammaglobulinemia. These individuals invariably develop a greater sensitivity to infection. The various types of this disorder can be distinguished by the lack of specific peaks in the electrophoretic patterns of serum proteins.

3. Specific Proteins

The production of many specific proteins is associated with disease. Hundreds of these plasma proteins have already been identified. They can be grouped into four major categories: carrier proteins,[31,97] immunoglobulins,[57,82] complement proteins, and acute phase reactants.[165,176] In many instances a plasma protein can be put into more than one group. The most frequently occurring specific problems will be discussed below.

4. Carrier Proteins

The primary role of these proteins is to carry a substance from one location in our body to another in an organized and controlled pattern.[45,166] In some instances, when the carrier protein is loaded with the ligand, the whole complex very rapidly disappears from circulation, such as protease inhibitors complexed with an enzyme,[98] haptoglobin bound hemoglobin molecule. Some carrier proteins also provide storage until the substance bound to them is released, such as carrier proteins participating in the transport of digested substances in the intestinal wall. Some carrier proteins may load and unload the respective substance repeatedly during their existence, some others link with substances more permanently.

The synthesis of carrier proteins is related. The demand or burden of the substance to be removed, such as β-lipoproteins, are increased when lipid supply is increased; on the other hand, transferrin levels are elevated when iron in the circulation is decreased. Elevated iron stores are associated with reduced transferrin synthesis. In many cases changes in carrier protein levels represent an adaptive response to a disease process and removal of a substance represents secondary action of synthesis.

a. Prealbumin

This carrier protein is rich in tryptophan and composed of four subunits with unique dual transport functions.[121,129] It binds thyroxine and on separate sites vitamin A (retinol). It plays a minor role in thyroxine metabolism and a major role in transporting retinol in complex form through the vascular space to cell surfaces where the vitamin A-prealbumin complex is anchored and vitamin A becomes bound to cell membrane.

The level of prealbumin is very low at birth and rises rapidly to adult level within the first few weeks of life. Its concentration is relatively high in the cerebrospinal fluid. Malnutrition and inflammatory processes adversely affect serum levels. Negative nitrogen balance generally, whether due to malignancy-associated cachexia, hepatic disease, chronic infection, or malnutrition causes a decreased level of prealbumin and retinol binding. Due to its relatively short life-span, a reduction of prealbumin in the blood is often an early indication of hepatitis, even before other problems are modified. Reversal of the disease process is often accompanied by a rapid synthesis and release of prealbumin and retinol. Corticoids stimulate prealbumin production.

b. Transferrin

Transferrin, or siderophilin, is a single protein responsible for the transport of iron from the site of absorption to the bone marrow where it is incorporated into the heme molecule.[46,78,102,136,167] Apoferritin is colorless; iron binding produces a pink color which contributes to the color of serum. Transferrin removes absorbed or released iron and makes it available for hemoglobin synthesis. Excess iron is stored bound to ferritin in the liver.

The concentration of transferrin is mainly influenced by the dietary intake of iron. Nutritionally deficient diets lacking in iron result in low serum iron and high transferrin levels. Parasitism intensifies iron deficiency through malabsorption or loss of blood. Transferrin plays an important role in bacteriostasis. Fe^{3+} is essential for the survival and replication of bacteria. Individuals with low iron and raised transferrin levels effectively suppress bacterial growth, the administration of iron on the other hand worsens the infection. In contrast, patients with low transferrin deal poorly with bacterial infection; congenital atransferrinemia leads to sepsis and death.

Severe liver diseases such as cirrhosis cause marked reduction of plasma transferrin levels. Similarly, protein-losing disorders such as nephrotic syndrome are also associated with reduced transferrin. Elevated levels may occur in severe iron deficiency or due to depletion of body iron depots.

c. Immunoproteins

These are not generally grouped as specific proteins. The role of various immunoglobulins in disease has been discussed earlier.

d. Cryoglobulin

In many diseases of the reticuloendothelial system the serum often contains cryoglobulin. These include multiple myeloma, rheumatoid arthritis, lupus erythematosus, leukemia, cirrhosis, lymphosarcoma, and polycythemia vera. In the bone marrow a rise in plasma cell, or cells resembling them, have been described and it is likely that these cells are the site of cryoglobulin production. The special feature of this protein is that it precipitates when the temperature is decreased below 37°C and becomes soluble again when the temperature is raised to 37°C. Cryoglobulins are found in the γ-globulin fraction, but are sometimes located separately between β- and γ-globulins. In many cases, when the serum cryoglobulin content is high characteristic clinical symptoms occur. These include purpura, ulceration of the skin, and peripheral gangrene. This protein precipitates in vivo in the small blood vessels of the extremities on cooling, which is responsible for the major manifestation of the disease:

intolerance to cold exposure. The appearance of the symptoms is dependent on serum cryoglobulin concentration, interaction with other proteins, and temperature reduction needed for the precipitation of the amount present in the blood. Precipitation of proteins in blood vessels is accompanied by an inflammatory response. The conditions, therefore, can be improved by the administration of adrenocorticotropic hormone, which decreases the serum cryoglobulin level.

In the cold precipitation of cryoglobulin, the protein sulfhydryl groups seem to play an essential role. Penicillamine and other mercapto compounds depolymerize the macromolecular aggregate. These drugs have been applied with some success in reducing the quantity of plasma cryoglobulin and controlling the disease. The principal cause of death is linked with the production of an abnormal serum constituent, but actual death may occur by the precipitation and blockage of the blood vessels of vital organs such as heart, lung, brain, and kidney.

e. Acute Phase Reactants

The concentration of these proteins increases or decreases in connection with inflammatory processes.[110] Most acute phase reactant proteins can be located in the α and β bands by electrophoresis. Some occur in relatively high concentrations, such as various mucoproteins, haptoglobin, α-antitrypsin, fibrinogen, ceruloplasmin, and C-reactive protein. Most of these proteins exert a specific function and their increased production and release during inflammation represents the action of the body biochemical organization to seminate and repair the injuring process. The biochemical mechanism responds similarly in sepsis, rheumatoid arthritis, and neoplasia suggesting a probable common pathway regulating the production and control of these proteins.

Many components have been proposed as mediators of the inflammatory response, such as kinins, clotting and fibrinolytic factors, and complements. The immune system may also be involved in initiating complement formation. Immunosuppression can reduce the inflammatory response. The mechanism of the production of acute phase reactants is not known. Individual components associated with tissue damage, such as lysosomal enzymes and prostaglandins, can stimulate the hepatic production of these glycoproteins.[33]

f. Mucoproteins

These are mainly present in the α_1-globulin fraction.[3] The determination of the serum mucoprotein level often provides means to diagnose some disease. In the acute phase of the inflammatory process and in proliferative diseases serum mucoproteins are enhanced, probably due to their release from the tissues. Many infections and traumatic and neoplastic processes cause elevated levels of mucoproteins. However, only in some conditions has a relationship been established between a particular disease and an increased mucoprotein level. In lymphosarcoma, enhanced serum mucoprotein and low γ-globulin levels are observed; in tuberculosis, characteristically high levels can be used for diagnosis. Some conditions modify the need for serum mucoproteins, such as the increased level that occurs in uncomplicated inflammatory conditions. If it is accompanied by liver damage or insufficiency of the adrenal system, the increase may be compensated and normal level is noted although the diseased state is still present. Hepatic cell injury due to toxic or viral hepatitis or cirrhosis usually lowers the serum mucoprotein level. Adrenal insufficiency is also accompanied by a reduction of mucoproteins; in contrast in Cushing's disease cortisone or corticotropin therapy causes an increase.

g. Haptoglobin

The serum level of another specific protein, haptoglobin, is also associated with disease conditions.[85,135] Haptoglobin is synthetized in the liver and it is found chiefly in α_2-globulin fraction. The main characteristic of this protein is its specific binding of hemoglobin which may be present in the plasma. The removal of the haptoglobin-hemoglobin complex from the serum is very fast so normally the serum level is zero. Very low haptoglobin concentrations, therefore, suggest hemolysis. The importance of the formation of the haptoglobin-hemoglobin complex is that it prevents considerable loss of iron through urinary hemoglobin elimination.

At birth little or no haptoglobin is present in the blood. The rate of synthesis is stimulated by inflammatory processes. The increase is most drastically observed in tissue necrosis, particularly if it is due to malignancy. Haptoglobin level is also elevated in bacterial and viral infections and tuberculosis and in patients with extensive tissue damage from burns and abscess. Increase occurs in many cases of cancer with metastases associated with fever and necrosis. Hypohaptoglobulinemia is connected with diseases complicated with severe hepatic involvement.

h. Paraproteins

The presence of specific proteins in the serum and urine characterizes several lymphoproliferative diseases.[16] Multiple myeloma or myelomatosis represents a tumor of widespread locations, mainly in the bone marrow of skull, spine, sternum, ribs, shoulders, and occasionally in other bones. The clinical signs of this disease include the development of soft areas in the skull and elsewhere, causing skeletal deformity and spontaneous fractures. The tumor is composed of abnormal plasma cells and its proliferation destroys the neighboring bone tissue. These cells produce unusual globulins without any demonstrable biological function, called paraproteins. These proteins show structural differences in each individual being and are eventually excreted through the kidney. In multiple myeloma the urine of most patients contains an unusual protein, called Bence-Jones protein, which was the first recognized tumor product. It precipitates at acid pH by heating between 45 and 60°C. Bence-Jones protein is often detectable in the urine in other lymphoproliferative disorders such as lymphomas, soft tissue plasmacytoma, lymphatic leukemia, and Waldenström's macroglobulinemia. Bence-Jones proteinuria has clinical significance: patients with frankly invasive malignant lymphomas have a higher incidence of the synthesis of this abnormal protein. However, once the patient develops renal sufficiency, it can also be demonstrated in the serum.

i. Bence-Jones Protein

This protein represents a heterogeneous group of plasma proteins, being a mixture of monomers and dimers of the light chains of immunoglobulins, with molecular weights varying between 20 to 45 $\times$ 10^3 daltons.[146] It is usually synthesized as monomers of molecular weight 22 $\times$ 10^3 but these undergo postsynthetic modification to form a dimer through disulfide linkages. On electrophoresis, it mainly migrates into the region of γ-globulins but also may be found between the β and γ bands or in the α_2- or γ-globulin region. The different character of Bence-Jones protein as compared with normal globulins can also be demonstrated by chemical and immunologic methods. Amino acid sequencing studies have shown that, in the basic paraprotein part of this molecule, the amounts of glutamic and aspartic acid terminals are significantly different from those of normal serum globulins.

Many clinical changes found frequently in multiple myelomia include neuropathy, amyloidosis, anemia, and nephropathy. Neuropathy may be developed from the compression of the nerve or the cord from the expanding bone tumor associated with demyelination or amyloid involvement of the peripheral and other nerves. Anemia maybe due to hemolysis related to the synthesis of abnormal serum proteins, or hemorrhage from the gastrointestinal tract affected by multiple myeloma. It may be directly associated with a suppression of the normal erythrocyte production in the bone marrow infiltrated by abnormal plasma cells, or indirectly to the myeloma renal disease causing impairment of erythropoiesis. Many patients with multiple myeloma develop nephropathy with proteinuria; due to diminished kidney function, casts containing uric acid or calcium phosphate deposits are formed and nitrogen retention progresses. The cause of death is often related to uremia in addition to infection and hemorrhage.

j. α_2*-Macroglobulin*

This is a large protein and important in hormone transport.[27,78,104] Values are high in newborns and rise during childhood, followed by a fall at puberty. In adults the serum level is sex-dependent, being higher in females than in males. This indicates that the suppression is connected with sex hormones, testosterone being more effective than progesterone. The normal serum level of α_2-macroglobulin is not influenced by most diseases. Significantly high values may, however, occur during diseases that affect glomerular filtration such as membranoproliferative glomerulonephritis and nephrotic syndrome. In the syndrome of Waldenström's macroglobulinemia the proliferation of lymphocytoid plasma cells in the bone marrow results in the excessive production of a α_2-macroglobulin. This protein often represents more than 20% of total serum protein. Bence-Jones protein is present occasionally in the urine. The proliferation of these abnormal cells leads neither to tumor formation nor to bone lesions. The clinical symptoms include anemia associated with bleeding from mucous membranes, liver and spleen enlargement, lymphadenopathy, microangiopathy, and hyperviscosity syndrome.

k. α_1*-Acid Glycoprotein*

This acute phase protein carries about 10% of the protein-bound carbohydrate of the serum. Children have significantly lower levels than adults.[164] α_1-Acid glycoprotein is enhanced in late pregnancy. Conditions with cell proliferation stimulate the synthesis of this protein. It has been applied to monitor malignancies in patients, although α_1-acid glycoprotein is not a tumor maker.

l. α_1*-Antitrypsin*

It is also called α_1-trypsin inhibitor. At birth the serum concentration of α_1-antitrypsin is low, but rises rapidly. There is no significant change throughout life except in inflammatory conditions. Pregnancy and contraceptive steroids bring about an increase, which returns to normal after medication is discontinued. Due to polymorphism in heterozygous individuals associated with low levels of antitrypsin, severe lung or liver disease may develop.[24-26,55,75]

The primary action of antitrypsin is the inhibition of proteolytic enzymes by producing partially reversible complexes. Besides trypsin, it can block the activity of a wide range of enzymes such as chymotrypsin, collagenase, elastase, thrombin, and plasmin. During inflammatory conditions α_1-antitrypsin levels may rise to extreme heights. In mild inflammation following minor surgery, the level is increased within 2 to 3 days and remains elevated for a week or more. Bacterial infection, rheumatoid arthritis, vasculitis, and carcinomatosis also lead to enhanced levels provided the individual belongs to the proper genotype.

m. Ceruloplasmin

This is a blue copper-containing protein. Its level is low at birth. Pregnancy and the administration of drugs cause an increase.[31] The level of ceruloplasmin is also enhanced in many inflammatory disorders such as infection and malignant diseases.

n. Fibrinogen

This is one of the major plasma proteins and, as a precursor of fibrin, plays a key role in inflammation.[107] It is, however, not present in the normal serum. The erythrocyte sedimentation rate is connected with the plasma fibrinogen associated with inflammatory processes.

Fibrinogen is produced in the liver and its synthesis can be enhanced by hormones or by the onset of inflammation. The polymerization of the soluble fibrinogen to insoluble fibrin is the basis of hemostasis. The metastable fibrinogen is converted to a stable fibrillar protein through a series of changes. Thrombin enzymatically attacks fibrinogen, resulting in the release of two peptides, A and B. Dissociation of peptide A leads to the formation of a long intermediate polymer. Several polymers associate in a parallel fashion producing the fibril. Branching of the fibrils occurs in the presence of Factor XIII and Ca^{2+} to produce the stable network of the clot. Through cross linkages, further stabilization takes place between chains resulting in the clot retraction. The clot is not permanent. Plasmin, a potent proteolytic enzyme, cleaves the fibrin polymer and releases split products of varying molecular sizes. Usually, lysis of fibrin results in low-molecular-weight products.

o. Plasminogen

This is the precursor of plasmin, which is an active proteolytic enzyme involved in the removal of fibrin. The level of plasminogen at birth is close to the adult level. It is raised during pregnancy.[9,130]

Plasminogen is converted to plasmin by the action of several endogenous activators existing in many tissues, particularly in the vascular walls and in the urinary tract. The latter is called urokinase. Kallikrein can also induce the transformation. Some exogenous substances are also potent activactors, such as streptokinase. By the action of these activators, a selective cleavage of plasminogen produces the active proteolytic enzyme plasmin. Plasmin hydrolyzes fibrin into low-molecular-weight fragments no longer capable of polymerization. If this process is uninhibited bleeding disorders are produced, such as in thoracic and prostate surgery, leukemia, and obstetrical complications.

There are endogenous and exogenous plasmin inhibitors which control the plasmin-induced fibrinolysis. Strong protease inhibitors are α_2-antitrypsin and α_2-macroglobulin, less effective are α_2-antiplasmin, antithrombin III, and C1-inhibitor.

p. C-Reactive Protein

This is probably the most sensitive acute-phase reactant protein. Its normal blood level is very low; it is raised to very high levels during an illness.[42,47] C-reactive protein is able to bind pneumococcal C-polypeptide in the presence of Ca^{2+}. It may also interfere with the immune mechanism by activating the complement pathway and binding with certain T-lymphocytes and thus impairs their functions.

q. Complement Proteins

Complement proteins form a family named from C1 to C9.[2,87,100,144,162] They are linked functionally to each other and connected by the antigen-antibody reaction to the inflammatory response. In some cases, the complement action is associated with the entire complement chain sequence C1 to C9, such as the lysis of erythrocytes or bacteria. In contrast, phagocytosis only requires the enzymic hydrolysis of C3 to form C3b, which alone can induce the process. Complement proteins participate in the acute phase response to a small degree as compared to the action of the phase reactant proteins.

D. Interrelations between Blood and Tissue

If protein deficiency is of long duration, all tissues become involved and many functions deteriorate. Prolonged protein deficiency has many manifestations, being frequently associated with the lack of many essential body constituents, such as important cations like potassium and calcium and other minerals and vitamins. Deficiency of these important components often enhances the severity of symptoms caused primarily by the deprivation of proteins.

Low serum protein levels often cause edema affecting the peripheral organs and, occasionally, the lungs and intestines. In edema, the susceptibility of various organs involved in the synthesis and metabolism is increased. Edema formation is not always diagnosed easily in these circumstances because the low protein content is accompanied by a general loss of sodium and chloride ions and dehydration. However, when large quantities of fluids are given the edema becomes apparent. In delayed wound healing due to protein deficiency, the formation of edema and lack of vitamin C are the contributing factors.

Since the liver is the major site of plasma protein synthesis, during chronic protein deficiency liver damage develops and the hepatic cells become more susceptible to the toxic side effects of foreign compounds. The combination of chronic protein and vitamin deficiency in growing children may cause chronic hepatic disease.

Many other conditions involve increased susceptibility to shock due to reduced plasma volume associated with the low protein level. Increased susceptibility of infection is an often-occurring side effect because of the reduced concentration of immunoglobulins and several tissue immune factors of protein nature. Osteoporosis and delayed regeneration of fractures may also be the consequences of low protein and cacium levels. Anemia frequently develops in protein deficiency since proteins are also important in the synthesis of hemoglobin. Finally, the urine concentration ability of the kidney is often impaired because lower quantities of urea are produced and, therefore, less is available for the osmotic effect in the medullary interstitial cells.

E. Serum Enzymes

Enzymes are important functional units of the cell, catalyzing many biochemical reactions essential for the normal functions of a tissue or organ. Most enzymes occur as intracellular constituents and thus their release into the circulation represents damage or death of the cell.[10,16,32,44,67,91,92,126,149,153,157] Any alteration of enzyme levels in the blood or serum, therefore, may be a reflection of cellular injury and disease.[10,11] The normal concentrations of several enzymes vary within relatively narrow ranges and it is assumed that changes from normal levels indicate an alteration of enzyme production or metabolism in the specific tissues or organs where they originate. All serum enzymes have their origin in various cells. Some, such as aldolase and phosphohexoisomerase, are found in many tissues; others are concentrated in one or two specific tissues, such as ornithine carbamoyl transferase which is located almost exclusively in the liver; creatine phosphokinase is found in significant amounts only in skeletal muscle, heart, and brain.

Various blood cells also contain large numbers of enzymes, but only a few have any relationship with specific diseases.[63,79,122,182] In leukocytes it has been suggested that changes in alkaline phosphatase indicate certain types of leukemias; in erythrocytes the levels of glucose 6-phosphatase and glutathione reductase show limited association with some diseases. Apart from these cases, serum enzymes indicate tissue levels and any departure from normal values reflects a pathological event occurring in the processing tissue and organ or a disorder in the mechanism of excretion or disposal. The determination of these enzymes can be applied to the differential diagnosis of many disorders and certain clinical conditions although changes often do not represent specific effects. Moreover, an increase in the serum level of an enzyme with ubiquitous distribution gives a less specific indication of the site of injury than elevated levels of an enzyme normally found in one or two tissues.

Serum enzymes can be categorized into three groups: (1) enzymes with specific functions, (2) enzymes produced by specific cells, and (3) enzymes produced by secretory glands. Enzymes with specific functions are formed mainly in the liver cell, and are continuously released into the circulation to maintain an effective concentration. Accordingly, their level is highest in the plasma. Such plasma-specific enzymes include various protein factors of blood coagulation (prothrombin, factor V, factor VII, etc.), lipoprotein lipase, pseudocholinesterase, several carrier protein enzymes, and metalloproteins. Any damage or disease of the organs producing these enzymes causes a reduction of their respective activities.

Some enzymes synthesized in special cells exert their effect in the cells of origin and are involved in vital metabolic processes within the cell where they are produced. They play no role whatsoever in the blood. Their presence in the blood, therefore, is the result of cell breakdown. Increased permeability of the cell membrane occurs whenever the cell is damaged, resulting in a leak of intracellular material through the cell wall. Most of these enzymes are localized in cell sap, lysosomes, or mitochondria of various tissues and are present in the serum as a consequence of normal turnover during the life span of these organelles, but at levels much lower than in the tissue. Furthermore, some of them are mostly inactive since the plasma does not contain adequate amounts of coenzymes or activators required for their full catalytic activity. Enzymes belonging to this group include various phosphatases, pancreatic lipases, and amino transferases. Some enzymes are produced by various tissues, such as acid phosphatase which is synthesized in liver, spleen, muscle, stomach, prostate, and erythrocyte. Some occur only in a few tissues, such as alkaline phosphatase which originates from the osteoblast cells of the bone and is also released partly into the blood stream via the hepatobiliary system or excreted into the bile; amylase is synthesized in the pancreas, breast, genital tissues, and parotid glands. Increased rate of release is clearly responsible for the high levels of hepatic, pancreatic, and myocardial enzymes in diseases that produce necrosis. The pattern of abnormality depends on the normal enzyme content of the tissue involved. Enzymes of the secretory glands are secreted into a special organ or duct where they exert their action. Amylase and pepsinogen are examples of this group.

Intrinsic cellular enzymes probably never leak out from the intact cell to appear in the blood stream. Nevertheless, cells die due to natural turnover and wear and tear. Therefore, even in the blood of normal persons, many endogenous enzymes are present in measurable amounts. However, cells may become injured and die as a result of various pathological processes. When the normal cell structure is altered and the cell wall starts disintegrating, many internal enzymes escape into the interstital fluid and pass into the circulation. When there is only slight damage, intracellular boundaries are destroyed and soluble enzymes flow out from the cell into the plasma. In the case of severe damage, cellular organelles become unstable and enzymes escape through their broken membranes. The injured cell may still function normally, but due to the injury some cellular enzymes are produced in excess and thus the serum levels become significantly elevated. The increased serum level is considered a sign of disease and the determination of several enzymes can be used for clinical diagnosis (Table 3). In some cases the augmented appearance of a specific enzyme can be related to a localized lesion or disease of the organ where it originates. For instance, the concentration of alkaline phosphatase is increased in the serum in a variety of diseases associated with enhanced osteoblastic activity. In obstructive jaundice alkaline phosphatase is elevated, probably due to a failure of its excretion via the bile. *De novo* synthesis of phosphatase in the liver may also be involved as shown in bile duct-ligated animals. Another example is amylase; in the normal serum only small amounts of this enzyme are present. Since serum amylase originates from some selected glandular organs, increased levels therefore indicate abnormalities in the function of one or more of these glands.

There are some conditions where nonspecific enzyme changes occur which cannot be linked with a specific disease. Physiological conditions may cause an increase in the level of certain enzymes.[69,126,155,177] Aspartate aminotransferase is raised in the serum of the newborn and alkaline phosphatase levels are increased in childhood until after puberty. During pregnancy, in the last trimester, the concentration of leucine aminopeptidase and alkaline phosphatase are enhanced. Aminotransferases are raised during and immediately after labor. Following major surgery and trauma, the serum level of many enzymes may be increased, such as aspartate transaminase, creatine phosphokinase, and lactate dehydrogen-

Table 3
DIAGNOSTIC VALUE OF SERUM ENZYMES

Enzyme	Conditions with elevated levels	
	Significant	**Moderate**
Acid phosphatase		
	Paget's disease	
	Malignant metastases	
	Gaucher's disease	
	Thrombocythemia (occasional)	
Aldolase	Muscular dystrophy	
	Cancer	
Alkaline phosphatase	Bone disease — osteomalacia, rickets	
	Primary hyperparathyroidism	
	Secondary bone ιoma	
	Liver disease	
	Cholestasis	Hepatitis
		Cirrhosis
Aspartate aminotransferase	Viral hepatitis	Cirrhosis
	Liver necrosis	Cholestatic jaundice
	Biliary tract disorders	Hepatic metastases
	Myocardial infarction	Muscular dystrophy
		Trauma
α-Amylase		
Pancreatic	Pancreatitis	Peptic ulcer (perforated)
	Severe uremia	Intestinal obstruction
		Abdominal trauma
		Cholecystitis
Salivary		Mumps
		Salivary calculi
Creatine phosphokinase	Myocardial infarction	Muscle injury
	Muscular dystrophy	Hypothyroidism
γ-Glutamyl transferase	Hepatic diseases	
Glutathione reductase	Cancer	
	Hepatitis	
Isocitric dehydrogenase	Hepatitis	
	Hepatic metastases	
Lactic dehydrogease	Myocardial infarction	Malignancy
	Hematological disorders	Skeletal muscle disease
	Pernicious anemia	Viral hemolysis
	Leukemia	Acute hemolysis
		Pulmonary embolism
Leucine aminopeptidase	Pancreatic diseases	
	Hepatic diseases	
	Cholestasis	
5′-Nucleotidase	Obstructive hepatic involvements	

Note: This table shows the major clinical applications of serum enzyme determinations. Significant elevations usually follow the acute phases of these diseases. Values over the normal level are dependent on the permeability of the cell wall, concentration of the enzyme in the affected cells, number of cells or extent of area damaged, and localization of the enzyme in cellular particles relative to their binding to the structure.

ase. In shock, the level of several enzymes is elevated — mainly liberated from the liver and skeletal muscle.

Several enzymes catalyzing the same chemical reactions may exist in different forms which can be distinguished by their different physicochemical properties and can be separated by chemical or physical methods (chromatography, electrophoresis, etc.). These are known as isoenzymes.[28,95,142,155] The distribution and relative proportions of various isoenzymes in the blood can be used for the diagnosis and localization of site of many diseases.

1. Alkaline Phosphatase

Phosphatases or phosphomonoesterases include two main types: alkaline phosphatase with an optimal activity at approximately pH 9 and acid phosphatase with a pH optimum of approximately 5. Both alkaline and acid phosphatases include several different isoenzymes.[28,59,69]

The bone is rich in alkaline phosphatase.[38,73] Elevated serum levels occur in patients with bone diseases characterized by increased osteoblastic activity. These include osteomalacia, rickets, osteitis deformans, primary and secondary osteoblastic bone tumor, healing fracture, and hyperparathyroidism. Growing children and pregnant women in the third trimester have physiological increases of serum alkaline phosphatase levels. In the latter case, placental alkaline phosphatase contributes 40 to 65% of total activity, which declines to normal during the first month after birth.

Elevated levels of serum alkaline phosphatase also occur in hepatic diseases; this fact can be applied to the differential diagnosis of jaundice. In obstructive (posthepatic) jaundice, when the obstruction is complete, levels are enhanced to three to eight times normal. Biliary obstruction resulting from carcinoma produces higher values than gallstone obstruction. Elevated serum alkaline phosphatase levels also occur in hepatocellular jaundice due to viral hepatitis, or with toxic actions. Increases may be present in nonjaundiced patients with hepatobiliary disease, such as metastatic or primary carcinoma of the liver, liver abscess, granulomatous disease, and amyloidosis.

Low serum alkaline phosphatase levels are found in malnourished patients and in patients with hypophosphatasia. An inborn error of metabolism or arrested growth of bone formation such as cretinism, vitamin C deficiency, and achondroplasia also exhibit low levels. Experimental conditions may induce alkaline phosphatasemia. Phenobarbital has been shown to increase the activities of this enzyme in the serum and liver. This has been attributed to hepatic enzyme induction. Since many other inducers of hepatic microsomal enzymes bring about no enhancement, the effect of the barbiturate may be considered as a specific stimulatory action on the production of alkaline phosphatase, most likely in the biliary system which is the major site of this enzyme. There is now much evidence that a phenobarbital-induced increase of alkaline phosphatase is related to subclinical osteomalacia.

2. Acid Phosphatase

The prostatic tissue contains this enzyme in high concentration.[20,29,101,152] Increased levels occur in patients with prostatic carcinoma that has metastasized. Erythrocytes and platelets contain different types of acid phosphatase which may also release this enzyme into the serum. Differential effects on various substrates or the action of various inhibitors are used to distinguish the origin of acid phosphatase elevation in the serum.

3. Amino Transferases

Alanine and aspartate aminotransferases are present in many tissues; they occur in the greatest amounts in liver, heart and skeletal muscle, and brain and kidneys.[68,143,151] Normally, only traces are detectable. However, when a small area of the heart muscle is damaged by occlusion of the coronary arteries, the aminotransferase from the injured cells is liberated into the blood and a sharp increase can be seen in its activity. In myocardial infarctions the enhanced serum aspartate aminotransferase is usually apparent after 24 hr following the onset of injury. This often can be measured before the electrocardiogram indicates the lesion. Usually the increase lasts for 2 days and the level returns to normal within 3 to 5 days. The extent and duration of these changes are proportional to the severity of cardiac injury. It

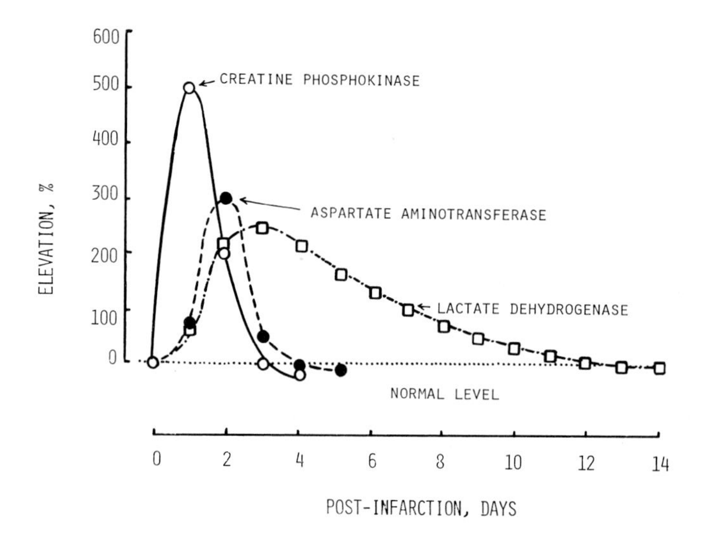

FIGURE 29. Aspartate aminotransferase, creatine phosphokinase, and lactate dehydrogenase activities after myocardial infarction. The increase of creatine phosphokinase level is the earliest, lactate dehydrogenase is the latest. Elevated lactate dehydrogenase levels persist longer than those of aspartate aminotransferase and creatine phosphokinase.

has been established that, in the diagnosis of myocardial infarction, determination of lactic dehydrogenase level and isoenzymes and creatine phosphokinase level and isoenzymes are sufficient. Aspartate aminotransferase levels are no longer measured in many laboratories. Elevation following myocardial necrosis varies: creatine phosphokinase rises within 24 hr, followed by aspartate aminotransferase, and the increase of the lactic dehydrogenase level is last (Figure 29).

Patients with congestive heart failure or marked tachycardia may have moderate increases of aspartate aminotransferase.[41] Diseases or injury producing inflammation or destruction of skeletal muscle, such as gangrene of the extremities or progressive muscular dystrophy, may also cause raised serum transaminase levels.

Various liver diseases are associated with striking elevations including acute hepatic necrosis (carbon tetrachloride intoxication, drug-induced side effects, viral hepatitis), posthepatic jaundice, and most cases of liver cirrhosis.[40] Alcoholic liver disease leads to moderate elevations; in metastatic carcinoma patients, only approximately half show increased serum levels. Similar observations have been reported in patients with cerebrovascular incidents. In acute pancreatitis aspartate transaminase levels may be increased.

In some conditions serum enzyme levels are reduced. Very low aminotransferase values are often found in patients undergoing long-term hemodialysis, vitamin B_6 deficiency, and prolonged isoniazid treatment. Chronic administration of isoniazid is frequently associated with liver disease and reduced enzyme levels. Phenothiazines and cephalosporin derivatives have also been found to markedly decrease serum aminotransferase activity. These effects are related to the reduction of pyridoxal phosphate, a coenzyme essential for the function of aminotransferase (Figure 6).

4. γ-Glutamyltransferase

This enzyme occurs in many tissues.[1,14,15,124,143] Kidney and, to a smaller extent liver, pancreas, and spleen are rich in γ-glutamyltransferase. This enzyme catalyzes the transfer of a γ-glutamyl group from a γ-glutamyl peptide to an amino acid or to another peptide.

The major clinical value of this enzyme is in hepatobiliary and pancreatic diseases. Its activity is highest when biliary stasis is apparent. Serum levels are also enhanced in hepatitis, but quite variable in cirrhosis. γ-Glutamyltransferase is increased in chronic alcoholism and during prolonged drug treatment, particularly with antidepressants, anticonvulsants, and barbiturates given to patients with epilepsy or psychiatric disease. Raised enzyme levels may represent a drug- or alcohol-induced microsomal enzyme synthesis or, in the case of chronic alcoholics, this may be the consequence of liver damage.

5. 5′-Nucleotidase

This enzyme is widely distributed in various tissues and endocrine organs; the vascular endothelium has the highest activities. 5′-Nucleotidase is applied for the differential diagnosis of hepatobiliary disease; it appears to be more specific than alkaline phosphatase. In this disease group, the highest values have been found in posthepatic jaundice, intrahepatic cholestasis, and in infiltrative lesion of the liver. 5′-Nucleotidase is relatively slightly elevated in hepatocellular disease. Patients with primary hepatic parenchymal disease also show a small rise. Raised serum 5′-nucleotidase occurs in metastatic liver disease. This often precedes the onset of jaundice, probably due to blockage of the bile ducts. In the treatment of liver cancer, measurements of 5′-nucleotidase levels provide a fairly good indication for the success of therapy. Serum 5′-nucleotidase values are normal in various bone diseases.

6. Leucine Aminopeptidase

Similarly to 5′-nucleotidase, this enzyme is also increased in the serum in most types of hepatobiliary disease. Highest levels have been reported in hepatitis, cirrhosis, obstructive jaundice, and metastatic carcinoma of the liver. Increased levels also characterize pancreatitis. Patients with carcinoma of the pancreas only show enhanced levels if obstructive jaundice or liver metastasis has developed. This enzyme is normal in osseous diseases. During the last trimester of pregnancy, increase in leucine aminopeptidase in the serum is derived from the placenta.

7. Lactic Dehydrogenase

This enzyme is increased in the serum in a variety of conditions.[16,173] Extreme elevations occur in patients with megaloblastic anemia associated with severe shock and hypoxia or with extensive carcinomatosis. Serum levels are moderately increased in myocardial infarctions, progressive muscular dystrophy, pulmonary infarctions, hemolytic anemia, acute or granulocytic anemia, and in infectious mononucleosis. Lactic dehydrogenase is relatively slightly enhanced in hepatitis, obstructive jaundice or cirrhosis, chronic renal disease, and myxedema.

Sometimes, the rises in the serum lactic dehydrogenase level cannot be applied to pinpoint specific changes since this enzyme is widely distributed in many types of cells and organs. In myocardial injury it escapes into the blood stream, but rises in the level of lactic dehydrogenase activity may also occur in many other diseases. Measurement of total lactic dehydrogenase elevation, therefore, does not indicate which organ is involved in the illness. Separation of isoenzymes provides a more specific method for diagnosis. However, the measurement of this enzyme can still be extremely useful as a routine, nonspecific test to reveal a serious disease in the same fashion as the erythrocyte sedimentation rate or leukocyte count.

Human lactate dehydrogenase occurs in many tissues and are found in the form of isoenzymes.[62] Different organs contain a characteristic proportion of lactic dehydrogenase isoenzymes (Figure 30). The relative amounts of these isoenzymes present in various tissues may also show variations in normal circumstances such as in the newborn and adult, but elevated levels of a particular isoenzyme can be used as a diagnostic tool for the identification of the organ from which it is mainly derived.

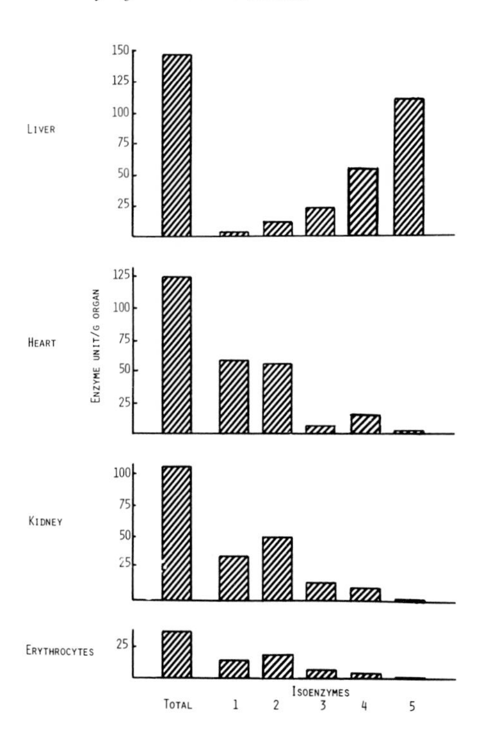

FIGURE 30. Lactic dehydrogenase activity of various human organs. The differentiation of the various isoenzymes can be made by separation on column chromatography or electrophoresis and by showing varying stability to heat inactivation.[200,201]

The lactic dehydrogenase molecule is composed of four genetically determined peptide chains. There are two different kinds of peptides, peptide M (skeletal muscle) and H (heart), and the tetramer chain of the active enzyme molecule is formed by the union of four peptide chains in the following fashion: HHHH, MHHH, MMHH, MMMH, and MMMM. These five possible combinations represent five possible lactic dehydrogenase molecules, LD_1, LD_2, LD_3, LD_4, and LD_5, since all these combinations possess enzyme activity. These isoenzymes differ from each other in their physicochemical properties and can be separated by electrophoresis. They are also different from each other in their ability to exert catalytic function under various environmental circumstances. One isoenzyme works better than the other in a relatively anaerobic environment. The fetal period may represent such a condition. It is therefore likely that in the fetus an isoenzyme which functions in relatively anaerobic conditions has a dominating role, and probably another form is responsible for the major lactic dehydrogenase action in the adult. Variations in different tissues in the relative ratio between isoenzymes may reflect differences in oxygen supply and metabolism requiring lactic dehydrogenase for function.

Predominance of isoenzymes occur at various stages of life. This indicates that although the synthesis of enzyme proteins is genetically regulated, the confirmation of the final molecule is markedly influenced by nongenetic factors. Since during various diseases lactic dehydrogenase isoenzymes are released into the circulation, beyond the absolute increase of enzyme level, the relative proportion of the individual isoenzymes can reveal the site of the pathological condition (Figure 31).

8. Cholinesterase

Low levels of cholinesterases also characterize certain diseases. There are two cholinesterases: one is synthesized in the liver and released in the blood and the other is formed

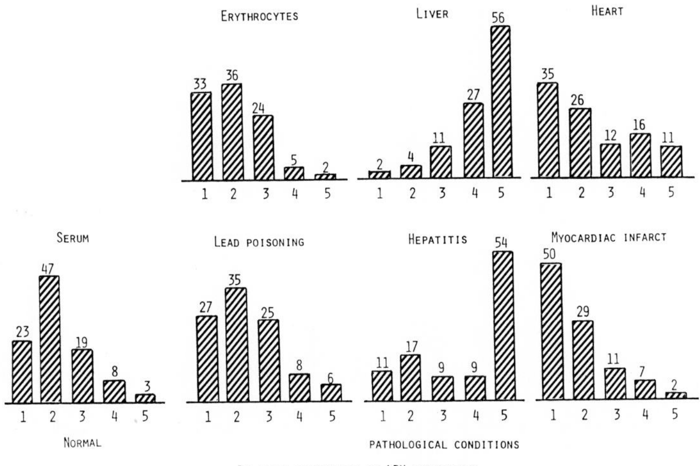

FIGURE 31. Relative proportion of lactic dehydrogenase isoenzymes in disease conditions. Membrane damage due to various diseases causes a leakage of cellular enzymes into the blood and elevated lactic dehydrogenase level in the serum. Lead poisoning affects erythrocyte membranes and the increased serum lactic dehydrogenase isoenzyme spectrum corresponds the distribution of the isoenzymes found in erythrocytes. Similarly, hepatitis and myocardiac infarct are associated with the pathological conditions resulting in changes in the distribution of lactic dehydrogenase isoenzymes existing in the liver or heart, respectively.

in the nervous tissue and erythrocytes.[83] The latter, acetylcholinesterase, is found in the serum only after extensive hemolysis. Serum cholinesterase is therefore called pseudocholinesterase to distinguish it from the true cholinesterase of nerve tissue and erythrocytes. Its level is reduced in parenchymatous liver disease including metastatic carcinoma, cirrhosis, viral and amoebic hepatitis, and hepatic congestion due to heart failure. The hepatic enzyme, pseudocholinesterase, is used as a liver function test. Cholinesterase levels return to normal after recovery, which may be used as an indication of prognosis. Persistent depression may be considered as a poor prognostic sign. Low levels occur in malnutrition, anemia, acute infections, and myocardial infarction. Cholinesterase changes show some parallelism with serum albumin levels and may reflect an impairment of protein synthesis in the liver.

Cholinesterase level is also decreased due to inherited deficiency, and as the result of toxic action by anticholinesterase drugs. Many endogenous factors may affect the level of serum enzymes, it can be reduced by the action of proteolytic enzymes or inhibitors, or by removal through the kidney, liver, and the reticuloendothelial system.

9. Creatine Phosphokinase

The concentration of this enzyme is very high in skeletal muscle and myocardium. Less is found in the brain; the liver tissue does not contain any creatine phosphokinase activity. Serum levels are high in myocardial infarction, progressive muscular dystrophy, alcoholic myopathy, and delirium tremens.[41,51] It is normal in various liver diseases. High levels in hypothyroidism may be connected with muscle changes. Elevation also occurs in pulmonary edema and infarction.[63] Strenuous exercise, acute psychotic reactions, and intramuscular injections raise creatine phosphokinase.

Creatine phosphokinase exists in three isomeric forms.[142] There are two different kinds of peptides, M (muscle) and B (brain). The active enzyme is assembled from two peptide units. The main component in the brain is CK_1(BB) and of the skeletal muscle is CK_3(MM).

Creatine phosphokinase is found in the myocardium which contains two components, CK_2 (MB) and CK_3. The increase of CK_2 isoenzyme in the serum is a good indication of myocardial infarction. Very high CK_3 levels characterize Duchenne muscular dystrophy. Creatine phosphokinase is increased in patients after epileptic seizures due to muscular activity rather than to the abnormal brain function. Similarly, patients with tetanus or with neurological disorders such as encephalitis, meningitis, and cerebral infarct show elevated creatine phosphokinase levels. The increase is mainly connected with the CK_3 isoenzyme, indicating that the skeletal muscle is the source of increase. After neurosurgery increased CK_1 levels have been reported in several cases. Recently three isoenzyme variants have been described. These may represent macroenzyme associations where several molecules of creatine phosphokinase units form an aggregate with each other or with serum proteins.

10. α-Amylase

The exocrine pancreas is the most important site of α-amylase synthesis.[145] The enzyme is secreted into the duodenum where it hydrolyzes glycogen or starch. Similar enzymes are produced in the salivary gland, kidney, small intestines, and testes. Serum α-amylase measurements are applied in the diagnosis of acute pancreatitis. There are certain conditions with enhanced α-amylase levels such as intestinal obstruction, cholecystitis, and perforated peptic ulcer. Mumps, orchitis, and parotitis also cause increased levels.

REFERENCES

1. **Albert, Z., Rzucidlo, Z., and Starzyk, H.,** Biochemical and histochemical investigations of the gamma-glutamyl transpeptidase in embryonal and adult organs of man, *Acta Histochem.*, 37, 74, 1970.
2. **Alper, C. A.,** Complement, in *Structure and Function of Plasma Proteins*, Vol. 1, Allison, A. E., Ed., Plenum Press, New York, 1974.
3. **Ambrus, C. M., Choi, T. S., Cunnanan, E., Eisenberg, B., Staub, H. P., Winetraub, D. H., Courey, N. G., Patterson, R. J., Jockin, H., Pickrin, J. W., Bross, I. D., Okhee, S. J., and Ambrus, J. L.,** Prevention of hyaline membrane disease with plasminogen: a cooperative study, *JAMA*, 237, 1837, 1977.
4. **Arias, J. M., Doyle, D., and Schimke, R. T.,** Studies on the synthesis and degradation of proteins of the endoplasmic reticulum of rat liver, *J. Biol. Chem.*, 244, 3303, 1969.
5. **Arias, J. and Jakoby, W. B., Eds.,** *Workshop on Glutathione*, Krok Foundation Conf., Raven Press, New York, 1976.
6. **Askonas, B. A. and Williamson, A. R.,** Balanced synthesis of light and heavy chains of immunoglobulin G, *Nature (London)*, 216, 264, 1967.
7. **Asofsky, R. and Thorbecke, G. J.,** Sites of formation of immune globulins and of a component of C_3. II. Production of immunoelectrophoretically identified serum proteins by human and monkey tissue in vitro, *J. Exp. Med.*, 114, 471, 1961.
8. **Baugh, C. M., Krumclieck, C. L., and Hershman, J. M.,** Synthesis and biological activity of thyrotropin releasing hormone, *Endocrinology*, 87, 1015, 1970.
9. **Beattie, A. G., Ogston, D., Bennett, B., and Douglas, A. S.,** Inhibitors of plasminogen activation in human blood, *Br. J. Haematol.*, 32, 135, 1976.
10. **Bengzon, A., Hippius, H., and Kanig, K.,** Veränderungen einiger Serumfermente während der psychiatrischen Pharmakotherapie, *Deutsch Med. J.*, 17, 217, 1966.
11. **Bernini, L., Latte, B., Siniscalco, M., Piomelli, S., Spade, V., Adinolfi, U., and Mollison, P. L.,** Survival of 51 Cr-labelled red cells in subjects with thalassaemia — trait or G6PD deficiency of both abnormalities, *Br. J. Haematol.*, 10, 171, 1964.
12. **Beutler, E.,** Glutathione reductase: stimulation in normal subjects by riboflavin supplementation, *Science*, 165, 613, 1969.
13. **Beutler, E. and Srivastava, S. K.,** Relationships between glutathione reductase activity and drug-induced haemolytic anaemia, *Nature (London)*, 226, 759, 1970.
14. **Binkley, F., Wiesemann, M. L., Groth, D. P., and Powell, R. W.,** Gamma-glutamyl transferase: a secretory enzyme, *FEBS Lett.*, 51, 168, 1975.

15. **Binkley, F. and Wiesemann, M. L.,** Minireview. Glutathione and gamma glutamyl transferase in secretory processes, *Life Sci.*, 17, 1359, 1975.
16. **Blatt, J., Spiegel, R. J., Lazaron, S. A., Magrath, I. T., and Poplack, D. G.,** Lactate dehydrogenase isoenzymes in normal end malignant lymphoid cells, *Blood*, 60, 491, 1982.
17. **Blundell, T. L., Dodson, E., Dodson, G., and Vijayau, M.,** The structure of a protein hormone insulin, *Contemp. Phys.*, 12, 209, 1971.
18. **Bodanszky, A., Bodanszky, M., Jorpes, E. J., Mutt, V., and Ondetti, M. A.,** Molecular architecture of peptide hormones optical rotary dispersion of cholecystokinin — pancreozymin bradykinin and 6-glycine bradykinin, *Experientia*, 26, 948, 1970.
19. **Bodanszky, A., Ondetti, M. A., and Mutt, V.,** Synthesis of secretin. IV. Secondary structure in a miniature protein, *J. Am. Chem. Soc.*, 91, 944, 1969.
20. **Bodansky, O.,** Acid phosphatases, *Adv. Clin. Chem.*, 15, 44, 1972.
21. **Boivin, P., Galand, C., Andre, R., and Debráy, J.,** Anémies hemolytiques congénitales avec déficit en glutathion synthetase, *Nouv. Rev. Franc. Hematol.*, 6, 859, 1966.
22. **Brady, A. H., Stewart, J. M., and Ryan, J. W.,** Optical activity and conformation of bradykinin and related peptides, *Adv. Exp. Med. Biol.*, 8, 47, 1970.
23. **Brady, R. O., Johnson, W. G., and Uhlendorf, B. W.,** Identification of heterozygous carriers of lipid storage diseases: current status and clinical applications, *Am. J. Med.*, 51, 423, 1971.
24. **Braxel, C., Versieck, J., Lemey, G., and Barbier, F.,** Alpha 1-antitrypsin in pancreatitis, *Digestion*, 23, 93, 1982.
25. **Carlson, J., Eriksson, S., and Hägerstand, I.,** Intra- and extracellular alpha-1-antitrypsin in liver disease with special reference to Pi phenotypes, *J. Clin. Pathol.*, 34, 1020, 1981.
26. **Bretzel, G. and Hochstrasser, K.,** Liberation of an acid stable proteinase inhibitor from the human inter-alpha-trypsin inhibitor by the action of kallikrein, *Hoppe-Seylers's Z. Physiol. Chem.*, 35, 487, 1976.
27. **Bridges, M. A., Applegarth, D., and Johannson, J.,** Assessment of protease binding by alpha 2-macroglobulin in multiple sclerosis, *Clin. Chim. Acta*, 121, 167, 1982.
28. **Bromhult, J., Fridell, E., and Sunblad, L,** Studies in alkaline phosphatase isoenzymes. Relation to γ-glutamyltransferase and lactate dehydrogenase isoenzymes, *Clin. Chim. Acta*, 76, 205, 1977.
29. **Bruce, A. W. and Mahan, D. E.,** The role of prostatic acid phosphatase in the investigation and treatment of adenocarcinoma of the prostate, *Ann. N.Y. Acad. Sci.*, 390, 110, 1982.
30. **Buckley, R. H., Sampson, H. A., Fisher, P. M., Becker, W. G., and Shirley, L. R.,** Abnormalities in the regulation of human IgE synthesis, *Ann. Allergy*, 46, 67, 1982.
31. **Burrows, S. and Pekala, B.,** Serum copper and ceruloplasmin in pregnancy, *Am. J. Obstet. Gynecol.*, 109, 907, 1971.
32. **Carson, P. E. and Frischer, H.,** Glucose-6-phosphate dehydrogenase deficiency and related disorders of the pentose phosphate pathway, *Am. J. Med.*, 41, 744, 1966.
33. **Carvalho, A. and Ellman, L.,** Hereditary antithrombin III deficiency. Effect of antithrombin deficiency on platelet function, *Am. J. Med.*, 61, 179, 1976.
34. **Chan, W. Y., Hruby, V. J., and Flouret, G.,** 4-Leucine-oxytocin: natriuretic, diuretic, and antivasopressin polypeptide, *Science*, 161, 280, 1968.
35. **Chase, L. R. and Aurbach, G. D.,** Renal adenyl cyclase: anatomicaly separate sites for parathyroid hormone and vasopressin, *Science*, 159, 545, 1968.
36. **Chasseaud, L. F.,** The nature and distribution of enzymes catalyzing the conjugation of glutathione with foreign compounds, *Drug Metab. Rev.*, 2, 185, 1973.
37. **Chasseaud, L. F.,** Glutathione S-transferases, *Z. Physiol. Chem.*, 354, 829, 1973.
38. **Chen, S. T.,** Alkaline phosphatase, in *Frontiers of Gastrointestinal Research*, Vol. 2, Van der Reis, L., Ed., S. Karger, Basel, 1976, 109.
39. **Chillemi, F. and Pecile, A.,** Solid-phase synthesis and bioassay by tibia test of monotetracontapeptide 81-121 and of ditracontapeptide 122-158 of human growth hormone, *Experientia*, 27, 385, 1971.
40. **Clermont, R. J. and Chalmers, T. C.,** The transaminase tests in liver disease, *Medicine*, 46, 197, 1967.
41. **Cohen, L. and Morgan, J.,** The enzymatic and immunologic detection of myocardial injury, *Med. Clin. North Am.*, 57, 105, 1973.
42. **Connell, E. B. and Connell, J. T.,** C-reactive protein in pregnancy and contraception, *Am. J. Obstet. Gynecol.*, 110, 633, 1971.
43. **Cusworth, D. C. and Dent, C. E.,** Renal clearances of amino acids in normal adults and in patients with aminoaciduria, *Biochem. J.*, 74, 550, 1960.
44. **Dacic, Z.,** Serum enzyme — Untersuchungen zur Frage der Leberschädigung durch oral Ovulationshemmer, *Med. Welt.*, 2, 108, 1966.
45. **Daiger, S. P., Schanfield, M. S., and Cavalli-Sforza, L. L.,** Group-specific component (Gc) proteins bind vitamin D and 25-hydroxy-vitamin D, *Proc. Natl. Acad. Sci. U.S.A.*, 72, 2076, 1975.
46. **Daigneault, R. and Vernet-Nyssen, M.,** Transferrin, in *Automated Immunoanalysis*, Ritchie, R. F., Ed., Marcel Dekker, New York, 1978, 253.

47. **Daniels, J. C., Larson, D. L., Abston, S., and Ritzmann, S. E.,** Serum protein profiles in thermal burns. II. Protease inhibitors complement factors and C-reactive proteins, *J. Trauma,* 14, 153, 1974.
48. **Davidson, D. C., McIntosh, W. B., and Forg, J. A.,** Assessment of plasma glutamyl transpeptidase activity and urinary D-glucaric acid excretion as values of enzyme induction, *Clin. Sci. Mol. Med.,* 47, 279, 1974.
49. **Debeljuk, L. and Rettori, V. B.,** Gonadal hormones and gonadotropin secretion, *Prog. Clin. Biol. Res.,* 87, 33, 1982.
50. **Doolau, P. D., Harper, H. A., Hutchin, M. E., and Shreeve, W. W.,** Renal clearance of 18 individual amino acids in human subjects, *J. Clin. Invest.,* 34, 1247, 1955.
51. **Doran, G. R. and Wilkinson, J. H.,** The origin of the elevated activity of creatine kinase and other enzymes in the sera of patients with myxoedema, *Clin. Chim. Acta,* 62, 203, 1975.
52. **Du Vigneaud, V., Ressler, C., Swan, J. M., Roberts, C. W., and Katsoyannis, P. G.,** The synthesis of oxytocin, *J. Am. Chem. Soc.,* 76, 3115, 1954.
53. **Eldjarn, L., Jellum, E., and Stokke, O.,** Pyroglutamic aciduria: studies on the enzymic block and on the metabolic origin of pyroglutamic acid, *Clin. Chim. Acta,* 40, 461, 1972.
54. **Erdös, E. G., Ed.,** *Bradykinin Kallidin and Kallikrein Handb. Exp. Pharmak.,* Vol. 25, Springer Verlag, Berlin, 1970.
55. **Eriksson, S.,** Studies in alpha-antitrypsin deficiency, *Acta Med. Scand.,* Suppl. 177, 432, 1965.
56. **Ette, S. I. and Laditan, A. A.,** Plasma amino acid ratios in Nigerian children suffering from marasmus and Kwashiorkor, *East Afr. Med. J.,* 58, 515, 1981.
57. **Fahey, J. L.,** Antibodies and immunoglobulins. I. Structure and function, *JAMA,* 194, 71, 1965.
58. **Feuer, G., Miller, D. R., Cooper, S. D., de la Iglesia, F. A., and Lumb, G.,** The influence of methyl groups on toxicity and drug metabolism, *Int. J. Clin. Pharmacol.,* 7, 13, 1973.
59. **Fishman, W. H.,** Perspectives on alkaline phosphatase isoenzymes, *Am. J. Med.,* 56, 617, 1974.
60. **Friedman, J. M. and Fialkow, P. J.,** Cell marker studies of human tumorigenesis, *Transplant. Rev.,* 28, 17, 1976.
61. **Frithz, G., Ronquist, G., and Hugossin, R.,** Perspectives of adenylate kinase activity and glutathione concentration in cerebrospinal fluid of patients with ischemic and neoplastic brain lesions, *Eur. J. Neurol.,* 21, 41, 1982.
62. **Fogh-Anderson, N., Sorensen, P., Moller-Petersen, J., and Ring, T.,** Lactate dehydrogenase isoenzyme I in the diagnosis of myocardial infarction, *J. Clin. Chem. Clin. Biochem.,* 20, 291, 1982.
63. **Galen, R. S.,** The enzyme diagnosis of myocardial infarction, *Hum. Pathol.,* 6, 141, 1975.
64. **Ganda, O. P.,** The role of gastrointestinal and neuronal peptides in the pathophysiology of diabetes mellitus, *Diabetes Care,* 4, 435, 1982.
65. **Giblett, E. R., Ed.,** *Genetic Markers in Human Blood,* Blackwell Scientific, Oxford, 1969.
66. **Glass, G. B.,** Gastrointestinal peptide hormones as modulations of bile secretion, *Prog. Liver Dis.,* 7, 243, 1982.
67. **Goldberg, D. M.,** Clinical enzymology, in *Progress in Medicinal Chemistry,* Vol. 13, Ellis, G. P. and West, G. B., Eds., North-Holland, Amsterdam, 1976, 1.
68. **Goldberg, D. M.,** Structural, functional and clinical aspects of γ-glutamyltransferase, *CRC Crit. Rev. Clin. Lab. Sci.,* 12, 1, 1980.
69. **Gorman, L. and Statland, B. E.,** Clinical usefulness of alkaline phosphatase isoenzyme determinations, *Clin. Biochem.,* 10, 171, 1977.
70. **Gratzer, W. B., Beaven, G. H., Rattle, H. W., et al.,** A conformational study of glucagon, *Eur. J. Biochem.,* 3, 276, 1968.
71. **Greenberg, E., Wollaeger, E. E., Fleisher, G. A., and Engstrom, G. W.,** Demonstration of gamma-glutamyl transpeptidase activity in human jejunal mucosa, *Clin. Chim. Acta,* 16, 79, 1967.
72. **Gross, R. T., Bracci, R., Rudolph, N., Schroeder, E., and Kochen, J. A.,** Hydrogen peroxide toxicity and detoxification in the erythrocytes of newborn infants, *Blood,* 29, 481 1967.
73. **Gutman, A. B.,** Serum alkaline phosphatase activity in diseases of the skeletal and hepatobiliary systems, *Am. J. Med.,* 27, 875, 1959.
74. **Hannaford, M. C., Leiter, L. A., Josse, R. G., Marliss, E. B., and Halperin, M. L.,** Protein wasting due to acidosis of prolonged fasting, *Am. J. Physiol.,* 243, E251, 1982.
75. **Hochstrasser, K., Bretzel, G., Feuth, H., Hilla, W., and Lempart, K.,** The inter-alpha-trypsin inhibitor as precursor of the acid-stable proteinase inhibitors in human serum and urine, *Hoppe-Seyler's Z. Physiol. Chem.,* 357, 153, 1976.
76. **Hoffman, K.,** Synthesis of melanocyte-stimulating hormone derivatives, *Am. N.Y. Acad. Sci.,* 88, 689, 1960.
77. **Hoffman, K., Yajima, H., Liu, F. Y., Yanaihara, N., Yanaihara, C., and Humes, J. L.,** Studies on polypeptides. XXV. The adrenocorticotropic potency of an eicosapeptide amide corresponding to the N-terminal portion of the ACTH molecule contribution to the relation between peptide chain-length and biological activity, *J. Am. Chem. Soc.,* 84, 4481, 1962.

78. **Horne, C. H. W., Weir, R. J., Howie, P. W., and Goudi, R. B.,** Effect of combined oestrogen-progestogen oral contraceptives on serum levels of α_2-macroglobulin, transferrin, albumin and IgG, *Lancet*, 1, 49, 1970.
79. **Jaffe, E. R.,** Hereditary hemolytic disorders and enzymatic deficiencies of human erythrocytes, *Blood*, 35, 116, 1970.
80. **Jellum, E.,** Multi-component analyses of human body fluids and tissue in health and disease, *J. Chromatogr.*, 239, 29, 1982.
81. **Jellum, E., Kluge, T., Börresen, H. C., Stokke, O., and Eldjarn, L.,** Pyroglutamic aciduria. A new inborn error of metabolism, *Scand. J. Clin. Lab. Invest.*, 26, 327, 1970.
82. **Johnson, K. J. and Ward, P. A.,** Biology of disease. Newer concepts in the pathogenesis of immune complex induced tissue injury, *Lab. Invest.*, 47, 218, 1982.
83. **Juul, P. and Leopold, I. H.,** Human plasma cholinesterase isoenzymes, *Clin. Chim. Acta*, 19, 205, 1968.
84. **Keiman, A. L. and Cohen, P. P.,** Ammonia detoxication in liver from humans, *Proc. Soc. Exp. Biol. Med.*, 106, 170, 1961.
85. **Kaur, H. and Shrivastava, P. K.,** An overview of haptoglobins, *Acta Anthropogenet.*, 4, 129, 1980.
86. **Khairallah, E. A. and Pitot, H. C.,** in *Symp. Pyridoxal Enzymes*, Yamada, M., Katsumama, N. H., and Wada, M. C., Eds., Maruzui, Tokyo, 1968, 159.
87. **Kim, Y., Friend, P. S., Dresner, I. G., Yunis, E. J., and Michael, A. F.,** Inherited deficiency of the second component of complement (C2) with membranoproliferative glomerulonephritis, *Am. J. Med.*, 62, 765, 1977.
88. **Kirsch, E., Ihme, A., Muller, P., and Krieg, T.,** Molecular defects in inborn disorders of collagen metabolism, *Enzyme*, 27, 239, 1982.
89. **Kolodny, E. H. and Cable, W. J.,** Inborn errors of metabolism, *Ann. Neurol.*, 11, 221, 1982.
90. **Konrad, P. N., Richards, F., II, Valentine, W. N., and Paglia, D. E.,** Glutamyl-cysteine synthetase deficiency. A cause of hereditary hemolytic anemia, *N. Engl. J. Med.*, 286, 557, 1972.
91. **Konttinen, A.,** Serum enzymes an indicators of hepatic disease, *Scand. J. Gastroenterol.*, 6, 667, 1971.
92. **Konttinen, A., Hupli, V., and Sulmenkivi, K.,** The diagnosis of hepatobiliary disease by serum enzyme analysis, *Acta Med. Scand.*, 189, 529, 1971.
93. **Kosower, E. M.,** A molecular basis for learning and memory, *Proc. Natl. Acad. Sci. U.S.A.*, 69, 3292, 1972.
94. **Kupchan, S. M.,** Novel natural products with antitumor activity, *Fed. Proc.*, 33, 2288, 1974.
95. **Latner, A. L. and Skillin, A. W.,** Isoenzymes, in *Biology and Medicine*, Academic Press, New York, 1968.
96. **Li, C. H. and Dixon, J. S.,** Human pituitary growth hormone 32. The primary structure of the hormone: revision, *Arch. Biochem. Biophys.*, 146, 233, 1971.
97. **Litwack, G., Ketterer, B., and Arias, J. M.,** Ligandin: a hepatic protein which binds steroids, bilirubin, carcinogens and a number of exogenous organic anions, *Nature (London)*, 234, 466, 1971.
98. **Laurell, C. B. and Jeppsson, J. O.,** Protease inhibitors in plasma, in *The Plasma Proteins*, Vol. 1, 2nd ed., Putnam, F. W., Ed., Academic Press, New York, 1975, 229.
99. **Lohr, G. W. and Waller, H. D.,** Eine neue enzymopenische hämolytische Anämie mit Glutathion reductase-Mangel, *Med. Klin.*, 57, 1521, 1962.
100. **Marden, R. J., Rent, R., Choi, E. Y. C., and Gewurz, H.,** Cl_q deficiency associated with urticarial-like lesions and cutaneous vasculitis, *Am. J. Med.*, 61, 560, 1976.
101. **Marshall, G. and Amador, E.,** Diagnostic usefulness of serum and β-glycerophosphatase activities in prostatic disease, *Am. J. Clin. Pathol.*, 32, 83, 1969.
102. **McFarlane, H., Reddy, S., Adcock, K. J., Adeshina, H., Cooke, A., and Akene, J.,** Immunity transferrin and survival in kwashiorkor, *Br. Med. J.*, 4, 268, 1970.
103. **McMurray, D. N.,** Cellular immune changes in undernourished children, *Prog. Clin. Biol. Res.*, 67, 305, 1981.
104. **McPhedran, P., Finch, S. C., Nemerson, Y. R., and Barnes, M. G.,** Alpha-2 globulin "spike" in renal carcinoma, *Ann. Intern. Med.*, 76, 439, 1972.
105. **Meister, A.,** Biochemistry of glutathione, in *Metabolism of Sulfur Compounds*, Greenberg, D. M., Ed., Academic Press, New York, 1975, 101.
106. **Meister, A., Tate, S. S., and Ross, L. L.,** in *Membrane-Bound Enzymes*, Vol. 3, Martonosi, A., Ed., Plenum Press, New York, 1975, 315.
107. **Menache, D.,** Abnormal fibrinogens: a review, *Thromb. Diath. Haemorrh.*, 29, 525, 1973.
108. **Mendenhall, H. W.,** Serum protein concentrations in pregnancy. I. Concentrations in maternal serum, *Am. J. Obstet. Gynecol.*, 106, 388, 1970.
109. **Mezick, J. A., Settlemure, C. T., Brierley, G. P., Barefield, K. P., Jensen, W. N., and Cornwell, D. G.,** Erythrocyte membrane interactions with menadione and the mechanism of menadione-induced hemolysis, *Biochim. Biophys. Acta*, 219, 361, 1970.

110. **Milford-Ward, A., Cooper, E. H., Turner, R., Anderson, J. A., and Neville, A. M.,** Acute-phase reactant protein profiles: an aid to monitoring large bowel cancer by CEA and serum enzymes, *Br. J. Cancer,* 35, 170, 1977.
111. **Mittman, C., Ed.,** *Pulmonary Emphysema and Proteolysis,* Academic Press, New York, 1972.
112. **Mohler, D. N., Majerus, P. W., Miniich, V., Hess, C. E., and Garrick, M. D.,** Glutathione synthetase deficiency as a cause of hereditary hemolytic disease, *N. Engl. J. Med.,* 283, 1253, 1970.
113. **Morely, J. S.,** Structure-function relationship in gastrin-like peptides, *Proc. R. Soc. B,* 170, 97, 1968.
114. **Mutt, V. and Jorpes, E.,** Hormonal polypeptides of the upper intestine, *Biochem. J.,* 125, 57, 1971.
115. **Nabel, G., Allard, W. J., and Cantor, H. A.,** A closed cell line mediating natural killer cell function inhibits immunoglobulin secretion, *J. Exp. Med.,* 156, 658, 1982.
116. **Nanji, A. A. and Halstead, A. J.,** Changes in serum anion gap and sodium level in monoclonal gammopathie, *Can. Med. Assoc. J.,* 127, 32, 1982.
117. **Necheles, T. F., Boles, T. A., and Allen, D. M.,** Erythrocyte glutathione-peroxidase deficiency and hemolytic disease of the newborn infant, *J. Pediatr.,* 72, 319, 1968.
118. **Nemazee, D. A. and Sato, W. L.,** Enhancing antibody: a novel component of the immune response, *Proc. Natl. Acad. Sci. U.S.A.,* 79, 3878, 1982.
119. **Newsholme, B. A.,** The interrelationship between metabolic regulation weight control and obesity, *Proc. Natr. Soc.,* 41, 183, 1982.
120. **Niall, H. D.,** Revised primary structure for human growth hormone, *Nature (London),* 230, 90, 1971.
121. **Nilsson, S. F., Rask, L., and Peterson, P. A.,** Studies on thyroid hormone-binding proteins. II. Binding of thyroid hormones retinol-binding protein and fluorescent probes to prealbumin and effects of thyroxin on prealbumin subunit self-association, *J. Biol. Chem.,* 250, 8554, 1975.
122. **Noble, A. and Ranney, H. M.,** in *The Human Red Cell in Vitro,* Greenwalt, T. J. and Jamieson, G. A., Eds., Grune & Stratton, New York, 1974, 91.
123. **Normansell, D. E. and Savory, J.,** Diagnosis of congenital dysproteinemias, *Ann. Clin. Lab. Sci.,* 12, 154, 1982.
124. **Novogrodsky, A., Tate, S. S., and Meister, A.,** γ-Glutamyl transpeptidase: a lymphoid cell-surface marker: relationship to blastogenesis differentiation and neoplasia, *Proc. Natl. Acad. Sci. U.S.A.,* 73, 2414, 1976.
125. **Pardridge, W. M.,** Transport of protein-bound hormones into tissues in vivo, *Endocrinol., Rev.,* 2, 103, 1981.
126. **Patel, S. and O'Gorman, P. O.,** Serum enzyme levels in alcoholism and drug dependency, *J. Clin. Pathol.,* 28, 714, 1975.
127. **Peters, T., Jr.,** Serum albumin, in *The Plasma Proteins,* Vol. 1, 2nd ed., Putnam, F. W., Ed., Academic Press, New York, 1975, 133.
128. **Peterson, G. M., McLean, S., Aldous, S., Von Witt, R. J., and Millingen, K. S.,** Plasma protein binding of phenytoin in 100 epileptic patients, *Br. J. Clin. Pharmacol.,* 14, 298, 1982.
129. **Peterson, P. A.,** Studies on interaction between prealbumin retinol-binding protein and vitamin A, *J. Biol. Chem.,* 246, 44, 1971.
130. **Peterson, R. A., Krull, P. E., Finley, P., and Ettinger, M. G.,** Changes in antithrombin III and plasminogen induced by oral contraceptives, *Am. J. Clin. Pathol.,* 53, 468, 1970.
131. **Post, R. M., Gold, P., Rubinow, D. R., Ballenger, J. C., and Goodwin, F. K.,** Peptides in the cerebrospinal fluid of neuropsychiatric patients: an approach to central nervous system peptide function, *Life Sci.,* 31, 1, 1982.
132. **Potts, J. T., Murray, T. M., and Peacock, M.,** Parathyroid hormone: sequence, synthesis, immunoassay studies, *Am. J. Med.,* 50, 639, 1971.
133. **Powell, L. W. and Halliday, J. W.,** Iron, ferritin and the liver, *Prog. Liver Dis.,* 7, 599, 1982.
134. **Price, R. G.,** Urinary enzymes, nephrotoxicity and renal disease, *Toxicology,* 23, 99, 1982.
135. **Putnam, F. W.,** Haptoglobin, in *The Plasma Proteins,* Vol. 2, 2nd ed., Putnam, F. W., Ed., Academic Press, New York, 1975.
136. **Putnam, F. W.,** Transferrin, in *The Plasma Proteins,* Vol. 1, 2nd ed., Putnam, F. W., Ed., Academic Press, New York, 1975.
137. **Ratnoff, O. D.,** Disordered homeostasis in liver disease, in *Diseases of the Liver,* 4th ed., Schiff, L., Ed., Lippincott, Philadelphia, 1975, 184.
138. **Reeds, P. J.,** Energy costs of protein and fatty acid synthesis, *Proc. Nutr. Soc.,* 41, 155, 1982.
139. **Refetoff, S.,** Syndromes of thyroid hormone resistance, *Am. J. Physiol.,* 273, E88, 1982.
140. **Richards, F., II, Cooper, M. R., Pearce, L. A., Cowan, R. J., and Spurr, C. L.,** Familial spinocerebellar degeneration, hemolytic anemia and glutathione deficiency, *Arch. Intern. Med.,* 134, 534, 1974.
141. **Ritzmann, S. E. and Daniels, J. C., Eds.,** *Serum Protein Abnormalities, Diagnostic and Clinical Aspects,* Little, Brown, Boston, 1975.
142. **Roberts, R. and Sobel, B. E.,** CPK isoenzymes in evaluation of myocardial ischemic injury, *Hosp. Pract.,* 11, 55, 1976.

143. **Rosalki, S. B.,** Gamma-glutamyl transpeptidase, *Adv. Clin. Chem.*, 17, 53, 1975.
144. **Ruddy, S., Gigli, I., and Austen, K. F.,** The complement system of man, *N. Engl. J. Med.*, 287, 1972.
145. **Salt, W. B., III and Schenker, S.,** Amylase — its clinical significance: a review of the literature, *Medicine (Baltimore)*, 55, 269, 1976.
146. **Scarpioni, L., Ballocchi, S., Bergonzi, G., and Cecchettin, M.,** Glomerular and tubular proteinuria in myeloma. Relationship with Bence-Jones proteinuria, *Contrib. Nephrol.*, 26, 89, 1981.
147. **Schindler, A. E.,** Hormones in human amniotic fluid, *Monogr. Endocrinol.*, 21, 1, 1982.
148. **Schlössman, S. F.,** The immune response. Some unifying concepts, *N. Engl. J. Med.*, 277, 1355, 1967.
149. **Schmidt, E. and Schmidt, F. W.,** Fundamentals and evaluation of enzyme patterns in serum, *Prog. Liver Dis.*, 7, 411, 1982.
150. **Schreiber, A. D.,** Immunohematology, *JAMA*, 248, 1380, 1982.
151. **Schulman, J. D., Goodman, S. T., Mace, J. W., Patrick, A. D., Tietze, F., and Butler, E. J.,** Glutathionuria: inborn error of metabolism due to tissue deficiency of gamma-glutamyl transpeptidase, *Biochem. Biophys. Res. Commun.*, 65, 68, 1975.
152. **Schumann, G. B., Badawy, S., Peglow, A., and Henry, J. B.,** Prostatic acid phosphatase. Current assessment in vaginal fluid of alleged rape victims, *Am. J. Clin. Pathol.*, 66, 6, 1976.
153. **Schwartz, M. K.,** Laboratory aids to diagnosis — enzymes, *Cancer*, 37, 542, 1976.
154. **Scott, E. M., Duncan, J. W., and Ekstrand, V.,** Purification and properties of glutathione reductase of human erythrocytes, *J. Biol. Chem.*, 238, 3928, 1963.
155. **Shugar, D.,** *Enzymes and Isoenzymes, Structure, Properties and Function*, Vol. 18, Academic Press, New York, 1970.
156. **Simpson, L. O.,** A hypothesis proposing increased viscosity as a cause of proteinuria and increased vascular permeability, *Nephron*, 31, 89, 1982.
157. **Skreder, S., Blomkoff, J. P., and Gjone, E.,** Biochemical features of acute and chronic hepatitis, *Am. Clin. Res.*, 8, 182, 1976.
158. **Staal, G. E., Visser, J., and Veeger, C.,** Purification and properties of glutathione reductase of human erythrocytes, *Biochim. Biophys. Acta*, 185, 39, 1969.
159. **Strauss, A. W. and Boime, I.,** Compartmentation of newly synthesized proteins, *CRC Crit. Rev. Biochem.*, 12, 205, 1982.
160. **Sveger, T. and Ekelund, H.,** Variations of protease inhibitors in foetuses, newborn infants and in some neonatal disorders, *Acta Paediatr. Scand.*, 64, 763, 1975.
161. **Taylor, J. M., Dehlinger, P. J., Dice, J. F., and Schimke, R. T.,** The synthesis and degradation of membrane proteins, *Drug Metab. Dispos.*, 1, 84, 1973.
162. **Tucker, E. S. and Nakamura, R. M.,** Abnormalities of the complement system, in *Serum Protein-Abnormalities Diagnostic and Clinical Aspects*, Ritzmann, S. E. and Daniels, J. C., Eds., Little, Brown, Boston, 1975, 265.
163. **Viberti, G. C., Hill, R. D., Jarett, R. J., Mahmud, U., and Keen, H.,** Microalbuminuria as a predictor of clinical nephropathy in insulin-dependent diabetes mellitus, *Lancet*, 1, 1430, 1982.
164. **van Oss, C. J., Bronson, P. M., and Border, J. R.,** Changes in the serum alpha glycoprotein distribution in trauma patients, *J. Trauma*, 15, 451, 1975.
165. **von Kaulla, E. and von Kaulla, K. N.,** Antithrombin III and diseases, *Am. J. Clin. Pathol.*, 48, 69, 1967.
166. **Wacker, W. E. C., Ulmer, D. D., and Vallee, B. L.,** Metalloenzymes and myocardial infarction, *N. Engl. J. Med.*, 255, 449, 1956.
167. **Walbaum, R.,** Deficit congenital en transferrine, *Lille Med.*, 16, 1122, 1971.
168. **Waldmann, T. A., Gordon, R. S., and Rosse, W.,** Studies on the metabolism of the serum proteins and lipids in patients with analbuminemia, *Am. J. Med.*, 37, 960, 1964.
169. **Waterlow, J. C. and Alleyne, G. A.,** Protein malnutrition in children: advances in knowledge in the last ten years, *Adv. Protein Chem.*, 25, 117, 1971.
170. **Weeke, E. O. B.,** Urinary serum proteins, *Protides Biol. Fluids*, 21, 353, 1973.
171. **Whaun, J. M. and Oski, F. A.,** Relation of red blood cell glutathione peroxidase to neonatal jaundice, *J. Pediatr.*, 76, 550, 1970.
172. **Wheatherall, D. J.,** Thalassemia revisited, *Cell*, 29, 7, 1982.
173. **Weidner, N.,** Laboratory diagnosis of acute myocardial infarct: usefulness of determination of lactate dehydrogenase, *Arch. Pathol. Lab. Med.*, 106, 375, 1982.
174. **Weiner, H. L. and Hauser, S. L.,** Neuroimmunology: immunoregulation in neurological disease, *Ann. Neurol.*, 11, 437, 1982.
175. **Weisinger, R. A., Gollan, J. L., and Ockner, R. K.,** The role of albumin in hepatic uptake processes, *Prog. Liver Dis.*, 7, 71, 1982.
176. **Weissmann, G., Ed.,** *Mediators of Inflammation*, Plenum Press, New York, 1974.
177. **Welch, G. R.,** The role of protein fluctuation in enzyme action: a review, *Prog. Biophys. Mol. Biol.*, 39, 109, 1982.

178. **Wellner, V. P., Sekura, R., Meister, A., and Larsson, A.,** Glutathione synthetase deficiency. An inborn error of metabolism involving the gamma-glutamyl cycle in patients with 5-oxoprolinuria (pyroglutamic aciduria), *Proc. Natl. Acad. Sci. U.S.A.*, 71, 2505, 1974.
179. **Wilkinson, J. H.,** *The Principles and Practices of Diagnostic Enzymology*, Edward Arnold, London, 1976.
180. **Wolf, P. L., Williams, D., and Von der Muehle, E.,** *Practical Clinical Enzymology*, John Wiley & Sons, New York, 1973.
181. **Wunsch, E. Z.,** Die Totalsynthese des Pankreas-Hormons Glucagon, *Z. Naturforsch.*, 22, 1269, 1967.
182. **Zimmerman, H. J. and Seeff, L. B.,** Enzymes in hepatic diseases, in *Diagnostic Enzymology*, Coodley, E. L., Ed., Lea & Febiger, Philadelphia, 1970, 138.
183. **Yoshida, A.,** Hemolytic anemia and G6PD deficiency, *Science*, 179, 532, 1973.
184. **Young, V. R.,** Protein metabolism and nutritional state in man, *Proc. Nutr. Soc.*, 40, 343, 1981.
185. **Rose, W. C., Johnson, J. E., and Haines, W. J.,** *J. Biol. Chem.*, 182, 541, 1950.
186. **Rose, W. C., Wixsom, R. L., and Lockhart, H. B., and Lambert, G. F.,** *J. Biol. Chem.*, 217, 987, 1955.
187. **Dudrick, S. J. and Roads, J. E.,** *Sci. Am.*, 227, 73, 1972.
188. **Orlowski, M. and Meister, A.,** *Natl. Acad. Sci. U.S.A.*, 67, 1248, 1970.
189. **Waller, H. D., Lohr, G. W., Zysno, E., Gerok, W., Voss, D., and Strauss, G.,** *Klin. Wchschr.*, 43, 413, 1965.
190. **Palekar, A. G., Tate, S. S., and Meister, A.,** *Biochem. Biophys. Res. Commun.*, 62, 651, 1975.
191. **Boyland, E. and Chasseaud, L. F.,** *Adv. Enzymol.*, 32, 173, 1969.
192. **Wood, J. L.,** *Metabolic Conjugation and Metabolic Hydrolysis*, Vol. 2, Fishman, W. H., Ed., Academic Press, New York, 1970, 261.
193. **Jerina, D. M. and Daly, J. W.,** *Science*, 185, 573, 1974.
194. **Putnam, F. W.,** *The Proteins*, Vol. 3, Neurath, H., Ed., Academic Press, New York, 1965.
195. **Goldstein, A., Aronow, L., and Kalman, S. M.,** *Principles of Drug Action*, Harper & Row, London, 1968.
196. **Fleischman, J. B., Porter, R. R., and Press, E. M.,** *Biochem. J.*, 88, 220, 1963.
197. **Nisonoff, A., Wissler, F. C., Lipman, L. N., and Woernley, D. L.,** *Arch. Biochem. Biophys.*, 89, 230, 1960.
198. **Weir, R. C., Porter, R. R., and Givol, D.,** *Nature (London)*, 212, 205, 1966.
199. **Thorpe, N. O. and Deutsch, H. F.,** *Immunochemistry*, 3, 329, 1966.
200. **Wieland, T. and Wachsmith, E. D.,** *Biochem. Z.*, 334, 185, 1961.
201. **Wroblewski, F.,** *Progr. Cardiovasc. Dis.*, 6, 63, 1963.

FURTHER READINGS

Anfinsen, C. B., Anson, M. L., Bailey, K., and Edsall, J. T., Eds., *Advances in Protein Chemistry*, Annu. Ser., Academic Press, New York, 1944—1976.

Arias, I. M. and Jakoby, W. B., Eds., *Glutathione: Metabolism and Function*, Kroc Found. Ser., Vol. 6, Raven Press, New York, 1976.

Bondy, P. K. and Rosenberg, L. E., Eds., *Metabolic Control and Disease*, W. B. Saunders, Philadelphia, 1980.

Evered, D. and Whelan, J., Eds., *Receptors, Antibodies, and Disease*, Pitman, London, 1982.

Fudenberg, H. H., Stites, D. B., Caldwell, J. L., and Wells, J. V., Eds., *Basic and Clinical Immunology*, 2nd ed., Lange Medical Publ., Los Altos, Calif., 1978.

Kawai, T., *Clinical Aspects of the Plasma Proteins*, Lippincott, Philadelphia, 1973.

Miescher, P. A. and Mueller-Eberhard, H. J., Eds., *Textbook of Immunopathology*, Vol. 1 and 2, Grune & Stratton, New York, 1976.

Motta, M., Zanisi, M., and Piva, F., Eds., *Pituitary Hormones and Related Peptides*, Academic Press, New York, 1982.

Pesce, A. J. and First, M. R., *Proteinuria*, Marcel Dekker, New York, 1979.

Scriver, C. R. and Rosenburg, L. E., Eds., *Amino Acid Metabolism and Its Disorders*, W. B. Saunders, Philadelphia, 1973.

Wilkinson, J. H., *The Principles and Practice of Diagnostic Enzymology*, Arnold, London, 1976.

Chapter 3

ABNORMALITIES OF LIPID SYNTHESIS AND METABOLISM

I. INTRODUCTION

Man produces many different lipids which are grouped into several major classes (triglycerides, sterols, phospholipids, glycolipids, waxes) and subclasses occurring in various molecular species depending on the structure of the fatty acid, phospholipid base, or polymer alcohol component (Figure 1). Lipids are present in all cells and serve two major functions: energy storage[52] and constitution of structural elements.[50,85,97,125] Energy can be stored in the long hydrocarbon chain of triglycerides and, by the oxidation of these molecules, the body receives a high yield of energy which provides part of the heat for the maintenance of body temperature. Phospholipids and cholesterol are important in various cellular structures in combination with proteins. The structural role of lipid comprises the formation of the complex architecture of membranes, on the one hand, and participation in the catalytic function of membrane-bound enzymes on the other. In the structural role, mainly the phospholipids and sterols are dominant.[98,103] The hydrophobic terminal of these latter lipids faces the aqueous milieu surrounding the cell, thus separating the cell from its environment. These lipids are also essential in the organization of subcellular compartments or organelles. Their presence is intricately associated with the conformation of the membranes necessary for their role as building units with biochemical functions. The lipid-protein complex structural-functional unit provides relatively stable areas in the aqueous cytoplasm where probably the sites containing fat-soluble substrates are bound. Since the presence of phospholipid is an absolute requirement for the catalytic activity of these membranes, it is likely that many enzyme processes are organized around these lipid-containing sites.[18,33,133,134]

The major portion of the normal lipid synthesis and metabolism is directed toward internal use of these components required by various organs and tissues.[13,71] There are, however, several disease conditions where lipid synthesis and metabolism are impaired, leading to alteration of the membrane structure and subsequent abnormal function or to an accumulation of lipids in various tissues.[72] Many diseases are associated with dilatation and fragmentation of the membranes and prolonged alteration due to toxic drug action,[89] chronic effects accompanied by fat infiltration, and accumulation such as phospholipidosis. Fatty streaks, high serum and tissue triglycerides, and cholesterol are associated with atherosclerosis;[3,42] lipid storage is connected with inborn errors.[8,15,17,19-21,26,27,45,51,56,107,110,132,163] These lipid disorders will be discussed in detail in their specific sections. This chapter describes only diseases related to changes in the basic mechanism of lipid synthesis and metabolism.

Representative examples of abnormal lipid accumulation in the cell are seen in Plates 1 through 5.

II. LIPID ABSORPTION AND DIGESTION

A. Lipid Uptake

The mechanism of lipid uptake consists of two steps: (1) transport by diffusion across an aqueous unstirred layer up to the cellular membrane, followed by (2) penetration through the membrane. Diffusion across the unstirred layer is a major determinant of fat uptake. Solubilization directly increases the flux of lipids by raising their concentration for diffusion. Penetration of the cell membrane is probably a monomolecular process, consisting of partition of lipid molecules from the aqueous side of the interface into the membrane followed by diffusion, and then partition from the membrane into the aqueous cytoplasm. Many features

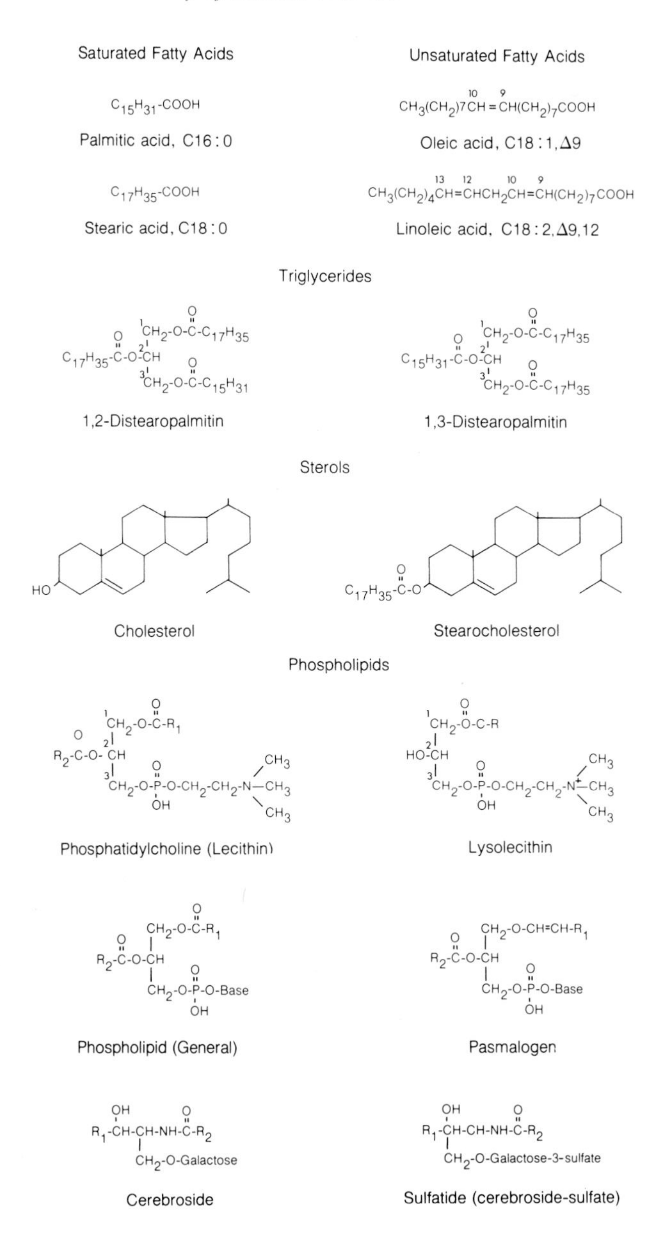

FIGURE 1. Major lipid classes found in human tissue. These include simple lipids: triglycerides, waxes, and sterols; complex lipids: phospholipids and glycolipids; and lipid derivatives: fatty acids, glycerol, fatty aldehydes, and ketone bodies. This figure shows only some representative examples. More relevant details are given in later chapters.

of the uptake of fatty acids and monoglycerides can be explained partly on a physicochemical basis. In the absorption of cholesterol, however, some other factors participate, such as the hydrolysis of esters.[63,128]

The efficient digestion of fats requires an adequate absorption which is connected with the rapid flux of poorly water soluble or insoluble molecules through a thick, unstirred aqueous coating on the mucosa of the small intestine. Some fats such as esters undergo hydrolysis by lipase and a nonspecific esterase and then micellar dispersion by bile acids.[5,108,137,165] Other insoluble lipids, such as cholesterol, are not altered during digestion but, for absorption, micellar solubilization is essential. Digestion occurs rapidly when fat

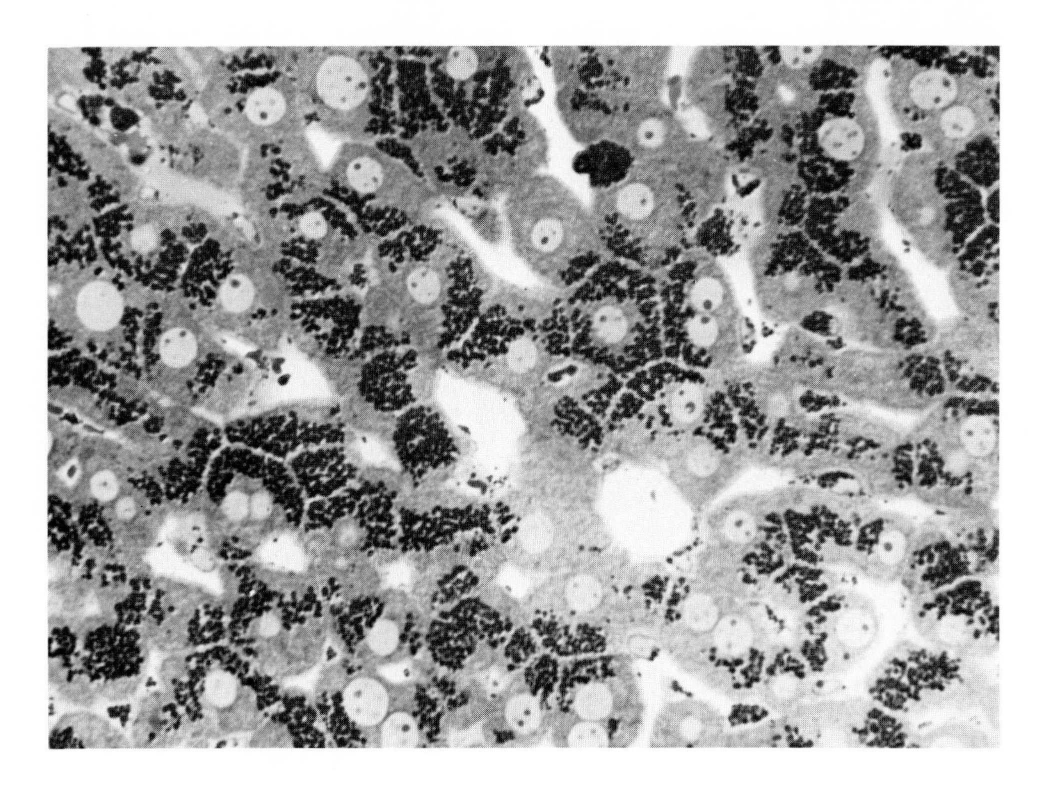

PLATE 1. Example of phospholipidosis in the liver of an experimental animal. The dark granules in the illustration represent abnormal phospholidipid accumulated in the lysosomes. This condition is a drug-induced side effect due to the intake of certain lipophilic compounds.

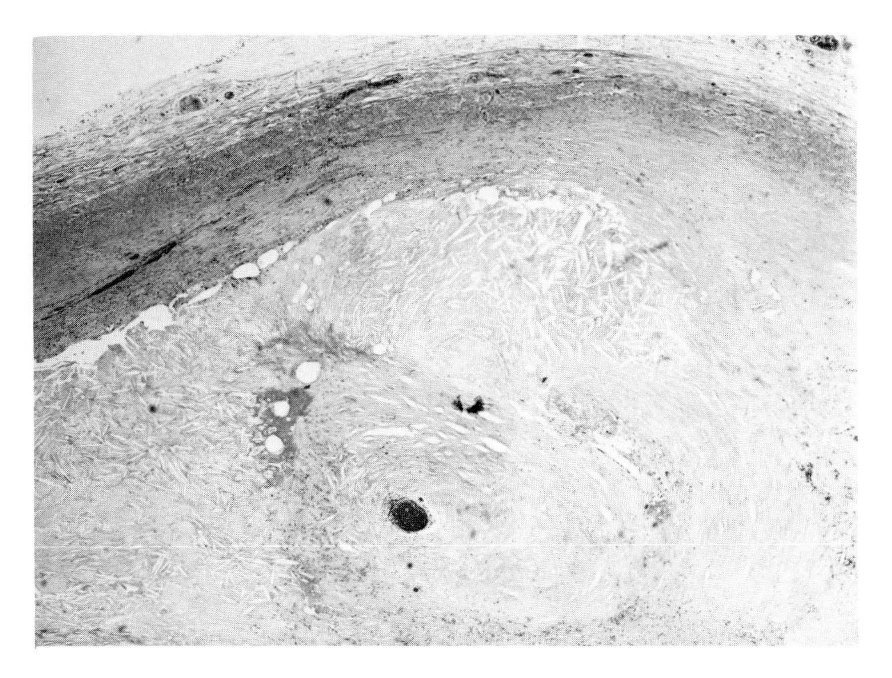

PLATE 2. Abnormal accumulation of cholesterol in tissues. This light micrograph shows a portion of a severely occluded branch of the coronary artery. The sharply defined cholesterol clefts are seen in the subintima of the vessel.

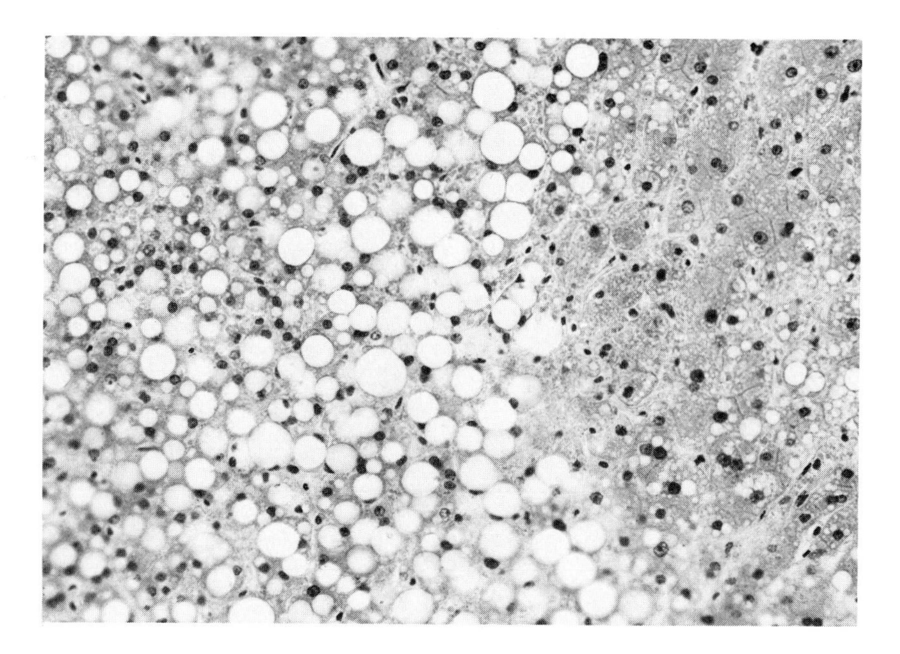

PLATE 3. Severe triglyceride accumulation in the liver of an obese patient. The hollow images correspond to cytoplasmic accumulation of hepatocytic neutral fat, lost during tissue processing.

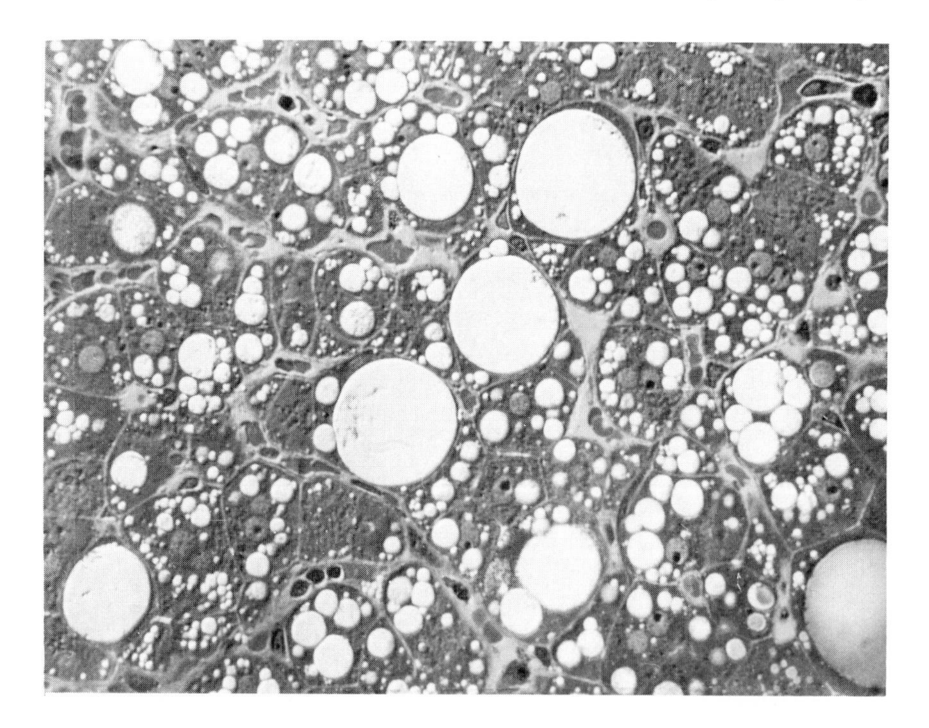

PLATE 4. Appearance of liver tissue in a patient with hyperlipoproteinemia type IIb. The liver cells show varying degrees of fat accumulation and in some instances, the fat droplets coalesce into larger ones.

enters the duodenum. Gall bladder contraction and pancreatic secretion occur almost instantaneously resulting in a rapid secretion of bile, pancreatic enzymes, and bicarbonate. The duodenal cells respond to the acidity of gastric contents and release secretin which triggers the secretion of bicarbonate from the pancreas. The duodenal and small-intestine

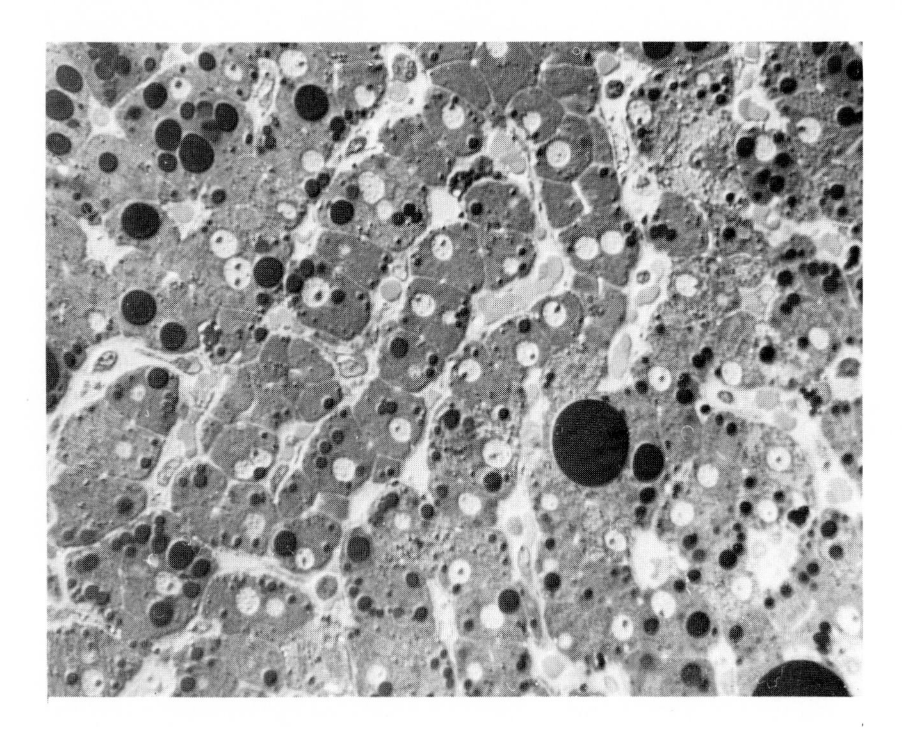

PLATE 5. Moderate fatty change in a liver biopsy of a patient with hyperlipoproteinemia type IV. The darkly stained fat droplets have taken up avidly the osmium fixative and indicate the degree of fat unsaturation, characteristic of triglyceride-rich lipoproteins.

contents are enzyme-rich micellar solutions which initiate a fast lipolysis. Two pancreatic proteins, lipase and colipase, are involved in the intraduodenal digestion of dietary fat.[16,28,48,96,127] The end result of this digestive process and micellar dispersion is the formation of a micellar phase containing lipolytic products. The micelle contains partially ionized fatty acids, monoacyl glyceride, cholesterol, and traces of other endogenous and exogenous fat-soluble compounds such as bilirubin conjugates, fat-soluble vitamins, and plant steroids. When all the lipid phase is transferred to the micellar phase, the concentration of fatty acids falls and fat absorption is decreased.[58]

Bile salts play several important roles during lipolysis. They stimulate the activity of colipase, probably by enhancing the binding of lipase with the cofactor. They may create a molecular template which provides a more accessible enzyme site. Calcium ions are also necessary for acceleration of the hydrolysis rate by improving the lipase-substrate binding.[29]

When deficiency occurs in bile salt production, lipolysis is still completed and fatty acids and monoglycerides are present in amounts relative to their saturation solubilities. However, in this condition the reserve buffering capability of the micelles is reduced and transport through the unstirred mucosal layer becomes rate limiting. As a consequence, the site of fat absorption is spread throughout the small intestine and steatorrhea may occur. Biliary obstruction may be associated with different degrees of fat malabsorption. Steatorrhea depends on the chain length and saturation of dietary fatty acid. It is minimal when triacylglycerol-containing medium-chain-length fatty acid is taken, and increases with the chain length and degree of saturation of the fatty acid components.

B. Lipid Absorption in Disease States

Clinical conditions often associated with lipase deficiency include: (1) decreased synthesis or output of the enzyme due to congenital deficiency, exocrine pancreatic insufficiency (pancreatitis, cystic fibrosis), vagal surgery, resection of the pancreas, impaired release and digestion, or (2) inactivation of the secreted enzyme due to bicarbonate deficiency, gastrinoma (Zollinger-Ellison syndrome), and chronic pancreatic disease. The overall consequence of pancreatic lipase deficiency is an impairment of fat digestion resulting in steatorrhea. There is a significant decrease of micellar fatty acid and monoglyceride absorption and increased amounts of fecal triglycerides.

Congenital lipase deficiency is linked with severe loss or absence of enzyme activity.[37] Although in these patients steatorrhea has been present since early life, growth and development have never been severely affected. Sometimes enzyme activity appears after the patient reaches adulthood. Patients with chronic inflammatory lesions of the pancreas such as in familial pancreatitis, idiopathic pancreatitis, and pancreatitis caused by alcoholism show lipolytic defects which cause steatorrhea. The degree of fatty stool production correlates well with the lipolytic enzyme deficiency in the intraduodenal juice. Cystic fibrosis, which occurs in children and has been detected in adults as well, is the disease commonly causing pancreatic insufficiency. The digestive deficiency often presents a wide spectrum depending on the severity of involvement of the organ.

Various types of pancreatic lesions and ligation of the pancreatic duct produce significant lipid and protein malabsorption and maldigestion.[143] The degree of impairment could be reduced by anastomosing the remaining portion of the pancreas to the small intestine. However, the degree of steatorrhea could be raised sharply by moderate increases of dietary fat suggesting that these patients readily decompensated in their ability to digest lipid. Defects of hepatic bile acid synthesis can also cause chronic diarrhea and steatorrhea. Acute and chronic parenchymal liver disease and various forms of cholestasis can cause lipid malabsorption.[152] In bile acid deficiency, the limiting step in fat assimilation is probably diffusion through the unstirred water layer, mucosal penetration, and decreased intracellular esterification.[69] Patients with extensive ileal resection also have steatorrhea.

Patients with steatorrhea usually exhibit diarrhea.[34] Jejunal studies in healthy men have shown that micellar solutions of oleic, ricinoleic, and 10-hydroxystearic acid progressively inhibited net water and electrolyte absorption and evoked net secretion of fluid. Ricinoleic acid, a monounsaturated, hydroxylated C_{18} fatty acid is the active principle of castor oil, a potent laxative. Large therapeutic doses of castor oil are incompletely absorbed and cause laxative actions. Small doses are well absorbed and produce no catharsis. It is probable that when the absorption of dietary fat is impaired, the presence of excess fat in the lumen acts as a cathartic agent.

III. LIPID METABOLISM

The significant lipids in the mammalian organism include triacylglycerols (triglycerides or neutral fat), phospholipids, and steroids together with their metabolic products such as long-chain free fatty acids, glycerol, and ketone bodies. Tissue lipids are not considered as inactive storehouses of calorigenic materials necessary in times of caloric shortage; the body fat is in a constant dynamic state.[1,58,95] Much of the dietary carbohydrate is converted to fat before it is utilized for the purpose of providing energy. As a result, fat is the major source of energy for many organs and may be used as fuel in preference to carbohydrate.[46,52] Besides this preference, fat has definite advantages over carbohydrate or protein. Its caloric value is more than twice as great as other food sources (lipids, 39 kJ/g; carbohydrates, 17 kJ/g; proteins, 17 kJ/g). When stored it is associated with less water, therefore it is a more concentrated potential source of energy.

As a source of energy, body fat can be replaced completely by carbohydrate or protein, although utilization of these foodstuffs for the purpose of providing fuel is much less efficient. However, a minimal amount of fat is required in the diet to provide adequate amounts of certain polyunsaturated fatty acids (so-called essential fatty acids) and of fat-soluble vitamins which cannot be synthesized in the human body in adequate amounts for optimal function. Fats also act as a carrier of these essential compounds and are necessary for their efficient absorption from the gastrointestinal tract.

A. Plasma Lipids

Extraction of serum with a lipid solvent and subsequent separation results in various classes of lipids including triacylglycerols, cholesterol and cholesteryl esters, phospholipids and small amounts of unesterified long-chain free fatty acids (Table 1).[63,67,87,139,141] Free fatty acid accounts for less than 5% of the total fatty acid present in the serum, but is metabolically the most active lipid. The various lipids occur in the serum in complex form. Since the serum presents an aqueous environment, the transport of large amounts of hydrophobic substances involves the production of an association of the more insoluble lipids such as triacylglycerols with polar molecules such as phospholipids and combinations with cholesterol and proteins to form a hydrophilic lipoprotein complex.[41,42] In this way, triacylglycerols derived from intestinal synthesis or released from the liver are carried in the blood as chylomicrons and very-low-density lipoproteins. Fat is released from adipose tissue in the form of free fatty acids which are transported in the plasma bound to albumin.

B. Lipid Catabolism and Biosynthesis

Triglycerides are hydrolyzed to their constituent fatty acids and glycerol before further degradation can occur.[54] This hydrolysis takes place mainly in the adipose tissue and the free fatty acid is released into the plasma and taken up by various tissues where they are oxidized. Many tissues such as liver, kidney, brain, heart, skeletal muscle, lung, testis, and adipose tissue can catabolize long-chain fatty acids. The breakdown of glycerol is associated with the function of glycerokinase which is present in great amounts in liver, kidney, intestine, and lactating mammary gland.

Several enzymes known as fatty acid oxidases are found in the mitochondrial matrix opposite the respiratory chain which is present in the inner membrane. These enzymes catalyze the β-oxidation of fatty acids containing an even number of carbon atoms finally to acetyl-CoA. The system is coupled with the formation of ATP from ADP (Figure 2). Fatty acids with an odd number of carbon atoms are oxidized by way of β-oxidation until a propionyl-CoA is produced. This compound is further converted to succinyl-CoA via the citric acid cycle. The reversal of the mitochondrial system with some modification of the β-oxidation sequence is responsible only for the elongation of fatty acids of moderate chain length. Mainly, a nonmitochondrial enzyme system catalyzes the complete synthesis of long-chain fatty acids from acetyl-CoA; an active microsomal enzyme system also produces chain elongation and desaturation.[23,120]

The mitochondrial system functions under anaerobic conditions and synthesizes mainly stearic and palmitic acid and some other acids with C_{20} and C_{14} chains by the incorporation of acetyl-CoA units. The incorporation of acetyl-CoA into the long-chain fatty acid is accomplished by addition to existing fatty acids rather than by *de novo* synthesis from acetyl-CoA units. This system requires ATP, NADH, and NADPH. ATP is probably required for the formation of acyl-CoA from endogenous fatty acids. The extramitochondrial system is found in the cytosol fraction of many tissues. ATP, NADPH, Mn^{2+}, and HCO_3^- are needed for the function of this system and the main product is palmitic acid. Another important difference from the mitochondrial enzyme is that during synthesis the acyl CoA derivatives are not real intermediates, the acyl moiety remains attached to the enzyme as an acyl-S-enzyme complex. The microsomal system is probably the main site for the elongation of existing fatty acid molecules. Acyl-CoA derivatives are converted to higher derivatives using malonyl-CoA as acetyl donor and NADPH as reducing cofactor. Intermediate CoA thioesters

Table 1
CHARACTERISTICS OF PRIMARY HYPERLIPOPROTEINEMIA

Lipoprotein phenotype	Clinical presentation	Lipoprotein elevation	Plasma		
			Appearance	Cholesterol	Triglyceride
I	Abdominal pain Eruptive xanthomas Hepatosplenomegaly Lipemia retinalis	Chylomicrons	Creamy layer above clear or slightly turbid infranate	Normal to moderately elevated	Markedly elevated
IIa	Accelerated atherosclerosis Juvenile corneal arcus Tendon and tuberous xanthoma Xanthelasma	LDL	Clear, occasionally increased yellow-orange color	Elevated	Normal
IIb	Similar to IIa	LDL VLDL	Slightly or moderately turbid	Elevated	Moderately elevated
III	Accelerated atherosclerosis Tendon and tubero-eruptive xanthomas	Abnormal B-VLDL LDL	Turbid occasionally creamy layer	Elevated	Moderately or markedly elevated
IV	Abnormal glucose tolerance Accelerated coronary vessel disease Hyperuricemia	VLDL	Turbid	Normal or slightly elevated	Moderately or markedly elevated
V	Abdominal pain Eruptive xanthomas Hepatosplenomegaly Hyperglycemia Hyperuricemia Lipemia retinalis	Chylomicrons VLDL	Creamy layer	Moderately elevated	Markedly elevated

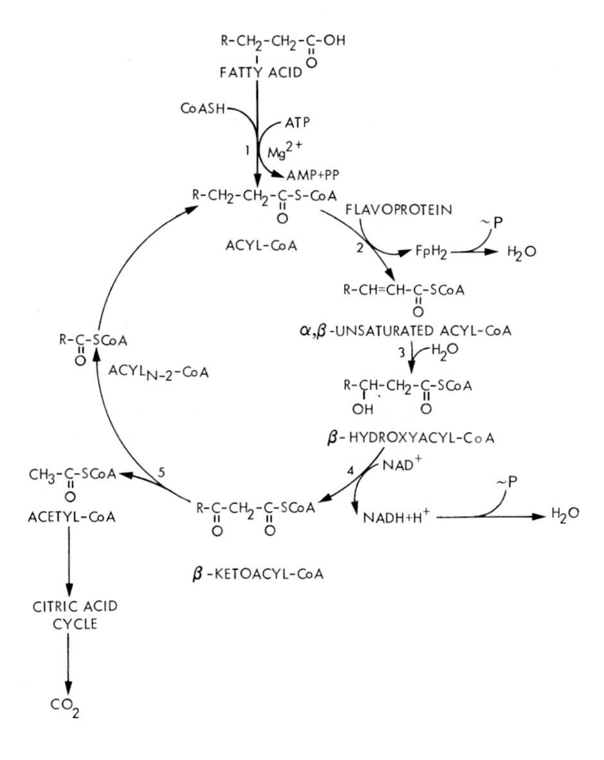

FIGURE 2. Metabolism of saturated fatty acids by β-oxidation. Thiokinase (1) catalyzes the conversion of fatty acid to an active fatty acid in the presence of ATP and CoA. Acyl-CoA is then reduced to α,β-unsaturated acyl-CoA by acyl-CoA dehydrogenase (2). Addition of water catalyzed by Δ^2-enoyl-CoA hydratase (crotonase, 3) saturates the double bond. The β-hydroxy derivative is further dehydrogenated by β-hydroxyacyl-CoA dehydrogenase (4) and finally the β-ketoacyl-CoA is split at the β-position by β-ketothiolase (5). The final products, acetyl-CoA and acyl-CoA, contain two carbon atoms less than the original fatty acid. The latter compound reenters the oxidation cycle and may be completely degraded to acetyl-CoA units.

are formed in the process. The acyl groups that are used as primer molecules include saturated acids between C_{10} and C_{16} and some unsaturated C_{18} fatty acids. The end product of microsomal action is the next higher homologue of the primer acyl-CoA molecule. Fasting and impairment of microsomal function largely abolishes chain elongation. The enzyme reactions involved in the oxidation of mono- and polyunsaturated fatty acids are partly identical with those normally responsible for β-oxidation until a Δ^3-*cis*-acyl-CoA or Δ^2-*cis*-acyl-CoA is produced, depending upon the position of the double bonds. Further enzymes convert the intermediates into the *trans* form, which is further metabolized by β-oxidation and hydration (Figure 3). The natural unsaturated fatty acids are *cis* isomers; however, partially hydrogenated vegetable oils such as margarine contain *trans*-unsaturated fatty acids.[71] These compounds do not possess essential fatty acid properties. They are metabolized more like the saturated than the *cis*-unsaturated fatty acids. This may be due to their similar straight chain conformation. However, these fatty acids are fairly stable; at autopsy up to 15% of tissue fatty acid has been found in the *trans* configuration. The long-term effect of these *trans*-unsaturated fatty acids in man is not known, but the high tissue level may raise the question of their safety as food constituents.

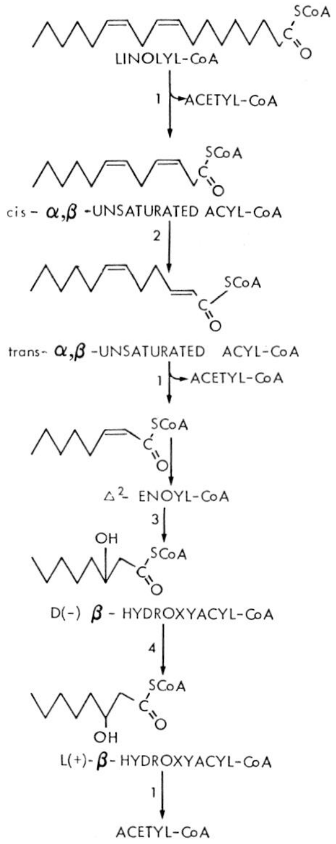

SCoA
LINOLYL-CoA
1 ACETYL-CoA
SCoA
cis- α,β -UNSATURATED ACYL-CoA
2
SCoA
trans- α,β -UNSATURATED ACYL-CoA
1 ACETYL-CoA
SCoA
Δ^2- ENOYL-CoA
3
OH
SCoA
D(-) β - HYDROXYACYL-CoA
4
SCoA
OH
L(+)- β- HYDROXYACYL-CoA
1
ACETYL-CoA

FIGURE 3. Metabolism of unsaturated fatty acids by oxidation. The CoA derivatives of these acids are degraded by β-oxidation (1) until either Δ^3-*cis*-acyl CoA or Δ^2-*cis*-acyl CoA derivative is formed, depending upon the position of the double bonds. These derivatives are isomerized by Δ^3-*cis*-Δ^2-*trans*-enoyl-CoA isomerase (2) to Δ^2-*trans*-CoA derivative, which is oxidized and hydrated by Δ^2-enoyl-CoA hydratase (3) to the β-hydroxy-acyl-CoA derivative. This is epimerized by β-hydroxyacyl-CoA epimerase (4) and further β-oxidation results finally in acetyl CoA.

C. Regulation of Lipogenesis and Lipolysis

Man takes his food intermittently and needs to store much of the energy for use between meals. In the process of lipogenesis, glucose and several intermediates such as acetyl-CoA and pyruvate are converted to fat.[62] The lipogenesis is controlled by the nutritional state of the organism and tissues. The rate is high when the diet contains a high percentage of carbohydrate. It is decreased on a high-fat diet, under conditions of restricted calorie intake, or when there is an insulin deficiency as in diabetes mellitus. Among these circumstances plasma free fatty acid concentration is enhanced. The elevated free fatty acid exerts a control on lipogenesis. Greatest inhibition occurred when the free fatty acid level ranged 30 to 80 μmol/dℓ plasma; this is the level free fatty acids increase during transition from a fed to a starved state. As little as 2.5% dietary fat causes a measurable reduction of hepatic lipogenesis. In the regulatory mechanism the microsomal enzyme system exerts a stimulatory effect. Its effect is probably associated with the removal of the feedback inhibition of acyl-CoA on acetyl-CoA carboxylase.

Insulin stimulates lipogenesis by multiple action.[47] It increases the transport of glucose into the cell, such as into the adipose tissue, and therefore the availability both of α-glycerophosphate and pyruvate. Insulin may serve as an activator of pyruvate dehydrogenase; it may convert the inactive form to the active enzyme. Insulin also depresses the level of intracellular cyclic AMP, which cause a reduction of lipolysis and hence reduces the concentration of long-chain acyl-CoA, which consequently leads to an increased lipogenesis.

Carnitine (β-hydroxy-γ-trimethylammonium butyrate) plays a part in the regulation of fatty acid metabolism by mitochondrial oxidation. This compound is widely distributed in various tissues and is particularly abundant in the muscle. It stimulates the oxidation of long-chain fatty acids. The activation occurs in microsomes or on the outer membranes of mitochondria, then carnitine transfer and replacement takes place on the outer and inner surface of the inner mitochondrial membrane (Figure 4). Long chain acyl-CoA does not penetrate mitochondria and is oxidized only if carnitine is present. Activation of shorter fatty acids may occur within the mitochondria, independent of carnitine.

The regulation of lipogenesis is linked with the metabolism of adipose tissue and mobilization of fat from these depot cells.[41,46,52,129,153] The triglyceride stores in adipose tissue are continually undergoing hydrolysis and reesterification. These opposite processes are not forward and reverse phases of the same mechanism (Figure 5). Different pathways and various enzymes are connected with these routes and many metabolic, nutritional, and hormonal factors which regulate the metabolism of adipose tissue act either on lipolysis or esterification. The results of these processes and the effect of regulatory factors determine the level of the free fatty acid pool in adipose tissue which, in turn, is the source and determinant of the amount of circulating free fatty acid in the blood. The level of free fatty acid in the blood has marked effect on the metabolism of other tissues, particularly of liver and muscle; the various factors which regulate the production of free fatty acids in adipose tissue and their outflow exert a profound influence on many tissues.[146]

In adipose tissue, triacylglycerol is synthesized from α-glycerophosphate and acyl-CoA. The hydrolysis of this compound to free fatty acid and glycerol is catalyzed by a hormone-sensitive lipase. Glycerol is not utilized in the adipose tissue, it diffuses out and is carried to the liver and kidney where it is metabolized. The free fatty acid formed by lipolysis can be resynthesized in the adipose tisse to acyl-CoA and reesterification with α-glycerophosphate produces triglycerides. There is a continuous cycle within the adipose tissue. However, when the rate of reesterification is low, as compared to lipolysis, more free fatty acid is produced; the excess diffuses into the plasma where it increases the level of free fatty acids.

When nutritional intake is adequate, or when the utilization of glucose by the adipose tissue is increased, the outflow of free fatty acid decreases and plasma free fatty acid decreases. In contrast, when the availability of glucose in adipose tissue is decreased as in diabetes mellitus or in starvation, less α-glycerophosphate is formed and due to the reduced precursor level the rate of lipolysis exceeds the rate of esterification. Subsequently free fatty acids accumulate and their release into the circulation is enhanced. It seems that when the available carbohydrate is high the adipose tissue utilizes glucose for energy production and to esterify free fatty acids. When the supply of carbohydrate is short, it conserves glucose via α-glycerophosphate for esterification and uses fatty acid for energy production. There are observations indicating that more than one free fatty acid pool exists within the adipose tissue. Unsaturated fatty acids are not incorporated very rapidly into the depot fat, but appear first in various other compartments.

D. Hormone Action on Lipid Metabolism

Many hormones influence the rate of release of free fatty acids from adipose tissue, either by affecting the rate of esterification or the rate of lipolysis.[23] Administration of insulin causes a fall in the circulating plasma level of free fatty acids. Insulin enhances lipogenesis, inhibits the release of free fatty acids from adipose tissue, and increases glucose oxidation to carbon dioxide. All these actions are dependent on the presence of glucose. The insulin effect, therefore, is connected with an enhanced uptake of glucose into the fat tissue. Insulin also inhibits the activity of the hormone-sensitive lipase, thus decreases free fatty acid and glycerol release. Adipose tissue may represent the major site of insulin action since it is the tissue most responding to insulin. Prolactin has a similar effect to that of insulin on adipose tissue, but only in large doses.

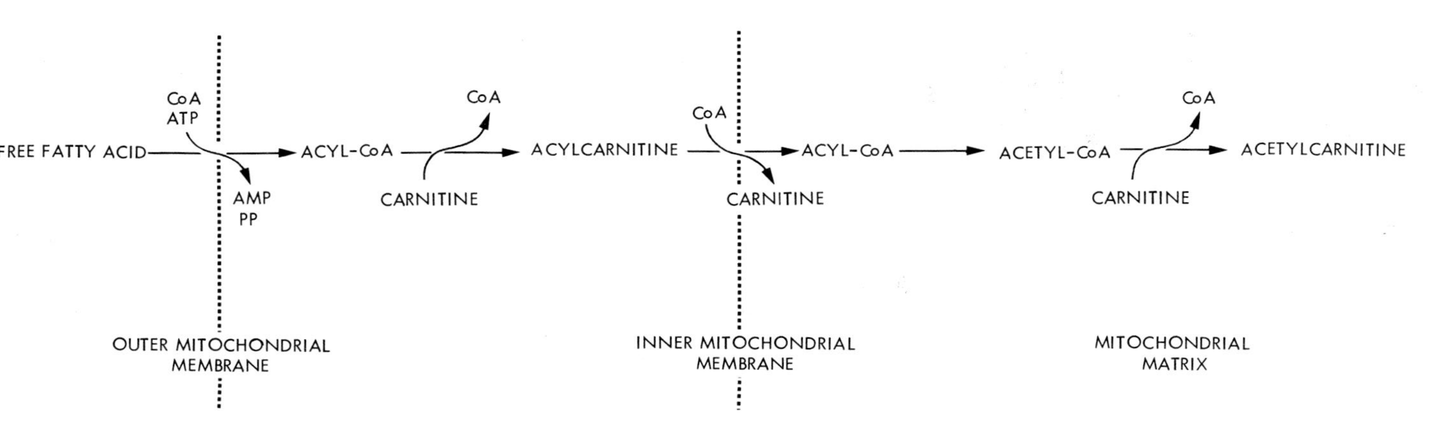

FIGURE 4. Role of carnitine in the metabolism of fatty acids. Long-chain fatty acids are activated to acyl-CoA by thiokinase (1) located in microsomes and the outer mitochondrial membrane, then acyl carnitine derivative is formed by carnitine-palmityl acyl transferase (2) located on the outer surface of the inner mitochondrial membrane. This binding is essential in the transfer of long-chain acyl groups through the membrane, where membrane-bound carnitine acyltransferase (3) reversibly converts it to acyl-CoA which is degraded by β-oxidation (4). Shorter-chain acyl groups can be transferred from CoA to carnitine by carnitine acetyl transferase (5).

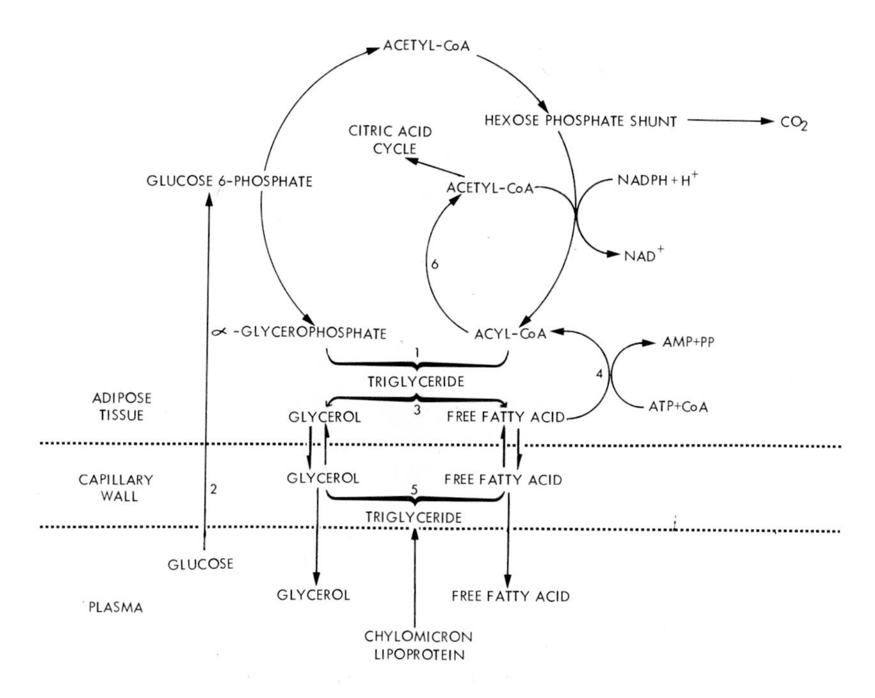

FIGURE 5. Regulation of fat metabolism in adipose tissue. Triglyceride is synthesized from acyl CoA and α-glycerophosphate (1); the latter is formed from glucose 6-phosphate. The entry of glucose into adipose tissue is promoted by insulin (2). Triglyceride is hydrolyzed by a hormone-sensitive lipase (3) to form free fatty acid and glycerol. This enzyme is activated by adrenocorticotropic hormone, glucagon, thyrotropic hormone, vasopressin, epinephrine, norepinephrine, and probably by thyroid hormone inhibited by insulin and prostaglandin E_1. Free fatty acid is partly metabolized to acyl-CoA and triglyceride is formed by reesterification (4). It is partly transported into the plasma. Glycerol also diffuses into the plasma and is carried into other organs where it is metabolized. The triglyceride content of chylomicrons and very low-density lipoproteins undergoes hydrolysis in the capillary wall by lipoprotein lipase (5). Excess acyl-CoA is degraded by β-oxidation (6).

Other hormones increase the release of free fatty acids from adipose tissue and enhance the plasma free fatty acid level by raising the rate of lipolysis of triglyceride stores. These hormones include thyroid-stimulating hormone, adrenocorticotropic hormone, α- and β-melanocyte-stimulating hormones, growth hormone, vasopressin, glucagon, epinephrine, and norepinephrine.[147] Many hormones stimulate the activity of the hormone-sensitive lipase and raise glucose utilization. Adipose tissue contains many lipases; one of them is the hormone-sensitive triacylglycerol lipase. Lipolysis is controlled largely by the amount of cyclic AMP present in tissue. This compound stimulates a protein kinase which converts the inactive hormone-sensitive lipase into the active form. The action of various hormones in promoting lipolysis is connected with the stimulation of adenylate cyclase, which catalyzes the transformation of ATP to cyclic AMP.[4] Cyclic AMP is degraded to 5′-AMP by the enzyme cyclic 3′,5′-nucleotide phosphodiesterase. This enzyme is inhibited by caffeine and theophylline. Drinking coffee, or the administration of caffeine, causes a pronounced and prolonged increase of plasma free fatty acid in man. In contrast, nicotinic acid and prostaglandin E_1 inhibit the synthesis of cyclic AMP. The various hormonal controls of adipose tissue lipolysis is illustrated (Figure 6).

E. Ketosis

There are certain metabolic conditions which are linked with a high rate of fatty acid oxidation and with the formation of considerable amounts of ketone bodies. These compounds are acetoacetic acid and β-hydroxybutyric acid which diffuse into the blood, and acetone

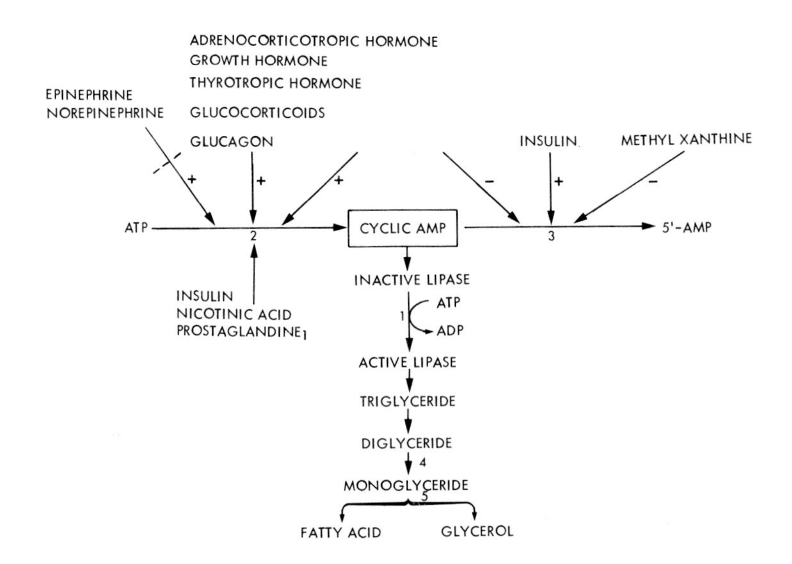

FIGURE 6. Hormonal and pharmacological control of lipolysis in adipose tissue. The hormone-sensitive triacylglycerol lipase is enhanced by protein kinase (1) which is stimulated by cyclic AMP (2). The enzyme is activated by various hormones. Some of these actions can be inhibited by β-adrenergic blockers. The activity of diacylglycerol lipase (4) and monoacylglycerol lipase (5) is not influenced by hormones. Adenylate cyclase is blocked by insulin, prostaglandin E_1, and nicotinic acid. Cyclic AMP is degraded to 5′-AMP by phosphodiesterase (3) which is activated by insulin and inhibited by methylxanthines and probably by thyroid hormones.

which is formed by the spontaneous decarboxylation of acetoacetic acid. The concentration of these ketone bodies is normally less than 1 mg/dℓ acetone equivalent in man, and loss through the urine is usually less than 1 mg/24 hr. Higher quantities may present in the blood or urine, resulting in ketonemia or ketonuria, respectively. The overall condition is ketosis. Acetoacetic and β-hydroxybutyric are both strong acids and are buffered when present in blood or the tissues. However, when excessive amounts are produced the neutralization of these acids causes some loss of buffering capacity which progressively depletes the alkali reserve causing ketoacidosis. This condition is fatal in uncontrolled diabetes. Ketosis occurs in simple form during starvation. This involves the depletion of the available carbohydrate source, coupled with excessive mobilization of free fatty acids. Pathological conditions may develop in the exaggerated form of ketosis. In milder form, ketosis is found under conditions of high dietary fat consumption and after severe exercise.

The production of ketone bodies takes place in the liver by active enzymatic mechanism; and since the utilization of these compounds is low in hepatic tissue, there is a flow of ketone bodies to extrahepatic tissues where they are metabolized and eliminated. Enzymes responsible for the formation of ketone bodies are mainly located in mitochondria. The pathway of ketogenesis involves many steps (Figure 7). The major compound formed in this set of reactions is acetoacetate. The active mechanism which is responsible for the production of this compound in the liver is practically irreversible; acetoacetate, once formed, cannot be reactivated and reutilized by the hepatic tissue. Acetoacetate, however, may be converted to β-hydroxybutyrate, which quantitatively is the predominant ketone body present in the circulation and urine in ketosis. β-Hydroxybutyric acid may be activated directly in extrahepatic tissues and the acetoacetyl-CoA formed by these reactions is split to acetyl CoA which is further oxidized in the citric acid cycle.

Ketone bodies are metabolized in extrahepatic tissues proportionately to their blood level. Ketonemia is associated with an increased hepatic production of these substances rather than

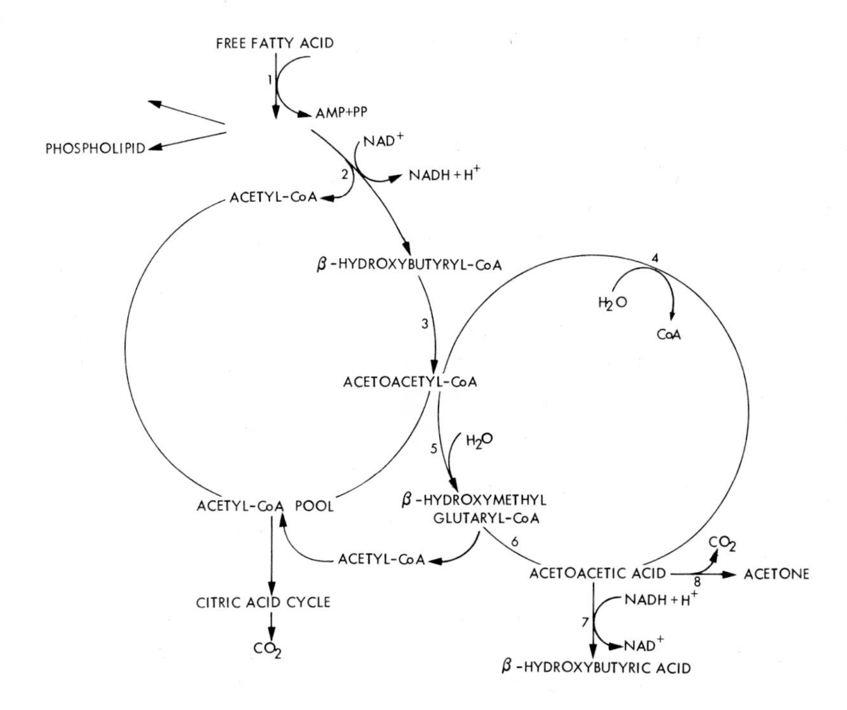

FIGURE 7. Production of ketone bodies in the liver. Free fatty acids undergo β-oxidation catalyzed by thiokinase (1) and the fatty acid oxidase system (2). Reduction processes produce β-hydroxybutyryl-CoA which is converted to acetoacetyl-CoA by β-hydroxyacyl-CoA dehydrogenase (3). Acetoacetate is formed by the enzyme acetoacetyl-CoA deacylase (4). This compound is converted to β-hydroxybutyrate by β-hydroxy-β-methylglutaryl-CoA synthetase (5), β-hydroxy-β-methylglutaryl-CoA lyase (6), and β-hydroxybutyrate dehydrogenase (7) enzymes. Acetone is formed by spontaneous decarboxylation (8).

with a deficiency of extrahepatic degradation. In moderate ketonemia, the urinary loss of ketone bodies is variable, but amounts to only a low percentage of total ketone production and utilization by the body. There is a threshold-like effect by the kidney which shows wide individual variations. Generally, ketosis occurs if there is a rise in the concentration of free circulating fatty acids. Severe ketosis is often accompanied by very high levels of free fatty acids in the blood, with the exception of certain congenital defects of gluconeogenesis. Esterification of the free fatty acids reduces the degree of ketosis and this process represents an essential antiketogenic mechanism. Ketosis occurs as a result of an imbalance between esterification and lipolysis in adipose tissue and is related to a deficiency of available carbohydrates.The release of free fatty acids provide the substrates for hepatic formation of ketone bodies. The availability of α-glycerophosphate and the hormonal state of the liver influence this release. Hormones affect the balance between esterification and oxidation of free fatty acids. Ketosis may be due to starvation, pregnancy toxemia, and uncontrolled diabetes mellitus. In the form of the milder disease, glycogen is present in the liver in variable amounts and free fatty acid levels are not high, accounting for the less severe ketosis under these conditions.

In diabetes mellitus the lack or relative absence of insulin affects adipose tissue, which is extremely sensitive to this hormone. As hypoglycemia develops, associated with low hepatic glycogen levels, lipolysis is enhanced, resulting in a pronounced release of free fatty acids which gives rise to high plasma free fatty acid levels and severe ketosis.

IV. DISEASES OF LIPID ACCUMULATION

A. Obesity

This is the most common disorder of triglyceride metabolism in man.[11,31,41,140,155] Obesity is closely associated with hyperlipidemia, atherosclerosis, and gallstone formation. Due to this link, obesity may not represent a single disease, rather the clinical manifestation of a group of disorders. Obesity is defined arbitrarily, since there is no clear-cut borderline between the obese and thin individual. Body weight is not always the best index of the relative proportion of adipose tissue, and obesity is usually considered to be that condition when the body weight is over 20% above the ideal average weight related to body size. Abnormal metabolic and physiological changes and pathological impairments, however, appear parallel with deviation from normal body weight. Furthermore, there is a continually progressive mortality rate associated with the increasing degree of overweight.

Many possible factors are involved in the pathogenesis of obesity: excessive lipid deposition, decreased lipid metabolism, and decreased lipid utilization.[12,13,60,67,69,95,149] Excessive lipid deposition may be connected with increased food intake, adipose cell hyperplasia, hyperlipogenesis, or hypothalamic lesions. Diminished lipid mobilization may be due to decreased synthesis or function of lipolytic hormones, defective adipose cell lipolysis, or some abnormality of autonomic innervation. Decreased lipid utilization may be associated with aging processes, defective lipid oxidation and subsequent degradation, defective thermogenesis, or inactive life style. Irrespective of these manifold factors associated with obesity, there are generally only two types. In lifelong obesity the abnormality almost starts at birth. Although the birth weight of these individuals is generally normal, there is a trend, as children, to become heavier from early school grades and a larger weight gain occurs during puberty; in females, each successful pregnancy is also connected with irreversibly increased body weight. All attempts in these patients to lose weight by dietary means or all available methods of caloric restriction will lead only to temporary weight loss and a gradual return to the prereduction level of overweight. These individuals are grossly obese and their weight is more than 150% of the ideal average body weight. In lifelong or grossly obese individuals, there is a characteristic increase in adipose cell numbers as well as adipose cell size. After weight reduction the size of the adipose cells shrinks, but the hyperplasia related to the increased cellular number remains unchanged.

Adult-onset obesity is more common.[72,73] These individuals are generally thin and gain weight between 20 and 40 years of age. This weight gain is associated with a more sedentary life style and other environmental factors. One possible explanation for the weight gain during adult life is the slight decline in basal oxygen consumption with age, although the proportionally raised body fat and reduced lean body mass representing muscle and bone compensate for the demand for oxygen consumption. Adult-onset obesity may be the simple reflection of imbalanced caloric intake and utilization, the consequence of which is that the altered body composition with age has not adjusted itself with the reduced calorie consumption. In contrast to the hypercellularity of lifelong obesity, the adult-onset type is characterized predominantly by adipose cell hypertrophy with a minimal increase in cell number. The number of adipose cells is determined early in life and the effect of the feeding pattern during infancy and early childhood affects the development of lifelong obesity. The rapidity of weight gain in infancy provides a good correlation with the risk of becoming overweight in adult years. There also appears to be a genetic influence on obesity.

The etiologic basis of obesity is not well known. In rare instances, damage of the ventromedial hypothalamic nucleus due to tumor or trauma leads to adipose cell hyperplasia. Obesity is cured when the tumor is removed. The hypothalamic center regulates the deposition of triglycerides in adipose tissue. It has been thought that this center is the appetite or satiety center, but recent experiments have shown that hypothalamic lesions alter lipogenesis and insulin level independent of food intake. Emotional and cerebral influences and eating pattern also play a role in obesity. Habit and environment have an effect on the eating pattern by influencing the appetite.

The metabolic consequences of obesity are independent of the cause or type. Acquired resistance to the action of insulin on glucose translocation in fat and muscle cells is directly related to the increased size of the fat cells.[47,59] One of the consequences of this resistance is a feedback compensatory hyperinsulinism. The beta cells of the pancreatic islets are stimulated to produce more insulin and these cells eventually become hypertrophic. Increased circulating insulin levels are directly related to the degree of adiposity.[76]

The degree and duration of obesity influences the development of adult-onset diabetes mellitus. Prolonged hyperinsulinism may lead to the exhaustion of the beta cells in genetically susceptible individuals; with weight reduction, glucose intolerance is reversed. Fatty acid mobilization is less affected by insulin resistance connected with obesity. The hypertriglyceridemia associated with adiposity may be partly due to hyperinsulinism. Serum cholesterol levels are less closely linked with this condition.

B. Hyperlipidemia

This disorder is characterized by an excessive accumulation of one or more of the major lipids in the plasma connected with a severe anomaly in their metabolism. Hyperlipidemia manifests as hypertriglyceridemia, or hypercholesterolemia, or both.[10,25,44,102,142] In these diseases, the lipid-protein molecular aggregates, the lipoproteins, are elevated. The increased plasma levels of lipids and lipoproteins often produce overt signs and symptoms such as xanthoma, lipemia retinalis, and acute abdominal crisis. These conditions are also associated with an increased risk of atherosclerosis. Hyperlipidemic individuals often develop atherosclerotic complications.[106,109] In this section only changes associated with elevated plasma lipid levels will be mentioned. Further associations will be seen in Volume 3.

Excessive accumulation of lipid in the plasma may be derived from either its defective removal or excessive endogenous production, or from both conditions. These abnormalities may be primary or may occur as the consequence of other diseases such as endocrine disorders (diabetes or hypothyroidism), or chronic drug administration during therapy, or with certain environmental pollutants.

Decreased removal of triglycerides may be associated with lipoprotein lipase abnormalities or with the defective uptake of very low-density lipoprotein remnants. Among the lipoprotein lipase abnormalities, the primary forms of hyperlipidemia can be divided into two classes: familial with genetic disposition, and sporadic. In these forms as well as in secondary hyperlipidemias the metabolic abnormalities, symptoms, and laboratory diagnosis are similar.[152] The differentiation between primary and secondary hyperlipidemia is, therefore, very difficult. In hypertriglyceridemia, alterations in fat transport are often parallel with mild glucose intolerance and clinical diabetes. The elevated plasma triglyceride level appear to be related to insulin. This hormone plays a role in both the production of triglyceride-rich lipoproteins and in the removal from the plasma. Insulin deficiency in uncontrolled diabetes, thyroxine deficiency, and nephrotic syndrome are related to lipoprotein lipase abnormalities. Defective uptake of very low-density lipoprotein may be due to primary causes or connected with hypothyroidism or diabetes.

Increased endogenous production of triglycerides may be associated with increased synthesis of precursors. These can be connected with elevated fatty acid production caused by stress, increased glucose levels, or alcoholism.[108] Enhanced insulin secretion can also cause triglyceridemia. This can be related to obesity, estrogen content of contraceptives, or abnormal production of glucocorticoids. Increased triglyceride production and decreased removal can be connected with diabetes and renal disease. Causes of hypertriglyceridemia are seen in Figure 8.

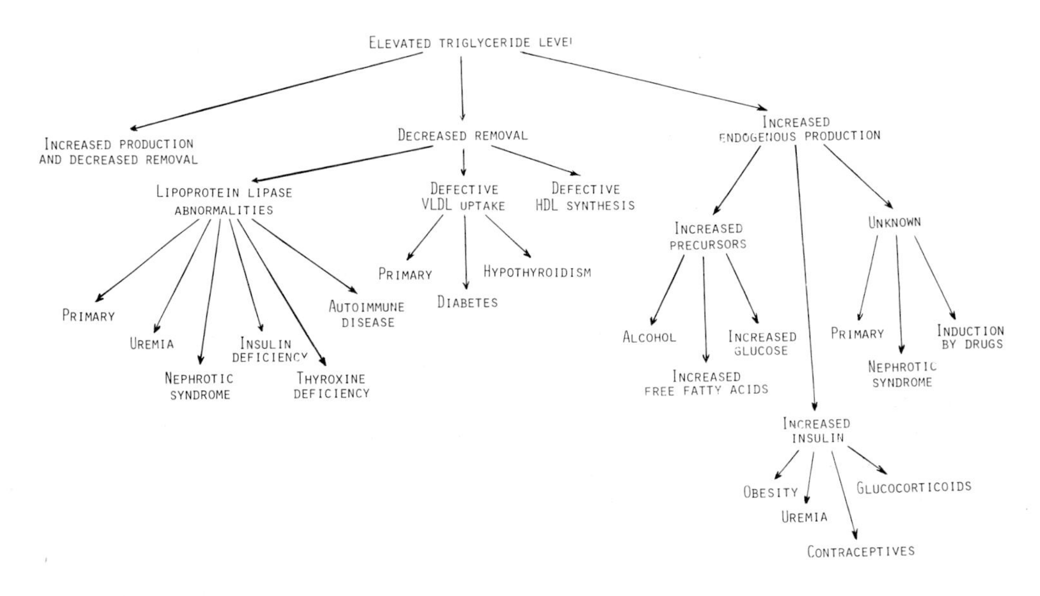

FIGURE 8. Causes of hypertriglyceridemia.

C. Hyperlipoproteinemia

Hypercholesterolemia or hypertriglyceridemia may represent the result of elevated lipoprotein level in the serum. There are primary and secondary hyperlipoproteinemias.[3,10,24,55,60,61,81,84,104,106,119,126,130,141] The primary disease is mainly connected with atherosclerosis of the coronary and peripheral vessels, xanthomas, and hepatosplenomegaly (Table 1); the secondary disease has many primary causes (Table 2). Both kinds of hyperlipoproteinemia are grouped into five types, each is characterized by various lipoprotein elevations.[68,100—102,114,161] The major biochemical parameters of these five lipoprotein phenotypes of hyperlipoproteinemia are presented in Table 1. Since hyperlipoproteinemias are connected with atherosclerosis and secondary changes are found in a variety of underlying disorders, a detailed discussion on this disease category will be dealt with in Volume 3.

D. Coronary Heart Disease

During recent years, strong evidence indicates that both constitutional and environmental factors influence the incidence of coronary heart disease.[2,64,77,80,86,138,141,150] Genetically determined inborn errors of metabolism of lipoproteins are responsible for early atherosclerosis in certain cases.

Among the major factors, elevated serum cholesterol and LDL cholesterol and reduced HDL cholesterol make a significant contribution to the risk.[64,66,88,144,145] Obesity, elevated plasma triglyceride, hyperglycemia, and hyperuricemia show weaker association.[67] Evidence for the association between high serum cholesterol and atherosclerosis include (1) the occurrence of hypercholesterolemia in groups of subjects with clinical manifestation of the disease, (2) epidemiologic studies of populations with differing serum cholesterol levels, (3) the mechanism and nature of atherosclerotic plaque formation, (4) the study of genetically determined hyperlipidemias associated with premature atherosclerosis, and (5) experimental production of atherosclerotic lesions in animals with high cholesterol-containing diets.

The relationship between cholesterol and coronary heart disease is connected with the distribution of cholesterol carrier lipoproteins. Serum LDL concentrations correlate closely with serum cholesterol, since normally 60 to 70% of total cholesterol is directly related to the risk of atherosclerosis; HDL cholesterol shows an inverse relationship. Further details on atherosclerosis will be given in Volume 3.

Table 2
CAUSES OF SECONDARY HYPERLIPOPROTEINEMIA

Lipoprotein phenotype	Causes
I	Dysglobulinemia
	Insulinopenic diabetes mellitus
	Lupus erythematosus
	Pancreatitis
II	Hypothyroidism
	Multiple myeloma
	Nephrotic syndrome
	Obstructive hepatic disease
	Porphyria
III	Dysgammaglobulinemia
	Hypothyroidism
IV	Alcoholism
	Diabetes mellitus
	Gaucher's disease
	Glycogen storage disease
	Nephrotic syndrome
	Niemann-Pick disease
	Oral contraceptives
	Pregnancy
V	Alcoholism
	Insulinopenic diabetes mellitus
	Myeloma
	Nephrotic syndrome
	Pancreatitis

E. Lipid Storage Diseases

Abnormal metabolism of sphingolipids often leads to accumulation, particularly in the central nervous system.[6-8,55,91,93,94,167] These various lipid disorders are called, variously, lipid storage diseases, lysosomal storage diseases, cerebral lipidoses, sphingolipidoses, or gangliosidoses.[19,21,27,51,83,92,105] Most lipid storage diseases are due to some inherited defect in the breakdown of sphingolipids causes by a lack of a specific enzyme (Figure 9). There are some instances where the enzyme may be present in an inactive form. This occurs in metachromatic leukodystrophy.

Sphingolipids are associated with cell membranes.[50] They may play a role in the receptor function of the membrane surface. The effect of certain glycoprotein hormones appears to be mediated by membrane-bound sphingolipids. Due to membrane association, lipid storage diseases are usually connected with membrane abnormalities.[70,98,99] This disorder is discussed in more detail in following chapters.

F. Gallstones

Gallstones occur in Western countries with increasing prevalence. Some gallstones are mainly composed of bile pigments or insoluble calcium salts formed from bile acids, but most contain 70 to 80% cholesterol — even as much as 98%. Cholesterol is insoluble in water, but solubility is enhanced in the bile due to the presence of bile acids and lecithin-producing micelles. When the bile is supersaturated, conditions of cholelithiasis are apparent.

Cholesterol-rich stones are formed when cholesterol excretion from the liver exceeds the normal solubilizing capacity of the bile, representing a disease in the regulation of cholesterol metabolism.

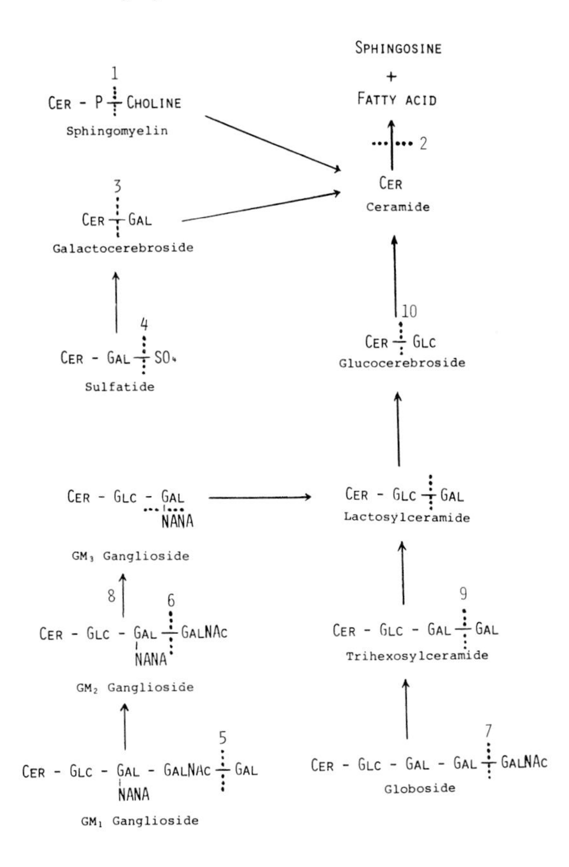

FIGURE 9. Schematic pathways of sphingolipid metabolism. Inborn errors in this pathway present as enzyme deficiency result in spingolipidoses. These include Niemann-Pick disease: sphingo-myelinase deficiency (1); Faber's disease (ceramide lipidosis, lipogranulomatosis): ceramide deficiency (2); Krabbe's disease (globoid cell leukodistrophy): cerebroside β-galactosidase deficiency (3); metachromatic leukodistrophy: sulfatide sulfatase (arylsulfatase A) deficiency (4); GM_1 gangliosidosis: β-galactosidase deficiency (5); Tay-Sachs disease (GM_2 gangliosidosis): hexosaminidase A deficiency (6); infantile variant of this disease is connected with hexosaminidase A and B (7) deficiencies; GM_3 gangliosidosis: deficiency of enzymes synthesizing GM_2 ganglioside from GM_3 ganglioside (8); Fabry's disease: trihexosyl ceramide α-galactosidase (9); and Gaucher's disease: cerebroside β-glucosidase deficiency (10).

The prevalence of gallstones increases with age, and they occur more frequently in females than in males. It has been suggested that the incidence of cholelithiasis is connected with diabetes and hypothyroidism. Both thyroid hormones and insulin affect the metabolism of steroids and untreated hypothyroidism and diabetes cases have shown high serum lipid levels.

V. LIPID SYNTHESIS IN THE SKIN

In contrast to the lipids produced for internal functions, we synthesize a much greater variety of lipids for external use. These lipids are formed in the skin, primarily in various types of sebaceous glands, and excreted externally.[148] The lipid products of these glands are different from the lipids which are formed and utilized internally by various organs. Production of these special lipids is localized with some particular areas of the skin. It separates the environment from the internal milieu and the synthesis of these specific molecules is probably the result of an adaptation to protect the organism from the unpredictable effects of the environment.

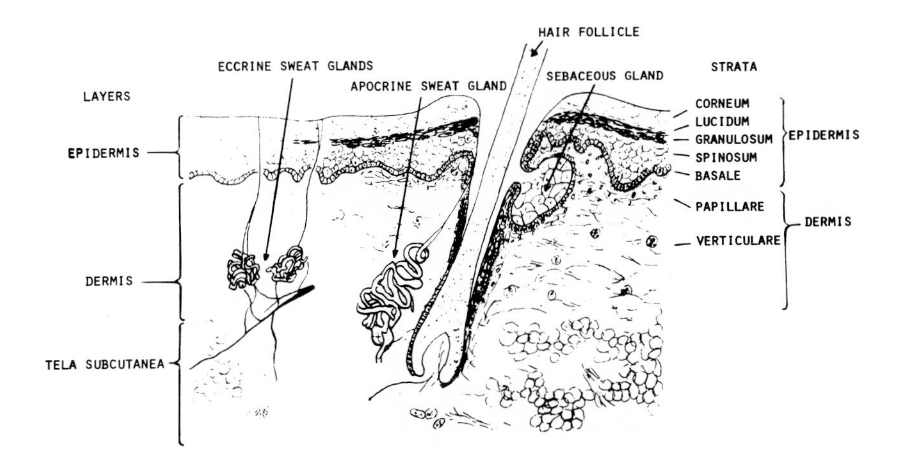

FIGURE 10. Section of the human skin composed of three major layers: epidermis, dermis, and subcutaneous fat. Epidermis is a superficial thin layer, divided into two sublayers: the stratum corneum or dead layer and living epidermis consisting of many different layers, some of which may be absent in parts of the body. The dermis forms a deeper, thick layer divided into two sublayers: the stratum papillare and a dense irregular connective tissue, the stratum reticulare. The dermis contains blood and lymphatic vessels, nerve endings, hair follicles, and various glands. Sebaceous glands are connected to the open space around hair follicles, where their product is excreted to the skin surface. Excreta of the sweat gland can reach the surface directly through capillary tubules.

Various lipids synthesized by the human skin come from the sebaceous glands and from the stratum corneum cells (Figure 10). The composition of these lipids depends on the number of sebaceous glands present at that particular part of the skin. They are found in most areas of the body, but their number varies from $900/cm^2$ on the scalp or face, to less than 40 on forearms, to none on palms or soles. The amount of sebum on the skin surface is not entirely related to the number of sebaceous glands since it can flow from one area to the other through the stratum corneum. There are two types of sweat glands in the skin: eccrine and apocrine. Eccrine sweat glands are distributed uniformly over the body surface, apocrine glands occur mainly in the axilla, areola and nipple, and pubic area and occasionally in other areas. Eccrine sweat glands are responsible for thermoregulation; the exact role of apocrine glands has not yet been established. Both types of sweat glands probably excrete lipids; however, the quantitative contribution of these sweat glands to the surface lipids is not known. It has also been demonstrated that small quantities of lipids originating in the microflora can modify the composition of skin surface lipids.

Lipid synthesis in the skin differs from that taking place in other organs in several ways. Many unusual compounds are formed solely for external excretion, such as a great variety of fatty acids and wax esters, and intermediates of cholesterol synthesis accumulate which do not occur normally in most internal organs.[164] The uniqueness of skin lipids is illustrated by the difference between the composition of the human skin and that of the plasma (Tables 3 and 4). Triglycerides and their derivatives constitute the highest percentage of the sebum and surface lipids, followed by various sterols and precursors; only glyco- and phospholipids occur in greater amounts in lipids derived from the epidermis.

Triglycerides are hydrolyzed to a variable extent by lipases in the duct of the sebaceous glands and on the skin surface resulting in free fatty acids and mono- and diglycerides. On the surface of human skin, these compounds are sometimes present in very large amounts. Unesterified fatty acids are toxic and they normally do not occur in free form in internal organs. In the blood the circulating free fatty acid level is low and they are bound to albumin in a complex form. The hydrolysis of triglycerides to various metabolites is characteristic

Table 3
LIPID COMPOSITION OF HUMAN PLASMA

Lipid	mg/dℓ
Triacylglycerol	80—180
Total phospholipid	123—390
Phosphatidylethanolamine	50—130
Phosphatidylcholine	50—200
Sphingomyelin	15—35
Total cholesterol	107—320
Free cholesterol (nonesterified)	26—106
Free fatty acid (nonesterified)	6—16
Total lipid	310—860

Note: Total fatty acids, expressed as stearic acid, range from 200—800 mg/dℓ; 45% is found in triacylglycerols, 35% in phospholipids, 15% in cholesterol ester, and about 5% in free fatty acids. The amount of triacylglycerol and free fatty acid varies with nutritional state. The determination of the phospholipid content is based on lipid phosphorus.

Table 4
LIPID COMPOSITION OF ADULT HUMAN SKIN

Component	Sebum	Surface lipids	Epidermis
Triglycerides	60	25	10
Di- and monoacyl glycerols	0	10	10
Fatty acids, free	0	25	10
Sterols, unesterified	0	1.5	20
Sterol esters	<1	2.5	10
Squalene	12	10	<0.5
Wax esters	23	22	0
Glyco- and phospholipids	0	0	30
Unidentified	5	4	10

Note: Numbers represent the percentage of total lipids.

of the human skin surface; most animals, including primates, synthesize very little of these breakdown products. Animal surface lipids contain mainly mono- and diester waxes, free sterols, and sterol esters.

A. Unusual Fatty Acids

The synthesis of the fatty acid chain shows most interestingly the specific characteristics of skin lipids.[122,123] The uniqueness is expressed in the number and kinds of carbon skeletons, containing an extremely wide range of chain lengths and unusual patterns of unsaturation. The variety of acyl groups occurring in human skin lipids is seen in many examples (Figure 11). The overall composition of the surface lipids differs significantly from that of any other organ or secretion (Table 5). Most internal tissues synthesize normal saturated C_{14}, C_{16}, and C_{18} fatty acids as well as oleic $C_{18}\Delta 9$ and linoleic $C_{18}\Delta 9,12$ acids (Table 6), where the double bond is located at carbon 9. These acids are also present in the lipids of the human skin in fairly large percentage, however, the remaining acids consist of more than 200

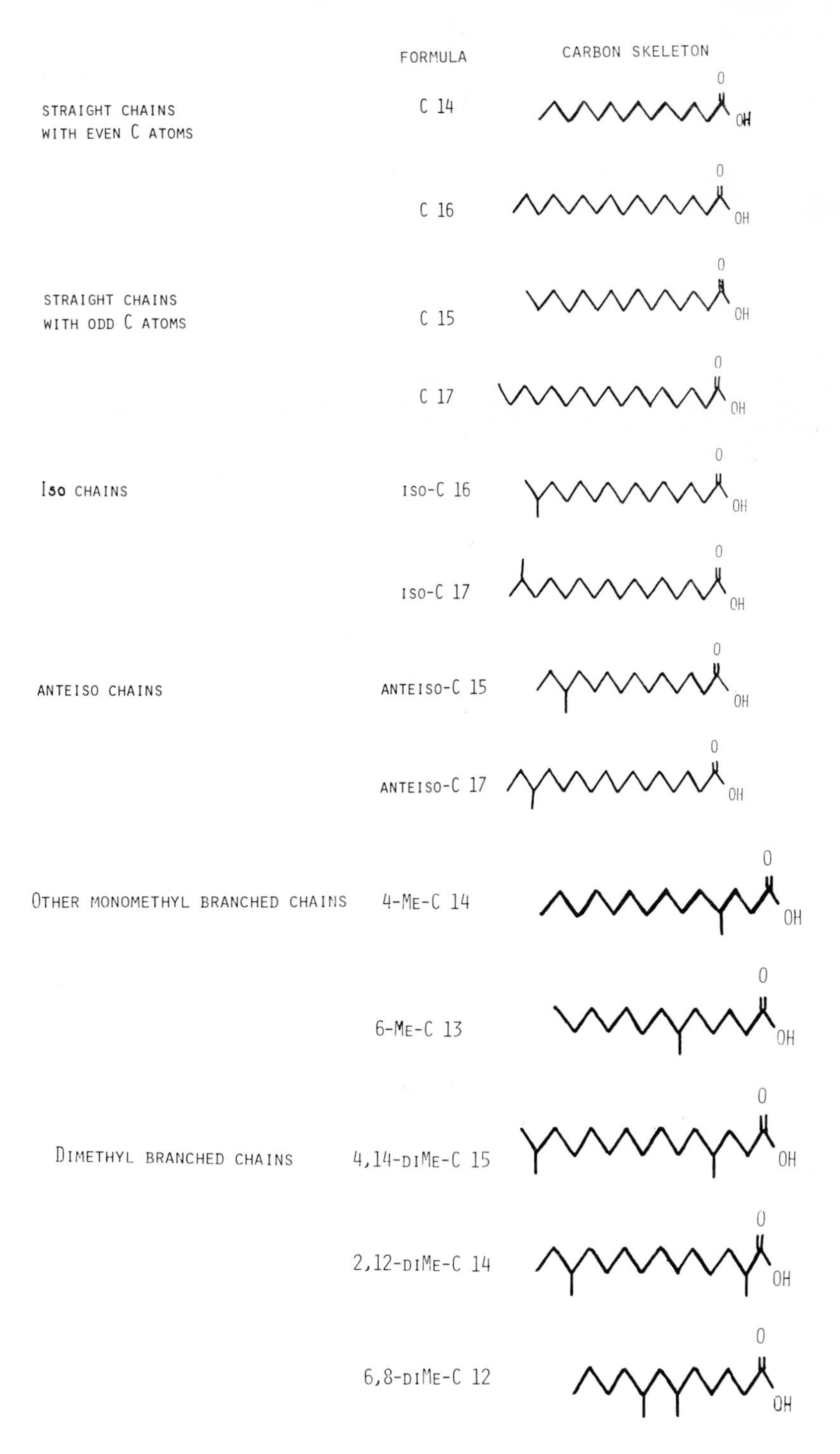

FIGURE 11. Examples of fatty acids in human surface lipids. The wide variation is related to the differences in the structure of acyl-CoA derivatives which are the starters of fatty acid synthesis.

Table 5
FATTY ACID COMPOSITION OF HUMAN SKIN, ORGAN FAT, BLOOD, AND BREAST MILK

Name	Formula	Skin	Fat	Serum	Milk
Butyric	C4				0.4
Caproic	C6				0.1
Caprylic	C8	present			0.3
Capric	C10	present			2.2
Lauric	C12	present	0.9		5.5
Myristic	C14	6.9	3.9	1.6	8.5
cis-Tetradeca-6-enoic	C14:Δ6	1.1	0.5		
4-Methyltetradecanoic	4Me-C14	0.7			
Pentadecanoic	C15	4.0			
12-Methyltetradecanoic	anteiso-C15	1.1			
Palmitic	C16	25.3	25.7	28.1	23.2
cis-Hexadeca-6-enoic	C16:Δ6	21.7			
cis-15-Methylpentadeca-6-enoic	iso-C16:Δ6	4.0			
Palmitoleic	C16:Δ9		7.6	7.6	
Heptadecanoic	C17	1.1			
cis-Heptadeca-6-enoic	C17:Δ6	1.3			
cis-14-Methylhexadeca-6-enoic	anteiso-C17:Δ6	0.8			
cis-Heptadeca-8-enoic	C17:Δ8	0.8			
Stearic	C18	2.9	5.2	3.7	6.9
Petroselenic (*cis*-Octadeca-6-enoic)	C18:Δ6	1.9			
cis-Octadeca-8-enoic	C18:Δ8	8.8			
cis-16-Methylheptadeca-8-enoic	iso-C18:Δ8	0.8			
Oleic	C18:Δ9	1.9	46.6	36.8	36.5
Sebaleic (Octadeca-5,8-dienoic)	C18:Δ5,8	1.1			
Linoleic	C18:Δ9,12	0.5	8.7	12.2	7.8
Linolenic	C18:Δ9,12,15			0.5	0.4
Arachidic	C20		0.6	0.6	
cis-Eicosa-10-enoic	C20:Δ10	0.5			
cis-Eicosa-7,10-dienoic	C20:Δ7,10	0.5			
	C20:3Δ positions			0.4	
Arachidonic	C20:Δ5,8,11,14		0.3	3.1	
Pentaene acids	C22:5Δ positions			1.2	

Table 6
MAJOR FATTY ACIDS IN HUMAN INTERNAL ORGANS

Saturated		Unsaturated	
Name	Formula	Name	Formula
Myristic	C14	Myristoleic	C14:Δ9
Palmitic	C16	Palmitoleic	C16:Δ9
Stearic	C18	Oleic	C18:Δ9
		Linoleic	C18:Δ9,12
		Linolenic	C18:Δ9,12,15
Arachidic	C20	Arachidonic	C20:Δ5,8,11,14
Behenic	C22		
Lignoceric	C24	Nervonic	C24:Δ15

Note: These acids are either the components of almost all normal tissue fats in amounts greater than 0.5 to 1%, or they are accumulated in disease states. Behenic and lignoceric acids constitute more than half of spleen cerebrosides in Gaucher's disease.

species. Among these, the branched acids form a large number of compounds even though they constitute a small portion of the total skin fatty acids.[157] Many of these unusual fatty acids, particularly those with an odd number of carbon atoms or which contain the iso or anteiso structures, occur only in very small quantities in other tissues.

The major site of the unusual fatty acid synthesis is the sebaceous gland.[40] The epidermis also forms these unique compounds occurring on the skin surface, but the pathways of these two syntheses are different. The living epidermis produces more than 80% fatty acids which are like the fatty acids synthesized internally, whereas in the dead layer this is reduced to below 60%. Furthermore, of the remaining 40% fatty acids made by the stratum corneum, a large amount contains odd numbers of carbon atoms, branched chains, and chains which are longer than C_{20}. The major difference between these two sites may be related to the dynamic state in which these two layers exist. Epidermal cells are viable, actively synthesizing keratin; the fatty acids produced are of the type also formed internally. In the final stages of keratinization the epidermal cells are somewhat separated from the dermis; subsequently, they are removed from their major source of nutrients which is essential to maintain the production of viable cellular organelles. These subcellular organelles start to deteriorate and fatty acid transformation leads to unusual compounds, too, thus contributing to the special composition of the surface lipids.

The consequence of changes in proteins, i.e., the keratinization process, is that significant differences occur in the chain length of synthesized fatty acids between skin and other organs. Internal tissues normally synthesize C_{16} and C_{18} fatty acids, which may be raised to a length of C_{24} in nervous tissue, producing nervonic acid as a component of membrane sphingolipids.[50] However, in the skin the chain lengths of fatty acids are extended far beyond this carbon number. Fatty acids have been identified in surface lipids of the adult human up to C_{30}. In the vernix caseosa, matter mainly originating from the fetal sebum which covers the skin of the newborn, significant amounts of fatty acids occur with an extremely long carbon skeleton, chain lengths up to C_{38} having been detected.[38] It is probable that short-chain fatty acids also are present; the presence of acids in the C_4, C_5, and C_7 range has been reported. The extreme range of the fatty acid chain length is a unique characteristic of the human surface lipids.

Another important feature of skin lipids lies in the structure of unsaturated acids, namely at the site of double bonds in the molecule. In most naturally occurring unsaturated fatty acids the double bond is in position $\Delta 9$. In human skin surface lipid a double bond is present at $\Delta 6$ in the monoenoic acid, which is otherwise very rare. It has also been identified in the sebaceous gland of some other animals. Unsaturated fatty acids with a $\Delta 9$ bond have been found in vernix caseosa and in small amounts in adult skin surface lipid. However, oleic acid $C_{18:1}\Delta 9$ which is the most frequently occurring monoenoic acid in tissues, is found only to a small extent in the skin. The main monounsaturated compounds are palmitoleic acid $C_{16}\Delta 9$ and its extension homologues. However, palmitoleic acid and its products are extremely rare minor components in internal tissues. In vernix caseosa the homologous chain is extended to form $C_{18:1},\Delta 11$, $C_{20:1}.\Delta 13$, $C_{22:1},\Delta 15$, $C_{24:1},\Delta 17$, $C_{26:1},\Delta 19$, $C_{28:1},\Delta 21$, and even greater chains are present in large quantities. In these monoenoic acids the double bond is located uniformly between the seventh and eighth carbon atom from the methyl end of the chain.

Monoenoic fatty acid families starting with $C_{16:1}\Delta 6$ and $C_{16:1}\Delta 9$ are the most common in human surface lipid, but many other homologous families are also present on the adult and fetal skin. Not only straight even chains, but straight odd chains, iso chains, and anteiso chains are frequent (Figure 12). Some of these unusual fatty acids may be formed from saturated fatty acids undergoing desaturation at the $\Delta 6$ or $\Delta 9$ position. Little is known about the special enzymes which catalyze the production of these fatty acids; however, they must possess some unique characteristics since some of the diene acids, e.g., sebaleic acid —

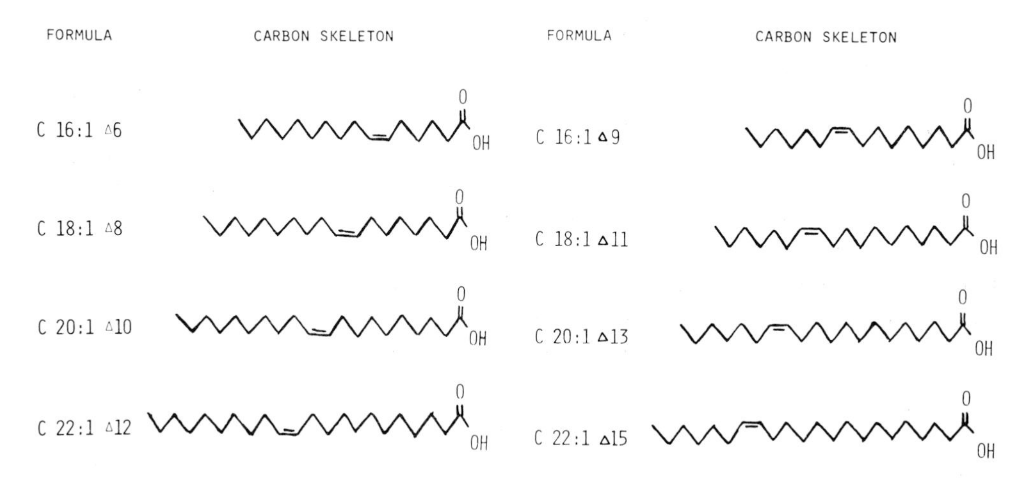

FIGURE 12. Some species of fatty acids in sebum. Within one species the position of the double bond always appears in the same position from the methyl end; the various acids differ from each other in structure only by an additional C2 unit.

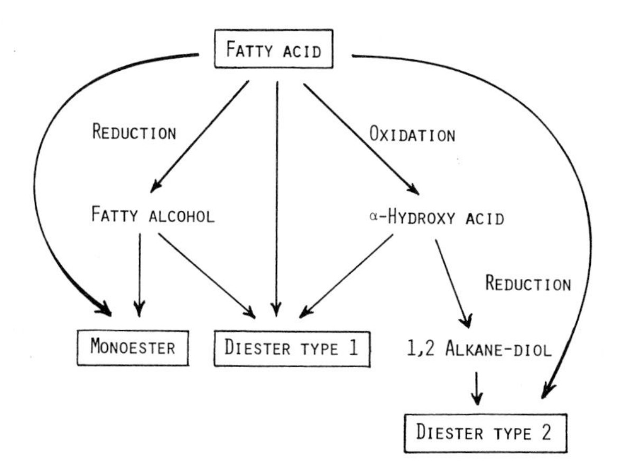

FIGURE 13. Scheme of wax synthesis in human skin.

cis $C_{18:2}\Delta 5,8$ — occur solely in the surface lipid of the human skin. Chain degradation or increased length produce an enormous variation in fatty acid types. Differences can also be related to a variety of starter molecules and to the many biochemical processes existing in the skin. Many fatty acid families may be produced through complex biochemical systems. These acids can be present in triglycerides and their degradation products, in wax esters of vernix caseosa (where 24 families representing 54 different molecular species have been identified), or in sterol esters (from which 28 families and 124 molecular species have been isolated).

B. Wax Esters

These compounds are entirely synthesized for external excretion and they serve partly as energy storage for many animals who live in water.[154] Waxes are derived from a long-chain fatty acid esterified to a long-chain alcohol. The mechanism of this process is depicted in Figure 13. In human sebum they occur mainly as wax monoesters; diesters are very rare. There are two major types of diesters: the hydroxy group of an α-hydroxy fatty acid can be esterified with another fatty acid and the carboxyl group with an alcohol or two fatty acids can form a diester with both hydroxyl groups of a long-chain alkane diol (Figure 14). In

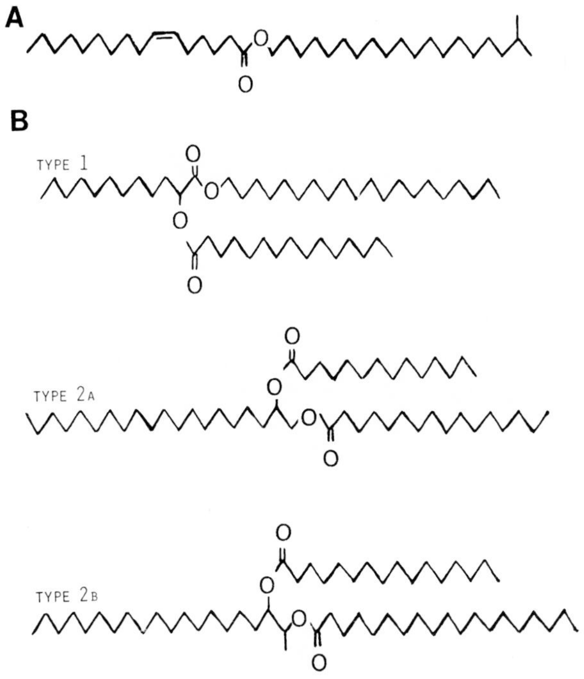

FIGURE 14. Waxes in human surface lipids. In decreasing amounts they contain monoesters (A) produced by the esterification of a long-chain fatty alcohol to a long-chain fatty acid; and diesters (B). The latter group can be formed from an α-hydroxy fatty acid, fatty alcohol and another fatty acid (type 1), or from 1,2-alkane diol (type 2a) or from 2,3-alkane diol (type 2b) esterified by two fatty acids.

man, trace amounts of diesters of 1,2-alkane diols and 2,3-diols occur only in the adult sebum, but they constitute 3% of the lipid of the fetal skin.

Fatty alcohols are produced by the reduction of fatty acids. In the sebum, the fatty alcohols show the same carbon skeleton variations as the fatty acids, indicating a common mechanism of formation. The major representative alcohols contain a double bond in position 10 from the methyl end, suggesting that they are mainly derived from the $C_{16:1}$,Δ6 family of acids. In the synthesis of wax diesters, α-hydroxy fatty acids are formed by hydroxylation of fatty acids; subsequent reduction leads to alkane diols.

C. Steroid Intermediates

During the formation of cholesterol in the human skin several intermediates have been described.[65] These intermediate products are rapidly converted to cholesterol in various internal tissues; however, in the skin they accumulate in fairly large quantities. In the adult dermis, squalene — and in the vernix caseosa, squalene and lathosterol — are the main components. The presence of small amounts of farnesol and geraniol has also been shown. The accumulation of these compounds is related to the hydrolysis of components occurring on the main pathway of cholesterol biosynthesis or else is due to a shift to a secondary route (Figure 15).

D. Secondary Processes

These are also important in the synthesis of various skin lipids, occurring both in the ducts of the sebaceous glands and on the skin surface. Hydrolysis of triglycerides takes

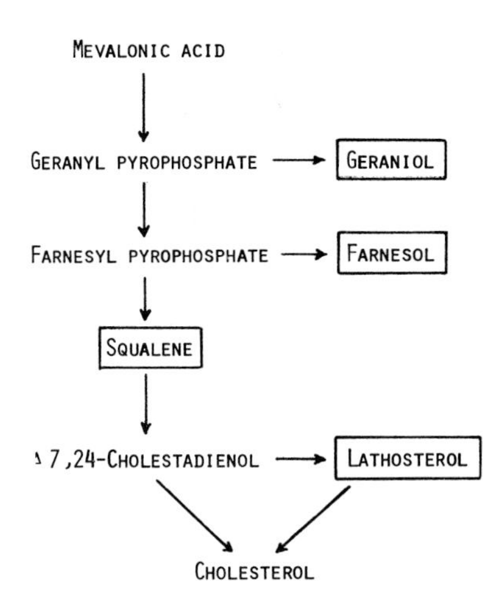

FIGURE 15. Formation of cholesterol intermediates in human skin.

place at these sites by lipases. These lipases probably derive entirely from bacteria and fungi that live on the skin. If the lipid remains on the surface longer, the hydrolytic action of the microflora increases. In the comedo, an accumulation of sebaceous material and dead cells is retained in the hair follicle and excretory duct of the sebaceous gland. The surface is covered by a dark crust associated with a hyperkeratosis. In these circumstances in the comedo, the hydrolysis of triglycerides is almost complete. This is the primary lesion of acne vulgaris and it has been suggested that in the pathogenesis of this skin disease the production of free fatty acids plays an important role. Another secondary process, the formation of cholesterol esters, is probably associated with the late processes of keratinization of the epidermis. Both epidermal and sebum type fatty acids participate in the esterification of cholesterol occurring in comedos. This indicates that a large fraction of the free cholesterol synthesized by the epidermis undergoes secondary processes.

E. Relation to Skin Disease

There are several suggestions as to the significance of the extraordinary amounts of lipid formation in the human skin. It has been proposed that the functional importance of this phenomenon is that various lipids serve as olfactory messages; they are the devices of chemical communications. Due to the great number of saturated and, in particular, unsaturated fatty acids and the complexity of mono- and dimethyl branched compounds it is very likely that each individual produces different substances in varying amounts related to variations of enzyme and cofactor concentrations, body temperature, pH, and ion milieu. This wide variation in skin fatty acids and other lipids may represent an individual chemical fingerprint which is the basis of the animal's recognition.

Each human being is loaded with bacteria and other infectious agents.[112,113] These also colonize our skin, by and large causing no harm. They are held in check by the body's natural defenses. The large number and variety of both saturated and unsaturated fatty chains synthesized by the human skin is probably responsible for part of this protection. Some substances present in these complexes prevent the growth of pathogens. Free fatty acids and other components determine the type of microorganism which can survive on the skin. This does not necessarily mean that these lipids are toxic to bacteria in general, rather that those which survive are compatible with the lipids of the normal healthy skin.

In particular, fatty acids with methyl side chains and fatty alcohols derived from these acids can be considered to be essential in the protective mechanism against microbial infection. This suggestion is based on a possible analogy with the antagonism between plants and insects through insect antijuvenile hormones which are produced by some plants for insect control. These compounds include farnesoic acid (3,7,11-trimethyldodec-2,6,10-trienoic acid), its methyl ester, and farnesol (alcohol derivative). The various methyl branched fatty acids occurring in the human surface lipid may be counterparts of the insect antijuvenile hormones. It may also be important to note that due to the presence of a methyl group in position 3 these compounds may be more resistant, since the infrequent biochemical process of α-oxidation is necessary for their degradation.

Hydrolytic products of triglycerides, mono- and diacyl glycerols, free fatty acids, and glycerol itself help in retaining moisture on the skin by forming a monomolecular film. This is probably an important medium in the balance attained by the process of natural selection. Sometimes, however, bacterial hydrolysis of lipids contributes to the development of skin disorders, such as acne vulgaris which represents a chronic inflammatory condition of the sebaceous structures.

It is feasible to consider that the accumulation of intermediates of cholesterol biosynthesis also provides some prevention of microorganisms from utilizing cholesterol which they must take up from the environment. Functionally all these lipids with unusual structures may inhibit the metabolism of potential pathogens and thus allow only the survival of the compatible microorganisms.

VI. STEROID METABOLISM IN THE SKIN

The metabolism of the normal skin is affected by a wide range of steroid hormones including androgens, estrogens, and corticosteroids.[115,166,168] The skin is considered to be a specific target organ to androgens, which is associated with their action on hair follicles, sebaceous, and apocrine glands. These hormones stimulate the activity and secretion of sebaceous glands. The effect is related to an increase in cell division and an enhanced intracellular lipid synthesis. They also promote hair growth in some parts of the human body. Particularly, hair follicles of the beard, axilla, and pubic regions are under androgenic control. Hirsutism is associated with an excessive androgen production; in contrast, hair growth is sparse when androgen biosynthesis is insufficient as in the case of feminizing adrenal tumors and Stein-Leventhal syndrome.

The action of testosterone involves an increase of holocrine activity and the elevated secretion is preceded by an increased cell replication. These effects are related to the transformation of testosterone to active metabolites (Figure 16).

It has been shown that the major synthesis of these sebotrophic hormones in normal adult men takes place predominantly in the testes, whereas in women they originate from the adrenal cortex and ovary. Some of the enzymes participating in the metabolic route are found in the skin. Testosterone, C_{17}, and many C_{19} steroids present in human blood are converted to 5α-dihydrotestosterone as a result of a high Δ^4-3-oxo steroid 5α-reductase activity. The activities of different enzymes involved in steroid metabolism show variations in distribution according to body site. In the scalp and facial skin the conversion of 17-hydroxyl steroids to 17-oxo steroids is much greater than the reverse process. In axillary and pubic skin the reaction proceeds in the other direction. 17 β-Hydroxysteroid dehydrogenases occur in two forms in human skin. One is a NADH- dependent microsomal enzyme which is the main constituent of facial skin, the other is NADPH-dependent and present in the soluble skin fraction; this is the main enzyme in axillary skin. These enzymes show a different rate of metabolizing testosterone and androstendione, which may represent the regulatory mechanism and relationship to skin disease. The metabolism of 17β-hydroxyandrosterone is increased in acne and hirsutism in women.

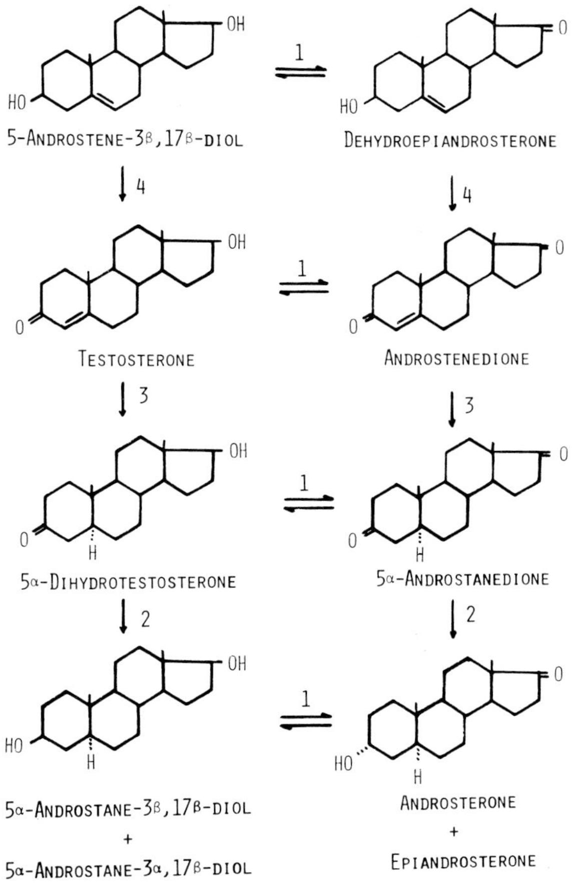

5-ANDROSTENE-3β,17β-DIOL ⇌(1) DEHYDROEPIANDROSTERONE

↓4 ↓4

TESTOSTERONE ⇌(1) ANDROSTENEDIONE

↓3 ↓3

5α-DIHYDROTESTOSTERONE ⇌(1) 5α-ANDROSTANEDIONE

↓2 ↓2

5α-ANDROSTANE-3β,17β-DIOL + 5α-ANDROSTANE-3α,17β-DIOL ⇌(1) ANDROSTERONE + EPIANDROSTERONE

FIGURE 16. Androgen metabolism in human skin. Four enzymes are responsible for the metabolic conversion: 17β-hydroxysteroid dehydrogenase, (1); 3α-hydroxysteroid dehydrogenase, (2); 5α-reductase; (3); and 3β-hydroxysteroid dehydrogenase $\Delta^{4,5}$ isomerase (4).

The importance of 5α-dihydrotestosterone in the human has been shown in the development of skin diseases, especially in women with acne vulgaris or idiopathic hirsutism. Alopecia is also an androgen-dependent condition. At puberty, the growth of sebaceous glands is often accompanied by acne. The clinical severity of the disease is well correlated with the secretory activity of these glands. The greater frequency of acne in women is associated with the excretion of abnormally high amounts of urinary androstanediol related to higher 5α-reductase activity of the skin. This excessive secretion may be the consequence of high plasma androgen or increased production of 5α-dihydrotestosterone in the skin. In males, acne is not connected with increased androgen concentrations. In pattern alopecia, a correlation was found between balding and the increased capacity of the skin to convert testosterone to 5α-dihydrotestosterone.

The response of the sebaceous gland to androgens is diminished by hypophysectomy, thus indicating a role of pituitary peptide hormones in these processes. After removal of the pituitary the response to testosterone can be restored by prolactin, growth hormone, and synthetic melanotropin. These preparations may act independently, but they also function synergistically with testosterone. It is possible that thyrotropin has a similar effect to these peptides, but since it directly affects the secretion of thyroid hormones influencing the activity of sebaceous glands the direct synergism between testosterone and thyrotropin cannot be established.

Specialized skin secretions responsible for the scent of animals are also steroid dependent. The mammalian pheromones are macrocyclic ketones and lactones. One such odorous sterol, 5α-androst-16-en-3-one, can be synthesized from progesterone, which raises the possibility of an interrelationship between scent production and hormone metabolism. The odor of the secretions depend further on bacterial action. The smell of axillary apocrine sweat in man is the result of microbial decomposition.

In contrast to the specific action of androgens on sebaceous and apocrine glands, estrogens stimulate the cutaneous tissue. The appearance and texture of the skin are important features of the female; after the menopause degenerative changes occur in these characteristics. Following ovariectomy the epidermal thickness is also decreased, but can be restored by oral treatment with estradiol succinate or valerate. The mechanism of action of estrogens is different from that of androgens. These steroids do not enhance cell mitosis; in pharmacological doses they decrease rather than stimulate sebaceous activity. Estrogens markedly reduce sebum production which indicates an inhibition of lipid synthesis within the sebaceous cells. The differences are also reflected in the various skin diseases.

The mechanism of action by adrenocortical hormones is not well established, although they are being used most effectively in dermatology in a variety of conditions. Corticosteroids can decrease many processes: the thickness of epidermis or the skin, capillary blood flow in inflammation, and mitosis. The latter actions will be dealt with later in appropriate chapters.

VII. ATYPICAL FATTY ACIDS IN BLOOD

Refsum's disease (heredopathia atactica polyneuritiformis) is caused by an inborn error of metabolism and is characterized by the accumulation of phytanic acid, (3,7,11,15-tetramethyl- hexadecanoic acid) in the plasma and many organs.[43,121,136,156,159] The symptoms of this disease include impairment of vision, night blindness, cerebellar ataxia dysfunction of the nervous system associated with a progressive degeneration of myelin and cardiac involvement, muscular atrophy, and bone lesions. The accumulation of this branch-chain fatty acid sometimes exceeds 20% of total plasma fatty acids. Phytanic acid is originated from plants and is present in the plasma in triglycerides (51%) and in its breakdown products, mono- and diacylglycerols and free fatty acids, and in phospholipids (25 to 50%). Cholesteryl esters contain only traces. Phytanic acid also accumulates in considerable amounts in various tissues. Its storage in myelin probably disrupts the bimolecular lipid layer due to steric hindrance associated with the presence of methyl groups. This process probably causes the demyelination of nerves and is responsible for many clinical manifestations of Refsum's disease.

The symptoms originate largely from defects in the peripheral nervous system, but some centrally derived disorders are also apparent. In the presence of phytanic acid myelin is formed, but it shows a reduced stability related to the poor fit of this fatty acid with its methyl branches. Due to the different architecture, the molecule tends to be wider and the spatial expansion of the side chains occupies the place of another fatty acid. As a consequence of the increased myelin sheath volume and fewer components, the structure becomes loose (Figure 17). The extended myelin is less stable, tends to break down, and elicits altered transport properties.

The presence of phytanic acid is unrelated to endogenous biosynthesis, since precursors are not converted to phytanic acid in these patients. However, from the corresponding α,β -monounsaturated alcohol, phytol, this compound is produced. Chlorophyll contains approximately 30% phytol and probably this is the endogenous natural source. Phytol is readily converted to phytanic acid by normal man and by patients with Refsum's disease. The accumulation is related to the patient's relative defect in the metabolic pathway for the

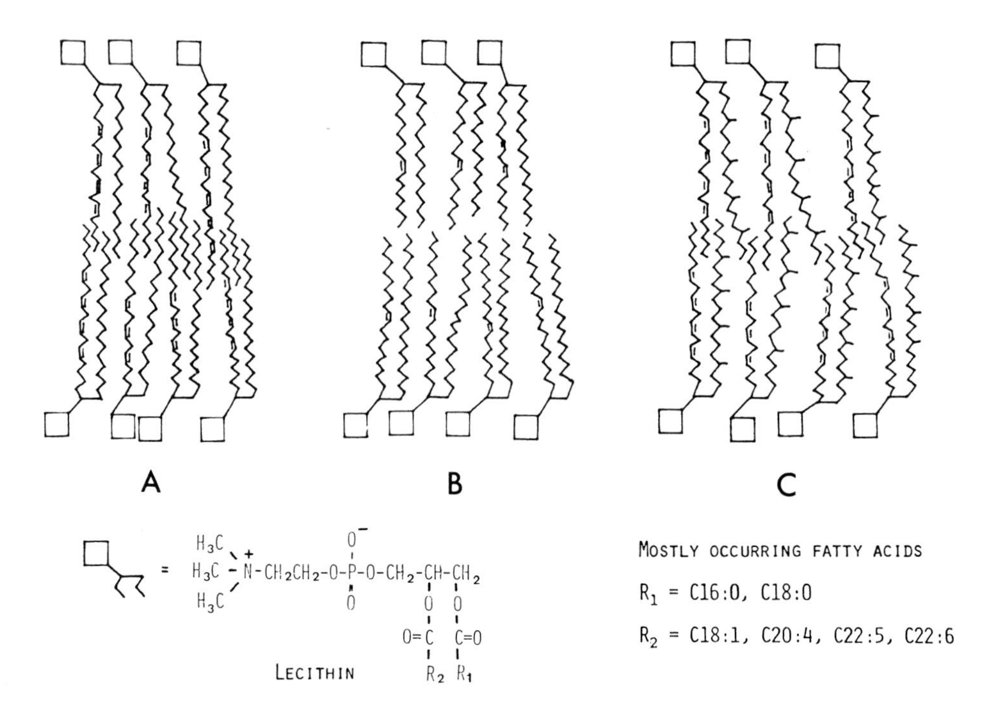

FIGURE 17. Scheme of structural defects in myelin fatty acids. Lipid bilayer structure in normal conditions, (A); in essential fatty acid deficiency, (B); and in Refsum's disease, (C). The lack of very long-chain unsaturated fatty acids causes a shortage of the lock between the nonpolar branches. Inclusion of branched-chain fatty acids increases the distance between the individual units and reduces density. In both cases, the stability of the membrane is decreased.

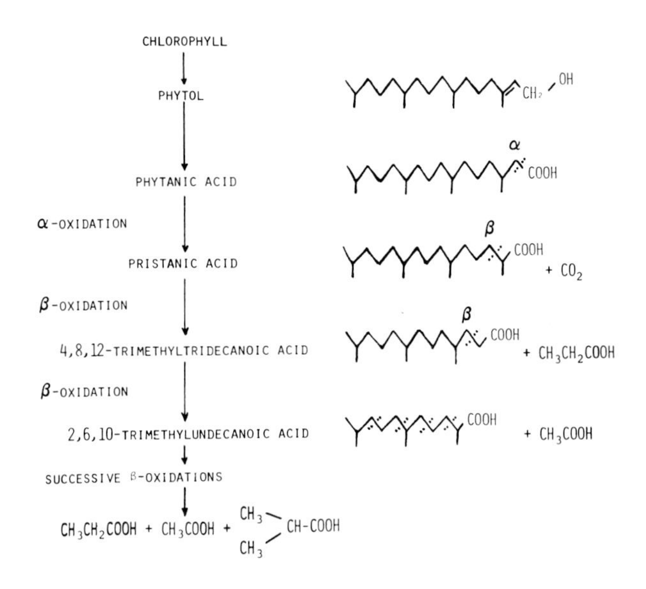

FIGURE 18. Scheme of phytanic acid metabolism.

oxidation of phytanic acid (Figure 18). Due to the methyl group in position 3 of the carbon chain β-oxidation is blocked. However, the α-oxidation of phytanic acid is very efficient in normal subjects, therefore phytanic acid normally occurs in the plasma only in trace amounts or not at all. Inborn error in the α-oxidation pathway is responsible for Refsum's disease.[14] In general, many abnormalities of this disorder are not specific; however, the presence of large amounts of phytanic acid in serum lipids can be decisively used in the diagnosis.

α-Hydroxylation is a particularly important process in the function of the central nervous system. Long-chain fatty acids are converted through this pathway to α-hydroxy fatty acids, which are essential constituents of many complex brain lipids. Polyneuritic changes similar to the symptoms of Refsum's disease are seen in other lipidoses. It may be, therefore, that the clinical manifestations represent some reflections of abnormalities in α-hydroxylation.

VIII. UNSATURATED FATTY ACIDS

Polyunsaturated fatty acids have many functions and they are essential constituents in the integrity of all tissues.[74,75,131,151] Man can synthesize long-chain saturated and monounsaturated fatty acids, but the synthesis of polyunsaturated fatty acids does not occur in mammalian tissues in an appreciable amount.[22] The intake of these polyunsaturated fatty acids is necessary, and inadequate levels in our diet lead to various essential fatty acid deficiencies. These deficiency diseases include many symptoms like reduced growth rate and increased susceptibility to bacterial infections.[57,116] The latter may be related to the enhanced permeability of the skin to water and greater penetration of microorganisms. More severe skin lesions, like parakeratosis, are probably associated with the damage to skin mitochondria. The lack of polyunsaturated fatty acids also alters the physiological functions of the liver and heart muscle mitochondria and platelet structure, and causes decreased prostaglandin biosynthesis, reduced myocardial contractility, abnormal thrombocyte aggregation, and other disorders.[32,36,49,117,118,124,135,158,160,162]

The function of polyunsaturated fatty acids in tissue metabolism may be related to the general physical properties of molecules containing many double bonds and to the production of specific derivatives of unsaturated fatty acids which are an integral part of phospholipids present in all biomembranes. In these phospholipids polyunsaturated fatty acids are esterified in position 2 of the glycerol moiety. This particular position seems to be essential in enzyme reactions when fatty acids are transferred to other molecules, e.g., cholesterol. The polyunsaturated structure may represent a fluidity and assure the liquid crystalline structure of lipids in biomembranes and may take part in the association of substrates with enzymatically active sites.[30]

The major synthesis of polyunsaturated fatty acids occurs from linoleic and linolenic acids.[53] However, in the absence of sufficient arachidonic and dihomo-γ-linolenic acids, due to lack of dietary intake of these compounds, oleic and palmitoleic acids can serve as the initial unsaturated precursors for the biosynthesis of other polyunsaturated fatty acids (Figure 19). The desaturation of the various fatty acids yields different polyunsaturated derivatives. Desaturation occurs in liver as well as in brain.[22] Linoleate exhibits an inhibition of the desaturation of oleate to 18:2, Δ6,9, thus showing important control of these processes. There is a general concept that there are separate desaturase enzymes for introducing double bonds at the 5, 6, or 9 positions. The presence of the original double bond determines the route and degree of double bond formation and evidently no cross reaction takes place between these reactions. The utilization of oleic and palmitoleic acids is an attempt of the body to compensate for the absence of arachidonic acid. Its replacement by other unsaturated fatty acids, however, is not satisfactory for the normal function of cellular membranes. Essential fatty acid deficiency increases the fragility of erythrocyte membranes and is associated with an increased degree of hemolysis. The stability of the membranes in hepatic mitochondria also depends on the presence of appropriate fatty acids. Synthetic and naturally occurring polyunsaturated fatty acids can correct abnormal swelling of liver mitochondria caused by essential fatty acid deficiency. There is an indication that the triene: tetraene ratio in membrane phospholipids is an essential feature of membrane stability.

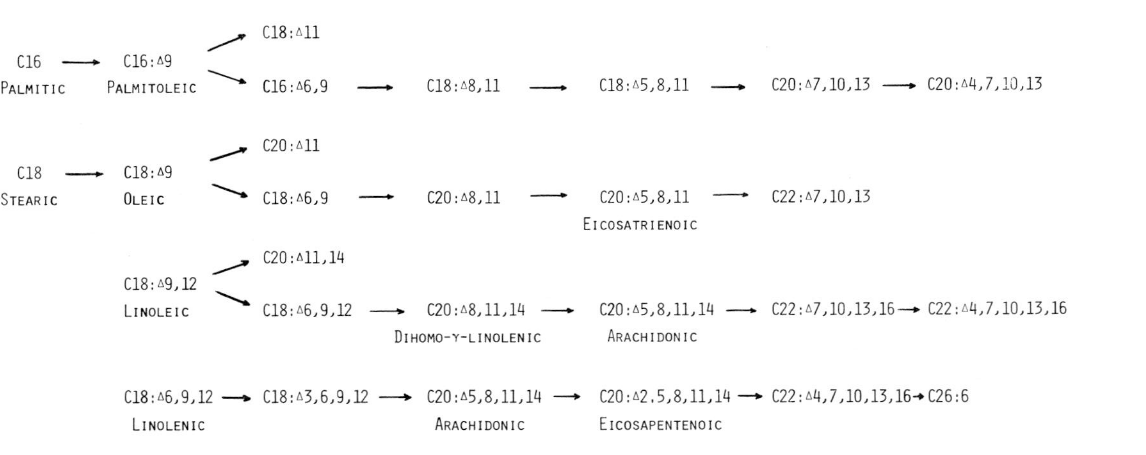

FIGURE 19. Biosynthesis of polyunsaturated fatty acids. This pathway consists of a series of alternate desaturation and chain elongation processes catalyzed by hepatic microsomes. In these enzymes actions, the alteration proceeds regularly and there is no crossover between the different reaction sequences. (From Mead, J. F., in *Progress in the Chemistry of Fats and Other Lipids*, Vol. 9, Holman, R. T., Ed., Pergamon Press, New York, 1971, 161. With permission.)

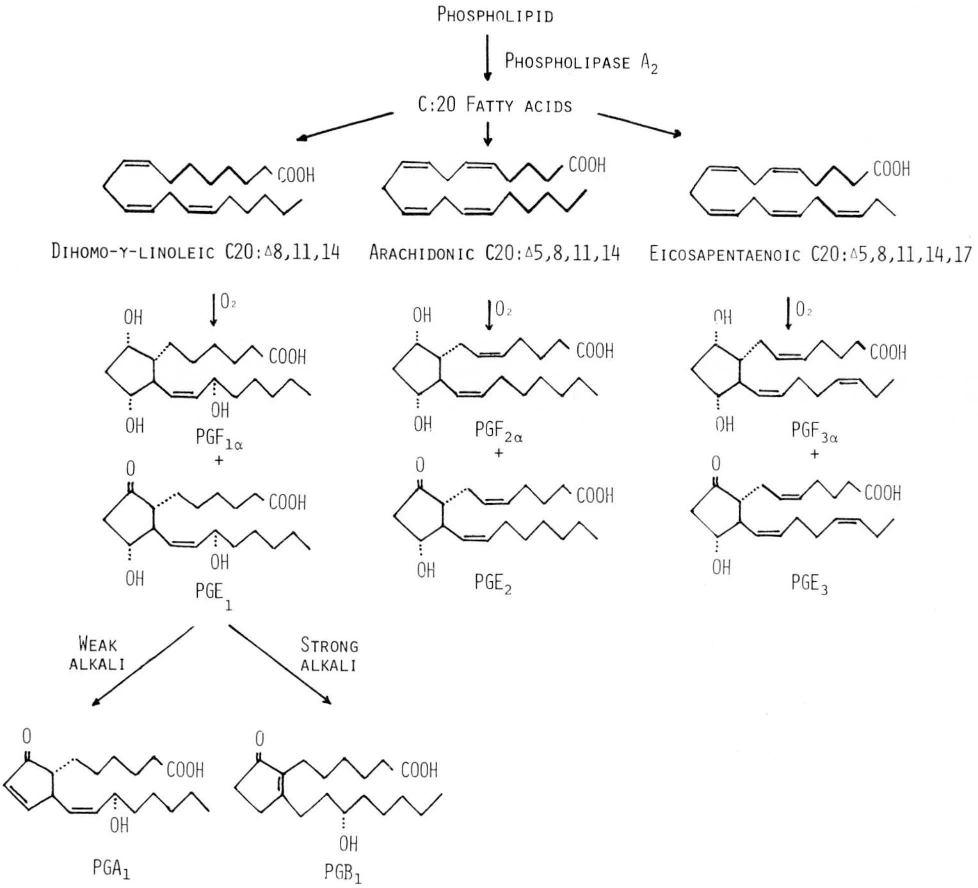

PHOSPHOLIPID

PHOSPHOLIPASE A_2

C:20 FATTY ACIDS

COOH COOH COOH

DIHOMO-γ-LINOLEIC C20:Δ8,11,14 ARACHIDONIC C20:Δ5,8,11,14 EICOSAPENTAENOIC C20:Δ5,8,11,14,17

O_2 O_2 O_2

OH COOH OH OH PGF$_{1\alpha}$ + OH COOH OH PGF$_{2\alpha}$ + OH COOH OH PGF$_{3\alpha}$ +

O COOH OH OH PGE$_1$ O COOH OH PGE$_2$ O COOH OH PGE$_3$

WEAK ALKALI STRONG ALKALI

O COOH OH PGA$_1$ O COOH OH PGB$_1$

FIGURE 20. Synthesis of primary prostaglandins and derivatives.

The activation of thrombocytes by stimuli like ADP and collagen is followed simultaneously by prostaglandin synthesis from arachidonic and dihomo-γ-linolenic acids.[78,79] This process affects the function of platelets which have a regulatory role in thrombus formation.[77,82,111] In essential fatty acid deficiency platelet function is abnormal and collagen-induced aggregation and serotonin release are reduced. In human skin prostaglandin is also produced, and its formation is inhibited by various unsaturated fatty acids and anti-inflammatory drugs.

Dietary linoleic acid is essential in prostaglandin synthesis.[35,54] Stimulation of cells and tissues results in the increased production of prostaglandins. The stimulus can be mechanical, hormonal, or neurological. Inflammation represents a major stimulation; during this process phospholipase A_2 is released from lysosomes which can cleave the unsaturated fatty acid moiety, arachidonic, dihomo-γ-linolenic, and eicosapentenoic acids, from phospholipids. These essential C_{20} acids undergo cyclization, molecular oxygen is incorporated by enzyme action, and they are finally coverted to prostaglandins (Figure 20). The synthesis takes place locally, therefore these compounds are produced in almost all tissues. The prostaglandins may serve as local regulators of cell functions. They possess a broad spectrum of pharmacological activities which involves an increase in cyclic adenosine 3′,5′-monophosphate concentration by the activation of membrane adenyl cyclase in a manner similar to the action of many hormones. Prostaglandins are involved in the inflammatory response and they have a stimulating influence on the coronary flow and contractile properties of the heart. The

Table 7
FATTY ACID CONTENT OF LECITHINS IN ADULT HUMAN BRAIN

	Fatty acid						
	C16:0	C18:0	C18:1	C20:1	C20:4	C22:5	C22:6
Gray matter	7	26	12	2	14	14	24
White matter	7	9	42	8	6	14	3

Note: The figures represent the percentage of major fatty acids. Unsaturated fatty acids mainly derive from position 2 of the glycerol nucleus. It is apparent that lecithins from the gray matter contain highly unsaturated fatty acids, whereas in the white matter oleic acid is the major constituent.

reduction of prostaglandin synthesis due to lack of linoleic acid may cause impairments in these physiological processes and in the regulation of the maintenance of normal health.

Polyunsaturated fatty acids lower the serum cholesterol level, indicating a role in normalizing cholesterol metabolism. There now seems to be considerable doubt about the role of polyunsaturated fatty acids in cholesterol metabolism and in prophylaxis of coronary disease. These compounds have been shown to reduce serum triglyceride levels.[64] They may, therefore, be important in reducing the occurrence of ischemic heart disease.

In the metabolism of the brain, polyunsaturated fatty acids play a special function.[9,32,39,70,160,169] They are present in various types of lipids, particularly in high concentration in the lecithin fractions of the gray matter (Table 7). Dietary unsaturated components are converted locally in the brain tissue to longer and more unsaturated derivatives. Some other organs such as the liver also produce these compounds, but the elongation of fatty acid beyond 20 to 22 carbon atoms is specific to the central nervous system; great amounts of fatty acids with chain lengths extended at least to C_{30} are found only in sphingolipids of brain tissue. The importance of these very long-chain fatty acids in the cerebroside and sulfatide fractions of the myelin is shown in animals. Abnormalities of microsomal enzyme systems responsible for the extension of the carbon chain of C_{20} fatty acids result in poor myelin formation associated with neurological symptoms. Prolonged essential fatty acid deficiency brings about impaired performance of learning in experimental animals.

Changes in the levels and relative proportions of dietary essential fatty acids modify the fatty acid composition of brain structural lipids which may cause a reduced membrane stability[70] (Figure 15). Tetraenoic acids derived from linoleic acid are continuously replaced by trienoic acids formed from oleic acid. Accumulation of trienes in the brain is followed by a depletion of body stores of polyunsaturated fatty acids. In deficiency, almost all lipids in various brain membranes are affected to a varying extent. The altered fatty acid pattern in the nervous tissue typical of essential fatty acid deficiency is reversible. On changing the diet, a continuous replacement takes place indicating the absence of a blood-brain barrier to essential fatty acids.

A stimulation of the central nervous system by convulsant drugs, electroshock treatment, anesthesia, or brain ischemia triggers the release of free fatty acids from phospholipids and selectively enhances the levels of arachidonic acid in the nervous tissue. The release of arachidonic acid leads to the endogenous formation of prostaglandins.[35] In this enzymatic process cyclic AMP is also involved. In the brain of essential fatty acid-deficient animals, the liberation of total arachidonic acid is diminished and lipid metabolism is disturbed.

Studies on intractable infant eczema and acrodermatitis enterohepatica have shown an association with low serum levels of unsaturated fatty acids and an abnormal proportion of other fatty acids, in particular, a low triene:tetraene ratio. In the latter condition, failure of zinc absorption represents a contributing factor. These symptoms are suspected to be caused by malabsorption related to an inadequate dietary intake or a metabolic disorder. There is a difference in the triene:tetraene ratio between breast milk (0.2 to 0.6) and cow's milk (0.6 to 1.4) and the replacement of cow's milk with breast feeding or proper supplementation of the essential fatty acids by intravenous infusion of fat emulsion has elicited a dramatic improvement in this skin condition.

In the past, prolonged intravenous perfusion of fat free preparations induced essential fatty acid deficiency in humans within weeks, particularly in infants. These preparations contained only glucose, amino acids, vitamins, and minerals. The onset of biochemical symptoms usually occurred after 1 to 3 months of intravenous alimentation. Fortunately, subsequent intake of normal food leads to rapid recovery. Infusion of lipid emulsions is now a standard procedure in prolonged intravenous alimentation.

Essential fatty acid deficiency facilitates microbiological invasion of wounds, thus reducing the resistance to infections. Surgical wounds do not heal normally in the absence of essential fatty acids; only when the deficiency is relieved by intravenous administration of fat emulsion containing these essential constituents does normal healing occur. The regrowth of tissue during wound healing and the elimination of infection require specific polyunsaturated fatty acids and if their precursors are not supplied in the diet then normal tissue cannot be formed.

REFERENCES

1. **Abumrad, N. A., Perkins, R. C., Park, J. H., and Park, C. R.,** Mechanism of long chain fatty acid permeation on the isolated adipocyte, *J. Biol. Chem.*, 256, 2183, 1981.
2. **Albrink, M. J., Meigs, J. W., and Mann, E. B.,** Serum lipids hypertension and coronary artery disease, *Am. J. Med.*, 31, 4, 1961.
3. **Alaupovic, P.,** Apolipoproteins and lipoproteins, *Atherosclerosis*, 13, 141, 1971.
4. **Amer, P. and Ostman, J.,** Importance of the cyclic AMP concentration for the rate of lipolysis in human adipose tissue, *Clin. Sci.*, 59, 199, 1980.
5. **Ammon, H. V., Thomas, P. J., and Phillips, S. F.,** Effect of lecithin on jejunal absorption of micellar lipids in man and on their monomer activity in vitro, *Lipids*, 14, 395, 1979.
6. **Assman, G., Fredrickson, D. S., Herbert, P., and Forte, T.,** An A-II lipoprotein particle in Tangier disease, *Circulation*, 50 (Suppl. 3), 259, 1974.
7. **Austin, J. H.,** Globoid (Krabbe) leukodystrophy, in *Pathology of the Nervous System*, Minkler, J., Ed., McGraw-Hill, New York, 1968, 843.
8. **Balint, J. A., Spitzer, H. L., and Kyriakides, E. C.,** Studies of red-cell stromal lipid in Tay-Sachs disease and other lipidoses, *J. Clin. Invest.*, 42, 1661, 1963.
9. **Bates, D., Fawcett, P. R. W., and Shaw, D. A.,** Polyunsaturated fatty acids in treatment of acute remitting multiple sclerosis, *Br. Med. J.*, 2, 1390, 1978.
10. **Beaumont, J. L., Carlson, L. A., and Cooper, G. R.,** Classification of hyperlipidaemia and hyperlipoproteinaemias, *Bull. WHO*, 43, 891, 1970.
11. **Berchtold, P., Berger, M., Greiser, E., Dolise, M., Irmscher, K., Gries, F. A., and Zimmerman, H.,** Cardiovascular risk factors in gross obesity, *Obesity*, 1, 219, 1977.
12. **Bjorntorp, P.,** Adipocyte precursor cells, in *Recent Advances in Obesity Research*, Vol. 3, Bjorntorp, P., Cairella, M., and Howard, A. N., Eds., John Libbey, London, 1981, 58.
13. **Bjorntorp, P. and Sjostrom, L.,** Adipose tissue cellularity, *Int. J. Obesity*, 3, 181, 1979.
14. **Blass, J. P., Avigan, J., and Steinberg, D.,** α-Hydroxy fatty acids in hereditary ataxic polyneuritis (Refsum's disease), *Biochim. Biophys. Acta*, 187, 36, 1969.
15. **Booth, D. A., Goodwin, H., and Cummings, J. N.,** Abnormal gangliosides in Tay-Sachs disease, Niemann-Pick's disease and gargoylism, *J. Lipid Res.*, 7, 337, 1966.
16. **Borgstrom, B. and Erleuson, C.,** Pancreatic lipase and co-lipase. Interactions and effects of bile salts and other detergents, *Eur. J. Biochem.*, 37, 60, 1973.

17. **Brady, R. O.,** The abnormal biochemistry of inherited disorders of lipid metabolism, *Fed. Proc.*, 32, 1660, 1973.
18. **Brady, R. O.,** Inherited metabolic diseases of the nervous system, *Science*, 193, 733, 1976.
19. **Brady, R. O., James, S. P., and Barranger, J. A.,** The liver in lipid storage disease. Biochemical basis of pathogenesis, *Prog. Liver Dis.*, 7, 331, 1982.
20. **Brady, R. O., Johnson, W. G., and Uhlendorf, W. F.,** Identification of heterozygous carriers of lipid storage disease, *Am. J. Med.*, 51, 423, 1971.
21. **Brady, R. O.,** Enzymatic defects in the sphingolipidoses, *Adv. Clin. Chem.*, 11, 1, 1968.
22. **Brenner, R. R.,** The oxidative desaturation of unsaturated faty acids in animals, *Mol. Cell Biochem.*, 3, 41, 1974.
23. **Brenner, R. R.,** Nutritional and hormonal factors influencing desaturation of essential fatty acids, *Progr. Lipid Res.*, 20, 41, 1982.
24. **Brewer, H. B.,** Current concepts of the molecular structure and metabolism of human apolipoproteins and lipoproteins, *Klin. Wochenschr.*, 59, 1023, 1981.
25. **Brown, M. S. and Goldstein, J. L.,** Familial hypercholesterolemia: a genetic defect in the low-density lipoprotein receptor, *N. Engl. J. Med.*, 294, 1386, 1976.
26. **Burton, B. K., Gerbie, A. B., and Nadler, H. L.,** Present status of intrauterine diagnosis of genetic defects, *Am. J. Obstet. Gynecol.*, 118, 718, 1974.
27. **Carmody, P. J., Rattazzi, M D., and Davidson, R. G.,** Tay-Sachs disease — the use of tears for the detection of heterozygotes, *N. Engl. J. Med.*, 289, 1072, 1973.
28. **Chapus, C., Sari, H., Sémérive, M., and Desmuelle, P.,** Role of colipase in the interfacial adsorption of pancreatic lipase at hydrophilic interfaces, *FEBS Lett.*, 58, 155, 1975.
29. **Cheung, W. Y.,** Calmodulin plays a pivotal role in cellular regulation, *Science*, 207, 19, 1980.
30. **Chio, K. S., Reiss, U., Fletcher, B., and Tappel, A. L.,** Peroxidation of subcellular organelles: formation of lipofuscin-like fluorescent pigments, *Science*, 166, 1535, 1969.
31. **Chumlea, W. C., Knittle, J. L., Roche, A. F., Siervogel, R. M., and Webb, P.,** Adipocytes and adiposity in adults, *Am. J. Clin. Nutr.*, 34, 1798, 1981.
32. **Colquhoun, I. and Bunday, S.,** A lack of essential fatty acids as a possible cause of hyperactivity in children, *Med. Hypotheses*, 7, 673, 1981.
33. **Cumings, N. J., Thompson, E. J., and Goodwin, H.,** Sphingolipids and phospholipids in microsomes and myelin from normal and pathological brains, *J. Neurochem.*, 15, 243, 1968.
34. **Cummings, J. H., James, W. P. T., and Wiggins, H. S.,** Role of the colon in ileal-resection diarrhoea, *Lancet*, 1, 344, 1973.
35. **Cunnane, S. C.,** Differential regulation of essential fatty acid metabolism to the prostaglandins, *Prog. Lipid Res.*, 21, 73, 1982.
36. **Darcet, P. and Mendy, F.,** Effect of a diet enriched with gammalinolenic acid on PUFA metabolism and platelet aggregation in elderly men, *Ann. Nutr. Aliment.*, 34, 277, 1980.
37. **DiMagno, E. P., Go, V. L., and Summerskill, H. J.,** Relations between pancreatic enzyme outputs and malabsorption in severe pancreatic insufficiency, *N. Engl. J. Med.*, 288, 813, 1973.
38. **Downing, D. T. and Green, R. S.,** Double bond positions in the unsaturated fatty acids of vernix caseosa, *J. Invest. Dermatol.*, 50, 380, 1968.
39. **Dworkin, R. H.,** Linoleic acid and multiple sclerosis, *Lancet*, 1, 1153, 1981.
40. **Ebling, F. J., Ebling, E., and Randall, V.,** The synergistic action of alpha-melanocyte-stimulating hormone and testosterone of the sebaceous, prostate, preputial, Harderian and lachrymal glands, seminal vesicles and brown adipose tissue in the hypophysectomized-castrated rat, *J. Endocrinol.*, 66, 407, 1975.
41. **Eisenberg, S.,** Lipoprotein metabolism and hyperlipemia, *Atheroscler. Rev.*, 1, 23, 1976.
42. **Eisenberg, S. and Levy, R. I.,** Lipoprotein metabolism, *Adv. Lipid Res.*, 13, 1, 1976.
43. **Eldjarn, L.,** Heredopathia atactica polyneuritiformis (Refsum's disease) -- a defect in the omega-oxidation mechanism of fatty acids, *Scand. J. Clin. Lab. Invest.*, 17, 178, 1965.
44. **Elkeles, R. S., Khan, S. R., Seed, M., and Wynn, V.,** High density lipoprotein subfraction cholesterol in hypertriglyceridaemia, *Atherosclerosis*, 43, 423, 1982.
45. **Evans, J. E. and McCluer, R. H.,** The structure of brain dihexosylceramide in globoid cell leukodystrophy, *J. Neurochem.*, 16, 1393, 1969.
46. **Faust, I. M.,** Nutrition and the fat cell, *Int. J. Obesity*, 4, 314, 1980.
47. **Felig, P., Marliss, E., and Cahill, G. F., Jr.,** Plasma amino acid levels and insulin secretion in obesity, *N. Engl. J. Med.*, 281, 811, 1969.
48. **Figarella, C., Negri, Ga., and Sarles, H.,** Presence of colipase in a congenital pancreatic lipase deficiency, *Biochim. Biophys. Acta*, 280, 205, 1972.
49. **Field, E. J. and Joyce, G.,** Effect of prolonged ingestion gamma linolenate by MS patients, *Eur. J. Neurol.*, 17, 67, 1978.
50. **Fishman, P. H. and Brady, R. O.,** Biosynthesis and function of gangliosides, *Science*, 194, 906, 1976.

51. **Fishman, P. M., Max, S. R., Tallman, J. F., Brady, R. O., MaClaren, N. K., and Cornblath, M.,** Deficient ganglioside biosynthesis: a novel human sphingolipidosis, *Science,* 187, 68, 1975.
52. **Flatt, J. P.,** Conversion of carbohydrate to fat in adipose tissue: an energy-yielding and therefore, self-limiting process, *J. Lipid Res.,* 11, 131, 1970.
53. **Flemming, C. R., Smith, L. M., and Hodges, R. E.,** Essential fatty acid deficiency in adults receiving total parenteral nutrition, *Am. J. Clin. Nutr.,* 29, 976, 1976.
54. **Fredholm, B. B.,** Local regulation of lipolysis in adipose tissue by fatty acids, prostaglandins and adenosine, *Med. Biol.,* 56, 249, 1978.
55. **Fredrickson, D. S.,** The inheritance of high density lipoprotein deficiency (Tangier disease), *J. Clin. Invest.,* 43, 228, 1964.
56. **Fredrickson, D. S., Altrocchi, P. H., Avioli, L. V., and Goodman, D. S.,** Tangier disease, *Ann. Intern. Med.,* 55, 1016, 1961.
57. **Gallin, J. I., Kaye, D., and O'Leary, W. M.,** Serum lipids in infection, *New Engl. J. Med.,* 281, 1981, 1969.
58. **Galton, D. J. and Wallis, S.,** The regulation of adipose cell metabolism, *Proc. Nutr. Soc.,* 41, 167, 1982.
59. **Garratt, C. J., Hubbard, R. I., and Ponnuduvai, T. B.,** Adipose tissue obesity and insulin, *Prog. Med. Chem.,* 17, 105, 1980.
60. **Garrison, R. J., Wilson, P. W., Casteli, W. P., Feinleib, M., Kannel, W. B., and McNamara, P. M.,** Obesity and lipoprotein cholesterol in the Framingham Offspring Study, *Metabolism,* 29, 1053, 1980.
61. **Ginsberg, H., Le, N. A., Mays, C., Gibson, J., and Brown, W. V.,** Lipoprotein metabolism in non-responders to increased dietary cholesterol, *Arteriosclerosis,* 1, 463, 1981.
62. **Glick, Z., Teague, R. J., and Bray, G. A.,** Brown adipose tissue: response increased by a single low protein high carbohydrate meal, *Science,* 213, 1125, 1981.
63. **Glomset, J. A. and Wright, J. L.,** Some properties of a cholesterol esterifying enzyme in human plasma, *Biochim. Biophys. Acta,* 89, 266, 1964.
64. **Gofman, J. W. and Jones, H. B.,** Obesity, fat, metabolism and cardiovascular disease, *Circulation,* 5, 514, 1954.
65. **Goodwin, P.,** The effect of corticosteroids on cell turnover in the psoriatic patient. A review, *Br. J. Dermatol.,* 94 (Suppl. 12), 95, 1976.
66. **Gordon, T., Castelli, W. P., and Hjortland, M. C.,** High density lipoprotein as a protective factor against coronary heart disease: The Framingham Study, *Am. J. Med.,* 62, 707, 1977.
67. **Gordon, T., Cateli, W. P., Hjortland, M. C., Kannel, W. B., and Dawber, T. R.,** Diabetes blood lipids and the role of obesity in coronary heart disease risk for women: The Framingham Study, *Ann. Intern. Med.,* 87, 393, 1977.
68. **Greenberg, B. H., Blackwelder, W. C., and Levy, R. I.,** Primary Type V hyperlipoproteinemia. A descriptive study in 32 families, *Ann. Intern. Med.,* 87, 526, 1977.
69. **Grey, N. and Kipnis, N.,** Effect of diet composition on the hyperinsulinemia of obesity, *N. Engl. J. Med.,* 285, 827, 1971.
70. **Gozzo, S. D. and D'Udine, B.,** Diet deprived of essential fatty acids affects brain myelination, *Neurosci. Lett.,* 7, 267, 1978.
71. **Hill, E. C., Johnson, S. B., and Holman, R. T.,** Intensification of essential fatty acid deficiency by dietary trans fatty acid, *J. Nutr.,* 109, 1759, 1979.
72. **Hirsch, J. and Batchelor, B.,** Adipose tissue cellularity in human obesity, *Clin. Endocrinol. Metab.,* 5, 299, 1976.
73. **Hirsch, J. and Knittle, J. L.,** Cellularity of obese and nonobese human adipose tissue, *Fed. Proc.,* 29, 1516, 1970.
74. **Holman, R. T.,** Essential fatty acid deficiency, in *Progress in the Chemistry of Fats and Other Lipids,* Holman, R. T., Ed., Pergamon Press, New York, 1966, 275.
75. **Holman, R. T. and Johnson, S. B.,** Linolenic acid deficiency in man, *Nutr. Rev.,* 40, 144, 1982.
76. **Homboll, P. and Olsen, T. S.,** Fatty changes in the liver: the relation to age, overweight and diabetes mellitus, *Acta Pathol. Microbiol. Immunol. Scand. A,* 90, 199, 1982.
77. **Hornstra, G.,** Dietary fats and arterial thrombosis, *Haemostasis,* 2, 21, 1974.
78. **Horrobin, D. F.,** Prosta glandins and schizophrenia, *Lancet,* 1, 706, 1980.
79. **Horrobin, D. F. and Manku, M. S.,** Possible role of prostaglandin E_1 in the affective disorders and alcoholism, *Br. Med. J.,* 1, 1363, 1980.
80. **Houstmuller, A. J., Van Hal-Ferwerda, J., and Zahan, K. J.,** Favourable influences of linoleic acid on the progression of diabetic micro and macro-angiopathy, *Nutr. Metab.,* 24, (Suppl. 1), 105, 1980.
81. **Hulley, S. B. and Rhoads, G. G.,** Plasma lipoproteins as risk factors, *Metabolism,* 31, 773, 1982.
82. **Jacono, J. M., Judd, J. T., and Marshall, M. W.,** The role of dietary essential fatty acids and prostaglandins in reducing blood pressure, *Progr. Lipid Res.,* 20, 349, 1982.
83. **Kampine, J. P., Brady, R. O., and Kanfer, J. N.,** Diagnosis of Gaucher's disease and Niemann-Pick disease with small samples of venous blood, *Science,* 155, 86, 1967.

84. **Kane, J. P., Richards, E. G., and Havel, R. J.,** Subunit heterogenicity in human serum beta lipoprotein, *Proc. Natl. Acad. Sci. U.S.A.*, 66, 1075, 1970.
85. **Kannagi, R., Nudelman, E., and Hakomori, S.,** Possible role of ceramide in defining structure and functions of membrane glycolipids, *Proc. Natl. Acad. Sci. U.S.A.*, 79, 9470, 1982.
86. **Kannel, W. B.,** Meaning of the downward trend in cardiovascular mortality, *JAMA*, 247, 877, 1982.
87. **Kannel, W. B. and Casteli, W. P.,** Prognostic implications of blood lipid measurements, in *Prognosis*, Fries, J. F. and Ehrujat, G. E., Eds., Charles Press, Maryland, 1980, 263.
88. **Kannel, W. B. and Dawber, T. R.,** Contributors to coronary risk implications for prevention and public health: The Framingham Study, *Heart Lung*, 1, 797, 1972.
89. **Kirby, J. D. and Munro, D. D.,** Steroid-induced atrophy in an animal and human model, *Br. J. Dermatol.*, 94, 111, 1976.
90. **Knittle, J. L.,** Obesity in childhood: a problem in adipose tissue cellular development, *J. Pediatr.*, 81, 1048, 1972.
91. **Kocen, R. S., Lloyd, J. K., Lascelles, P. T., Fosbrooke, A. S., and Williams, D.,** Familial α-lipoprotein deficiency (Tangier disease) with neurological abnormalities, *Lancet*, 1, 1341, 1967.
92. **Kolodny, E. H.,** Lysosomal storage diseases, *N. Engl. J. Med.*, 294, 1217, 1976.
93. **Krabbe, K.,** A new familial infantile form of diffuse brain sclerosis, *Brain*, 39, 74, 1916.
94. **Landing, B. H. and Rubinstein, J. H.,** Biopsy diagnosis of neurologic diseases in children with emphasis on lipidoses, in *Cerebral Sphingolipidoses: Symposium on Tay-Sachs Disease and Allied Disorders*, Aronson, S. M. and Volk, B. W., Eds., Academic Press, New York, 1962. 1.
95. **Landsberg, I. and Young, J. B.,** Diet induced changes in sympathoadrenal activity: implications for thermogenesis and obesity, *Obesity Metab.*, 1, 5, 1981.
96. **LaRosa, J. C., Levy, R. I., Windmueller, H. G., and Fredrickson, D. S.,** Evidence for two triglyceride lipases in post-heparin plasma, *J. Clin. Invest.*, 49, 55a, 1970.
97. **Laskarzewski, P., Khoury, P., Morrison, J. A., Kelly, K., and Glueck, C. J.,** Cancer cholesterol and lipoprotein cholesterol, *Prev. Med.*, 11, 253, 1982.
98. **Lee, R. E.,** The fine structure of the cerebroside occurring in Gaucher's disease, *Proc. Natl. Acad. Sci. U.S.A.*, 61, 484, 1968.
99. **Levine, A. S., Lemieux, B., Brunning, R., White, J. G., Sharp. H. L., Stadlan, E., and Krivit, W.,** Ceroid accumulation in a patient with progressive neurological disease, *Pediatrics*, 42, 483, 1968.
100. **Levy, R. I.,** Hyperlipoproteinemia. Concepts of diagnosis and management, *Curr. Probl. Cardiol.*, 1, 1, 1976.
101. **Levy, R. I. and Morganroth, J.,** Familial type III hyperlipoproteinemia, *Ann. Intern. Med.*, 87, 625, 1977.
102. **Levy, R. I.,** Hypercholesterolemia: genetic dietary and pharmacologic interrelationships, *Prog. Clin. Biol. Res.*, 67, 351, 1981.
103. **Levy, R. I.,** Consideration of cholesterol and noncardiovascular mortality, *Am. Heart J.*, 104, 324, 1982.
104. **Levy, R. I., Langer, T., Gotto, A. M., and Fredrickson, D. S.,** Familial hypobetalipoproteinemia: a defect in lipoprotein synthesis, *Clin. Res.*, 18, 539, 1970.
105. **Lowden, J. A., Rudd, N., Cutz, E., and Doran, T. A.,** Antenatal diagnosis of sphingolipid and mucopolysaccharide storage diseases, *Can. Med. Assoc. J.*, 113, 507, 1975.
106. **Ludwig, P. W., Hunninghake, D. G., and Hoidal, J. R.,** Increased leucocyte oxidative metabolism in hyperlipoproteinemia, *Lancet*, 2, 378, 1982.
107. **Luse, S.,** The fine structure of the brain and other organs in Niemann-Pick disease, in *Inborn Disorders of Sphingolipid Metabolism*, Proc. 3rd Int. Symp. Cerebral Sphingolipidoses, Aronson, S. M. and Volk, B. W., Eds., Pergamon Press, New York, 1967, 93.
108. **Malagelada, J. R., Pihl, O., and Linscheer, W. G.,** Impaired absorption of micellar long-chain fatty acid in patients with alcoholic cirrhosis, *Am. J. Dig. Dis.*, 19, 1016, 1974.
109. **Malinow, M. R.,** The reversibility of atheroma, *Circulation*, 64, 1, 1981.
110. **Malone, M. J.,** The cerebral lipidoses, *Pediatr. Clin. North Am.*, 23, 303, 1976.
111. **Manku, M. S., Oka, M., and Horrobin, D. F.,** Differential regulation of the formation of prostaglandins and related substances from arachidonic acid and from dihomogammalinolenic acid. I. Effects of ethyl alcohol, *Prostaglandins Med.*, 3, 119, 1979.
112. **Marples, R. R., Downing, D. T., and Kligman, A. M.,** Influence of *Pityrosporum* species in the generation of free fatty acids in human surface lipids, J. Invest. Dermatol., 58, 155, 1972.
113. **Marples, R. R., Kligman, A. M., Lantis, L. R., and Downing, D. T.,** The role of the aerobic microflora in the genesis of fatty acids in human surface lipids, *J. Invest. Dermatol.*, 55, 173, 1970.
114. **Mars, H., Lewis, L. A., Robertson, A. L., Jr., Butkus, A., and Williams, G. H., Jr.,** Familial hypo-β-lipoproteinemia, *Am. J. Med.*, 46, 886, 1969.
115. **Mauvais-Jarvis, P., Charrausol, G. S., and Bobas-Masson, F.,** Simultaneous determination of urinary androstanediol and testosterone as an evaluation of human androgenicity, *J. Clin. Endocrinol. Metab.*, 36, 452, 1973.

116. **McCarthy, D. M., May, R. J., and Maher, M.,** Trace metal and essential fatty acid deficiency during total parenteral nutrition, *Am. J. Dig. Dis.*, 23, 1009, 1978.
117. **McCormick, J. N., Neill, W. A., and Sim, A. K.,** Immunosuppressive effect of linoleic acid, *Lancet*, 2, 508, 1977.
118. **Mertin, J. and Stackpoole, A.,** Suppression by essential fatty acids of experimental allergic encephalomyelitis is abolished by indomethacin, *Prostaglandins Med.*, 1, 283, 1978.
119. **Miller, N. E.,** Coronary atherosclerosis and plasma lipoproteins: epidemiology and pathophysiologic considerations, *J. Cardiovasc. Pharmacol.*, 4 (Suppl. 2S), 190, 1982.
120. **Nervi, A. M., Peluffo, R. O., and Brenner, R. R.,** Effect of ethanol administration on fatty acid desaturation, *Lipids*, 15, 263, 1980.
121. **Nevin, N. C., Cumings, J. N., and McKeown, F.,** Refsum's syndrome: heredopathia atactica polyneuritiformis, *Brain*, 90, 419, 1967.
122. **Nicolaides, N., Kellum, R. E., and Woolley, P. V.,** The structures of the free unsaturated fatty acids of human skin surface fat, *Arch. Biochem. Biophys.*, 105, 634, 1964.
123. **Noren, B. and Odham, G.,** Antagonistic effects of *Myxococcus xanthus* on fungi. II. Isolation and characterization of inhibitory lipid factors, *Lipids*, 8, 573, 1973.
124. **Obi, F. O. and Nwanze, E. A. C.,** Fatty acid profiles in mental disease. I. Linolenate variations in schizophrenia, *J. Neurol. Sci.*, 43, 447, 1979.
125. **O'Brien, J. S.,** A molecular defect in myelination, *Clin. Res.*, 12, 276, 1964.
126. **Oftebro, H., Björkhem, I., Stormer, F. C., and Pedersen, J. I.,** Cerebrotendinous xanthomatosis: defective mitochondrial hydroxylation of chenodeoxycholic acid precursors, *J. Lipid Res.*, 22, 632, 1981.
127. **Oschry, Y. and Shapiro, B.,** Fat associated with adipose lipase. The newly synthesized fraction that is preferred substrate for lipolysis, *Biochim. Biophys. Acta*, 664, 201, 1981.
128. **Ostman, G., Arner, P., Engfeldt, P., and Kager, I.,** Regional differences in the control of lipolysis in human adipose tissue, *Metabolism*, 28, 1198, 1979.
129. **Patel, S. M., Owen, O. E., Goldman, L. I., and Hanson, R. W.,** Fatty acid synthesis by human adipose tissue, *Metabolism*, 24, 161, 1975.
130. **Patsch, W., Kuisk, I., Glueck, C., and Schonfeld, G.,** Lipoproteins in familial hyperalphalipoproteinemia, *Arteriosclerosis*, 1, 156, 1981.
131. **Paulsrud, J. R., Pensler, L., Whitten, C. F., Stewart, S., and Holman, R. T.,** Essential fatty acid deficiency in infants induced by fat-free intravenous feeding, *Am. J. Clin. Nutr.*, 25, 897, 1972.
132. **Philippart, M. and Menkes, J.,** Isolation and characterization of the main splenic glycolipids in Gaucher's disease: evidence for the site of metabolic block, *Biochem. Biophys. Res. Commun.*, 15, 551, 1964.
133. **Presentey, B.,** Peroxidase and phospolipid deficiency — a marker in population genetics, *Experientia*, 38, 628, 1982.
134. **Putnam, J. C., Carlson, S. E., DeVoe, P. W., and Barnes, L. A.,** Effect of variations in dietary fatty acids on the fatty acid composition of erythrocyte phosphatidylcholine and phosphatidylethanolamine in human infants, *Am. J. Clin. Nutr.*, 36, 106, 1982.
135. **Rapin, I.,** Progressive genetic-metabolic diseases of the central nervous system in children, *Pediatr. Ann.*, 5, 56, 1976.
136. **Refsum, S.,** Heredopathia atactica polyneuritiformis, *Acta Genet. (Basel)*, 7, 344, 1957.
137. **Reitz, R. C., Wang, L., and Schilling, R. J.,** Effects of ethanol ingestion on the unsaturated fatty acids from various tissues, *Progr. Lipid Res.*, 20, 209, 1982.
138. **Rogers, M. P.,** Plasma lipoprotein metabolism in relation to ischemic heart disease, *Biochem. Soc. Trans.*, 10, 159, 1982.
139. **Rose, G. and Shipley, M. J.,** Plasma lipids and mortality: a source of error, *Lancet*, 1, 523, 1980.
140. **Rössner, S. and Hallberg, D.,** Serum lipoproteins in massive obesity, *Acta Med. Scand.*, 204, 103, 1978.
141. **Rössner, S. J.,** Serum lipoproteins and ischemic vascular disease: on the interpretation of serum lipid versus serum lipoprotein concentrations, *J. Cardiovasc. Pharmacol.*, 4 (Suppl. 2S), 201, 1982.
142. **Sabesin, S. M.,** Cholestatic lipoproteins — their pathogenesis and significance, *Gastroenterology*, 83, 704, 1982.
143. **Sarles, H., Pastor, J., Pauli, A. M., and Barthelemy, M.,** Determination of pancreatic function: a statistical analysis conducted in normal subjects and in patients with proven chronic pancreatitis (duodenal intubation, glucose tolerance test, determination of fat content in the stools, sweat test), *Gastroenterologia*, 99, 279, 1963.
144. **Schaefer, E. J., Anderson, D. W., Brewer, H. B., Jr., Levy, R. I., Danner, R. N., and Plackwelder, W. C.,** Plasma triglyceride regulation of HDL-cholesterol levels, *Lancet*, 2, 391, 1978.
145. **Scontar, A. K.,** The metabolism of very low density and intermediate density lipoprotein in patients with familial hypercholesterolemia, *Atherosclerosis*, 43, 217, 1982.
146. **Seidel, D., Alaupovic, P., and Furman, R. H.,** A lipoprotein characterizing obstructive jaundice. I. Method for quantitative separation and identification of lipoproteins in jaundiced subjects, *J. Clin. Invest.*, 48, 1211, 1969.

147. **Shimazu, T. and Takahashi, A.,** Stimulation of hypothalamic nuclei has differential effects on lipid synthesis in brown and white adipose tissue, *Nature (London)*, 284, 62, 1981.
148. **Shuster, S. and Thody, A. J.,** The control and measurement of serum secretion, *J. Invest. Dermatol.*, 62, 172, 1974.
149. **Sims, E. A. H.,** Mechanisms of hypertension in the syndromes of obesity, *Int. J. Obesity*, 5 (Suppl. 1), 9, 1981.
150. **Sinclair, H. M.,** Dietary fats and coronary heart disease, *Lancet*, 1, 414, 1980.
151. **Sinclair, H. M.,** Prevention of coronary heart disease: the role of essential fatty acids, *Postgrad. Med. J.*, 56, 579, 1980.
152. **Smith, S. C., Scheig, R. L., Klatskin, G., and Levy, R. I.,** Lipoprotein abnormalities in liver disease, *Clin. Res.*, 15, 330, 1967.
153. **Smith, U., Hammersten, J., Bjorntorp, P., and Kral, J.,** Regional differences and effect of weight reduction on human fat cell metabolism, *Eur. J. Clin. Invest.*, 9, 327, 1979.
154. **Snyder, F. and Malone, B.,** Enzymic interconversion of fatty alcohols and fatty acids, *Biochem. Biophys. Res. Commun.*, 41, 1382, 1970.
155. **Sorlie, P., Gordon, T., and Kannel, W. B.,** Body build and mortality, *JAMA*, 243, 1828, 1980.
156. **Steinberg, D., Herndon, J. H., Jr., Uhlendorf, B. W., Mize, C. E., Avigan, J., and Milne, G. W. A.,** Refsum's disease: nature of enzyme defect, *Science*, 156, 1740, 1967.
157. **Tanaka, K. and Hine, D. G.,** Compilation of gas chromatographic retention indices of 163 metabolically important organic acids and their use in organic acidurias *J. Chromatography*, 239, 301, 1982.
158. **Ten Hoor, F., Van de Graaf, H. M., and Vergroesen, A. J.,** Effect of dietary erucic and linoleic acid on myocardial function in rats, in *Recent Advances in Studies on Cardiac Structure and Metabolism*, Vol. 3, Dhalla, N. S., Ed., University Park Press, Baltimore, 1973, 59.
159. **Try, K.,** The in vitro omega-oxidation of phytanic acid and other branched chain fatty acids by mammalian liver, *Scand. J. Clin. Lab. Invest.*, 22, 224, 1968.
160. **Vaddadi, K. S.,** The use of gamma-linolenic acid and linoleic acid to differentiate between temporal lobe epilepsy and schizophrenia, *Prostaglandins Med.*, 6, 375, 1981.
161. **Van Buchem, F. S. P., Pol, G., DeGier, J., and Bottcher, C. J. F.,** Congenital betalipoprotein deficiency, *Am. J. Med.*, 40, 794, 1966.
162. **Vergroesen, A. J.,** Dietary fat, a cardiovascular disease: possible modes of action of linoleic acid, *Proc. Nutr. Soc.*, 31, 323, 1972.
163. **Volk, B. W. and Schneck, L.,** Current trends in sphingolipidoses and allied disorders, in *Advances in Experimental Medicine and Biology*, Volk, B. W. and Schneck, L., Eds., Plenum Press, New York, 1976.
164. **Wilkinson, D. J. and Karasek, M. A.,** Skin lipids of a normal and mutant (asebic) mouse strain, *J. Invest. Dermatol.*, 47, 449, 1966.
165. **Williams, C. N. and Dickson, R. C.,** Cholestyramine and medium-chain triglyceride in prolonged management of patients subjected to ileal resection or bypass, *Can. Med. Assoc. J.*, 107, 626, 1972.
166. **Wilson, L.,** The clinical assessment of topical corticosteroid activity, *Br. J. Dermatol.*, 94 (Suppl. 12), 33, 1976.
167. **Wilson, J., Lake, B. D., and Dunn, H. G.,** Krabbe's leucodystrophy. Some clinical and pathogenetic considerations, *J. Neurol. Sci.*, 10, 563, 1970.
168. **Winter, G. D. and Burton, J. L.,** Experimentally induced steroid atrophy in the domestic pig and man, *Br. J. Dermatol.*, 94 (Suppl. 12), 107, 1976.
169. **Wolfgram, F., Myers, L., and Ellison, G.,** Serum linoleic acid in multiple sclerosis, *Neurology*, 25, 786, 1975.
170. **Mead, J. F.,** *Essential Fatty Acids*, Hawkins, W. W., Ed., University of Manitoba Press, Winnipeg, 1975.

FURTHER READING

Adachi, M., Schneck, L., and Volk, B., *Sphingolipidoses and Allied Disorders*, Eden Press, Westmount, Quebec, 1979.

Bremmer, W. F. and Third, J. L. H., *Hyperlipoproteinemias and Atherosclerosis*, Eden Press, Westmount, Quebec, 1979.

Conn, H. L., DeFilice, E. A., and Kuo, P. T., Eds., *Health and Obesity*, Raven Press, New York, 1983.

Eisenberg, S., Ed., *Lipoprotein Metabolism*, S. Karger, Basel, 1979.

Horrobin, D. F., *Clinical Uses of Essential Fatty Acids*, Eden Press, Montreal, 1982.

Horrocks, L. A., Ansell, G. B., and Porcellati, G., Eds., *Phospholipids in the Nervous System*, Raven Press, New York, 1982.

Paoletti, R. and Kritchevsky, D., Eds., *Advances in Lipid Research*, Academic Press, New York, 1976-1983.

Rifkind, B. M. and Levy, R. I., Eds., *Hyperlipidemia: Diagnosis and Therapy*, Grune & Stratton, New York, 1977.

Snyder, F., Ed., *Lipid Metabolism in Mammals,* Plenum Press, New York, 1977.

Stanbury, J. B., Wyngaarden, J. B., and Fredrickson, D. S., Eds., *The Metabolic Basis of Inherited Diseases,* 4th ed., McGraw-Hill, New York, 1978.

Stunkard, A. J., Ed., *Obesity,* W. B. Saunders, Philadelphia, 1980.

Thompson, G. A., *Regulation of Membrane Lipid Metabolism,* CRC Press, Boca Raton, Fla., 1980.

Chapter 4

ABNORMALITIES OF CARBOHYDRATE SYNTHESIS AND METABOLISM

I. INTRODUCTION

Carbohydrates account for a large proportion of the daily food intake. They exist in a great variety of basic foodstuffs; polysaccharides such as glycogen, starch, and dextrine make up approximately half of the usual carbohydrate intake. The other half contains disaccharides and monosaccharides such as sucrose, lactose, fructose, and maltose (Figure 1). Indirect sources provide carbohydrate or carbohydrate intermediates. About 10% of the fat and 50% of the proteins theoretically can be converted to carbohydrate. Conversely, much of the carbohydrate is transformed to fat and metabolized as fat. The major function of the carbohydrates is to provide energy by oxidation to other metabolic processes. Carbohydrates are utilized by all cells, mainly in the form of glucose. When the supply of carbohydrates or metabolites run low, due to faulty metabolism, widespread disorders occur such as in diabetes mellitus.[1,37,47,67,71,74] In addition, to provide fuel certain products of carbohydrate metabolism act as catalysts or promoters of oxidation processes. Carbohydrates are also used as starting material in the biosynthesis of other body constituents such as certain amino acids, fatty acids, and derivatives. They participate in the formation of biologically important complex molecules such as glycoproteins, nucleic acids, and glycolipids. Lack or deficiency of certain enzymes is responsible for the inborn errors of metabolism.[4,27,30,40,58]

Dietary sources provide polysaccharides and disaccharides which are converted to smaller units. Major monosaccharides are taken from the diet or produced from polysaccharides by digestion. These are hexoses such as glucose, galactose, and fructose. Fructose has a marked importance if large amounts of sucrose are taken in the food. Galactose has considerable significance when lactose is the major dietary carbohydrate. Both latter hexoses are readily converted to glucose by hepatic enzymes. Pentoses may be absorbed from the diet, but important representatives, D-ribose and D-2-deoxyribose are produced in the body; they are essential in the formation of nucleotides.

A large proportion of carbohydrate taken in the food is converted to fat and consequently metabolized as fat. The frequency of taking meals and the extent of carbohydrate conversion to fat has an influence on the development of certain diseases such as diabetes mellitus, atherosclerosis, and obesity.[35,37]

II. CARBOHYDRATE ABSORPTION AND DIGESTION

A. Carbohydrate Uptake

Carbohydrates are digested at all levels of the gastrointestinal tract.[17] Salivary amylase initiates the hydrolysis of starch and dextrin to maltose continued by the action of pancreatic amylase. The small intestines produce other enzymes including lactase, maltase, and invertase, which follow up the conversion of carbohydrates to monosaccharides. This process essentially completes digestion. However, some polysaccharides which are not digestible by these enzymes are broken down in the intestines by enzymes originating from the bacterial flora.

The various monosaccharides are absorbed from the intestines and transported into the liver in the portal circulation. The processes of digestion and absorption are gradual; following a carbohydrate meal the rate is about 1 g glucose per hour per kilogram of body weight. No appreciable amount of glucose is absorbed from the stomach and relatively small amounts are taken up from the colon by simple passive diffusion. The rate of absorption decreases

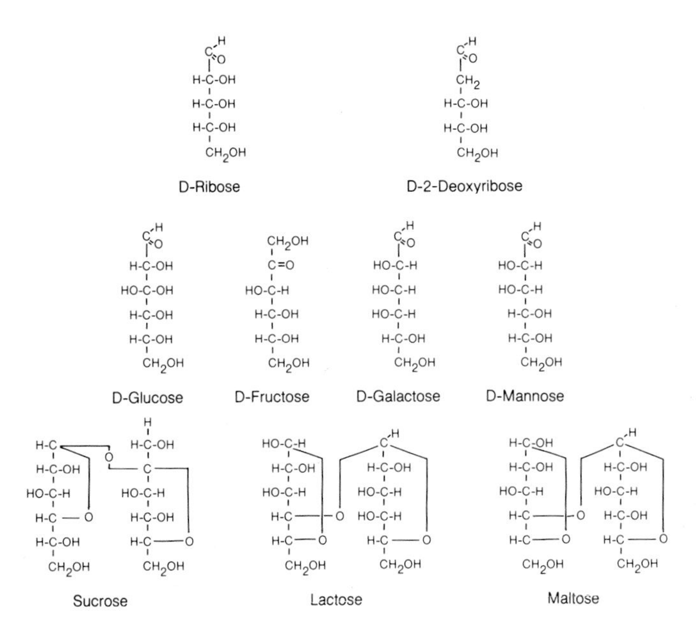

FIGURE 1. Major representatives of mono- and disaccharides in human tissue.

progressively as glucose passes through the jejunum and ileum. The intestinal mucosa absorbs various sugars by a highly selective action which is independent of molecule size. Two mechanisms are involved in the absorption of monosaccharides: (1) passive diffusion which is dependent on the sugar concentration gradient between the lumen of the intestines, mucosal cells, and plasma; (2) active transport which is affected by phosphorylating enzymes present in the epithelial cells of the small bowel. These enzymes act specifically on glucose, fructose, and galactose, resulting in a rapid removal of these sugar molecules at a speed which cannot be achieved by diffusion alone. From the intestinal lumen glucose diffuses into the epithelial cells and it is transformed to glucose-6-phosphate, catalyzed by glucokinase in the presence of adenosine triphosphate. This reaction facilitates the entry of more glucose molecules by diffusion. Within the cell, glucose 6-phosphate is hydrolyzed by nonspecific phosphatases and the free sugar passes into the intestinal fluid and portal capillaries. Fructose and galactose are phosphorylated similarly by fructokinase and galactokinase, respectively. These active processes need energy from adenosine triphosphate. Fructose 6-phosphate is converted to glucose 6-phosphate, which is further hydrolyzed to glucose. Therefore, part of fructose is transformed to glucose in the intestinal wall. In contrast is the intestinal transformation and active transport of hexoses; pentoses are absorbed by passive diffusion.

The hydrolysis of poly- and oligosaccharides is a fast process; the mechanisms responsible for the absorption of glucose, fructose, or galactose are rapidly saturated. Lactose represents an exception, absorption occurs at only half the rate of sucrose. When the lactose intake is elevated, hydrolysis of glucose proceeds at a slower rate due to lack of saturation of the transport mechanisms for glucose.

B. Carbohydrate Absorption in Disease

The localized absorption of hexoses from the intestines is influenced by the general condition of the organism. Functional or organic obstructions cause marked delays, or reduce the capacity, or even considerably prevent normal carbohydrate absorption. Moreover, diarrhea, gastrocholic fistula, or other conditions which increase intestinal mobility and accelerate the passage of foodstuff through the small bowel, or cause damage to the intestinal mucosa diminish the amount of sugar removed. Absorption may also be reduced in various structural and functional abnormalities of the mucous membrane such as inflammatory conditions in enteritis, edema as the consequence of congestive heart failure, and celiac disease. Anterior pituitary, adrenal cortical, and thyroid deficiencies interfere with the normal uptake of monosaccharides. It is delayed in hypothyroidism and accelerated in hyperthyroidism. Diminished absorption in adrenal cortical insufficiency is dependent upon the reduced sodium and chloride concentration in body fluids. Various vitamins such as pantothenic acid, thiamine, and pyridoxine interfere with the normal uptake. The exact mechanism of these various hormone actions or the effect of vitamin deficiency has not yet been established. It is likely that hormones and vitamins are essential for the activity of the cells which specifically absorb sugar molecules.

In contrast to the decreased absorption of hexoses by active transport, the uptake of sugars is increased by the production of excessive amounts of thyroid, adrenal cortical, and anterior pituitary hormones, or removal of the pyloric barrier by surgery.

III. CARBOHYDRATE METABOLISM

A. Metabolic Pathways

After absorption into the portal blood all carbohydrates are taken up by the liver and utilized via glucose through several mechanisms:[69,77,79,81] (1) after uptake various hexoses are metabolized to glucose, (2) part of the glucose is converted to glycogen for storage, (3) part of the glucose is oxidized for the production of energy, and (4) part of the glucose is converted to other compounds such as fatty acids and amino acids. In contrast to the utilization of glucose, opposite processes also take place which produce glucose. Glucose is formed and released from the liver into the circulation by (1) conversion of glycogen to glucose, (2) through the synthesis of glucose from other hexoses, and (3) glucose is also synthetized from noncarbohydrate sources. The amount of glucose present in systemic circulation at any moment is dependent upon the balance between these two kinds of opposing reactions mainly localized in the liver.[25,33,36,54] Glucose circulating in the blood is available for utilization by other tissues. However, the conditions of carbohydrate supply for extrahepatic tissues is dependent on the production of glucose by the hepatic tissue. The physiological state of the liver, therefore, exerts a significant influence on the carbohydrate metabolism of the whole body. It is expected, therefore, that major impairments of hepatic function may affect the normal activity of other organs.[56]

Glucose and other carbohydrate intermediates are also metabolized in the organism via several mechanisms — (1) glycolysis: representing the oxidation of glucose or glycogen to pyruvate or lactate; (2) glycogenesis: synthesis of glycogen from glucose; (3) gluconeogenesis: the formation of glucose or glycogen from noncarbohydrate precursors — principal substrates are glucogenic amino acids, glycerol, and lactic acid; (4) citric acid cycle: representing a common pathway of carbohydrate, protein, and fat oxidation. Through the formation of acetyl-CoA intermediates these various substrates are completely oxidized to carbon dioxide and water; (5) pentose monophosphate shunt; and (6) phosphogluconate oxidative pathway representing alternate routes to the citric acid cycle for the oxidation of glucose. Any impairment of these metabolic pathways may be associated with disease conditions (Figure 2).[16]

B. Blood Glucose

The amount of glucose in the systemic circulation is dependent on the various processes of carbohydrate metabolism.[9,43,75] The blood glucose level, therefore, represents the fate of the absorbed carbohydrate and the overall process of carbohydrate metabolism.

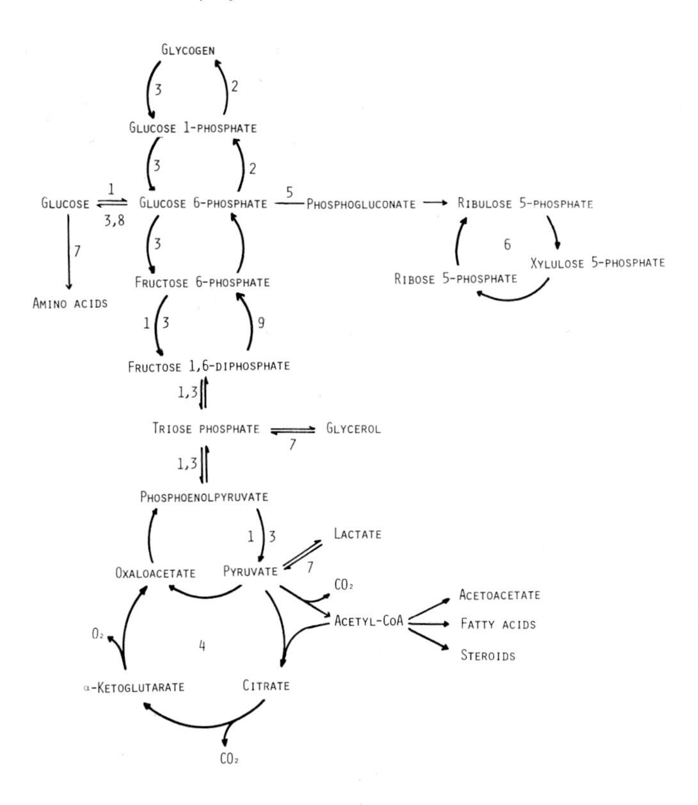

FIGURE 2. Major pathways of carbohydrate metabolism. These include oxidation of glucose or glycogen to pyruvate and lactate by glycolysis, termed the Embden-Meyerhof pathway (1); the synthesis of glycogen from glucose by glycogenesis (2); the breakdown of glycogen in the liver and muscle by glycogenolysis, producing glucose or pyruvate and lactate, respectively (3); the citric acid cycle or tricarboxylic acid cycle, termed the Krebs cycle (4). This cycle is a common pathway in the oxidation of carbohydrates, fats, and proteins through acetyl-CoA, ultimately yielding carbon dioxide and water. Direct oxidation exists through phosphogluconate by the hexose monophosphate shunt (5) and pentose phosphate cycle (6). Glucose or glycogen is formed from noncarbohydrate precursors. The principal sources for gluconeogenesis are glucogenic amino acids, glycerol, and lactate (7). Some gluconeogenic enzymes catalyze irreversible steps in this scheme such as glucose 6-phosphatase (8), fructose 1,6-diphosphatase (9), pyruvate carboxylase (10), and phosphoenolpyruvate carboxykinase (11).

The normal blood glucose level is 80 to 100 mg/dℓ (true glucose); the fasting level is lower. It is age dependent. After a meal (*post cibal*), glucose rises from 80 to 160 mg/dℓ. Major sources of glucose in blood are (1) diet: ingestion from intestines through an active glucose phosphorylation process; and (2) gluconeogenesis: 90% in liver, 10% in kidney, representing mainly glucose 6-phosphatase actions. Gluconeogenesis can be further subdivided: direct conversion from liver glycogen by glycogenolysis; partial breakdown products which are reconverted to glucose, such as lactate and glycerol (from fat breakdown); and amino acids and propionate.

The regulation of blood glucose represents a very sensitive mechanism. Liver cells are freely permeable to glucose, whereas extrahepatic cells require hormones for the transport of glucose across cell membranes. When blood sugar increases above certain levels, the kidney exerts a regulatory effect. Normally, glucose is filtered by the glomeruli but returned to the blood via the tubules in a process involving phosphorylation. There is a limit to the rate of reabsorption, termed the tubular maximum, for glucose. This occurs when blood sugar is in excess of 160 to 180 mg/dℓ, resulting in glucosuria.

IV. CARBOHYDRATE STORAGE AND UTILIZATION

The glucose taken up by the body or synthetized *de novo* is utilized for three purposes: (1) it is transformed immediately to special products if urgent physiological need exists for oxidative energy; (2) small amounts are converted to other sugars which are necessary for the synthesis of glycoproteins, nucleoproteins, phosphorylated intermediates, and glycolipids; and (3) excess glucose is deposited and stored in the liver, muscle, and other tissues as glycogen. This is the process of glycogenesis. From these stores, glycogen is transported to all tissues in the form of metabolites when demands rise.

The total store of glycogen in the body is small — about 300 to 500 g; two thirds is deposited in the muscle. The major endogenous carbohydrate source is the liver, which is directly available to the circulation. The glycogen stored in the muscle does not directly provide blood glucose. First, it is broken down to lactic acid during glycolysis and is then resynthesized to glucose and glycogen in the liver. Since the overall storage of glucose in the body is limited, excess quantities over the limit are transformed to fatty acids and stored as triglycerides in fat depots. There is no upper limit in the conversion of the surplus glucose to fat.

In all tissues glucose can be completely oxidized to water and carbon dioxide if there is a need for energy. In special circumstances, particularly in the muscle, glucose is only partially metabolized to lactic acid through the process of glycolysis. This product is further metabolized in various tissues, mainly in the liver. Small amounts of glucose are converted to various other sugars such as (1) galactose, which is the component of the disaccharide lactose secreted in the milk and also part of glycolipids; (2) glucuronic acid, involved in the mechanism of detoxication of endogeneous and foreign compounds, and a component of mucopolysaccharides; (3) glucosamine, galactosamine, and mannose, which take part in the synthesis of glycoproteins and mucopolysaccharides; and finally (4) glucose is also transformed to pentoses such as ribose and deoxyribose which form part of the nucleic acids.

There is an intimate interrelationship between carbohydrates, proteins, and lipids through the metabolism of glucose.[85] Certain nonessential amino acids receive their carbon skeleton from the metabolism of glucose; these are the glucogenic amino acids. Carbohydrate metabolites can also be converted to fatty acids. Fatty acids are produced from glucose when glycogen storage depots are exhausted. The conversion from glucose to fatty acid is rapid and to a large extent irreversible. Glucose forms the precursor of glycerol; this process is, however, reversible. Fragments from all classes of basic compounds finally form a common pathway through the components of the tricarboxylic acid cycle and they are ultimately converted to water and carbon dioxide while producing energy. The various sources and utilization of glucose and the interconversion between glucose, fatty acids, and amino acids are illustrated in Figure 3.

Acetyl-CoA is the most important key compound in the metabolism of carbohydrates and fatty acids. It is formed from pyruvate by oxidative decarboxylation, from fatty acids by oxidation, and directly from ketogenic amino acids through various reactions. In the liver the active acetate can take part in four main metabolic pathways: (1) condensation with oxaloacetate produces citrate and further metabolism of the latter compound in the tricarboxylic acid cycle finally leads to carbon dioxide and water; (2) fatty acid synthesis; (3) cholesterol synthesis; and (4) ketogenesis, through the condensation of two molecules into acetoacetate. These mechanisms are important in the synthesis of other body constituents and in the production and transfer of energy.

Certain hormonal mechanisms are involved in the maintenance of the energy supply. There is a balance between insulin and 11-oxysteroids plus pituitary growth hormone. Any impairment of the balance between these hormones leads to overall metabolic consequences. Absolute or relative insulin deficiency brings about a depression of carbohydrate utilization

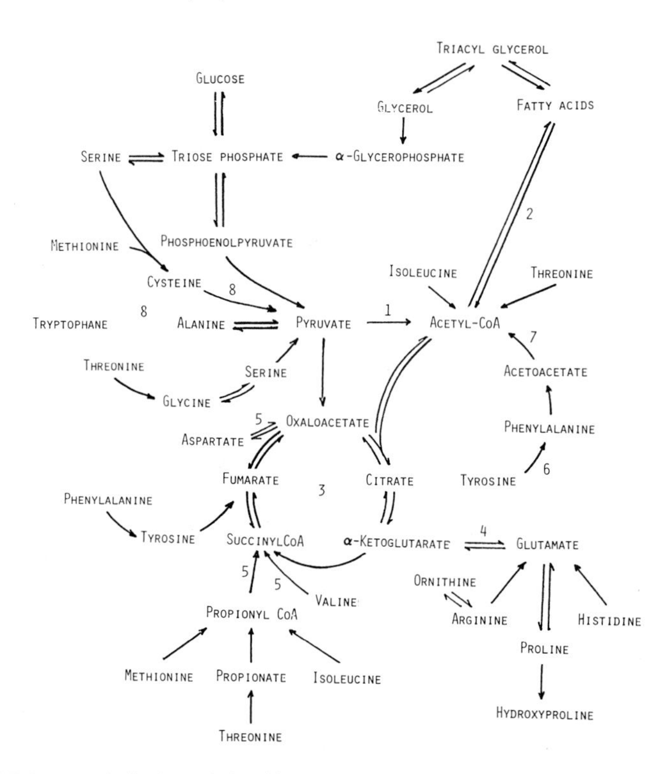

FIGURE 3. Metabolic interrelationships between the major classes of body constituents. In these interconversions the most important reaction is the transformation of pyruvate to acetyl CoA (1), which is the starting material in the synthesis of long-chain fatty acids (2). Nonessential amino acids can be produced from carbohydrates through the citric acid cycle (3) and transamination (4). By reversal of the transamination processes glycogenic amino acids produce carbon skeletons which are either members or precursors of the citric acid cycle (5). Ketogenic amino acids give rise to acetoacetate (6) which, in turn, can be metabolized to acetyl CoA in extrahepatic tissues (7). Essential amino acids are also converted to a carbon skeleton (8) but it is not possible to reverse the pathways of breakdown.

and a mobilization of fat from stores, resulting in hyperlipemia. Further consequences are excessive acetate formation, increased cholesterol synthesis, and increased catabolism of proteins. The latter process leads to negative protein balance.

V. REGULATORY MECHANISMS

The intake, storage, and metabolism of carbohydrates vary widely, still, the level of glucose in the blood is maintained within narrow limits.[8,54,77,78,85] This constant level is related to two opposite mechanisms. Processes are involved in the liver connected with glucose supply to the blood, and in various tissues connected with glucose utilization.[82] These processes collectively form a homeostatic regulatory mechanism. These processes are controlled by integration with each other and are also balanced with the metabolism of lipids and proteins which, again, are intrinsically connected with carbohydrate metabolism.

The regulation of carbohydrate metabolism occurs at two levels: (1) cellular and enzymatic processes; and (2) interaction between various organs affecting blood glucose. These two levels are interrelated, but for the sake of convenience will be discussed separately.

Gross effects in the cellular regulation of carbohydrate metabolism are dependent on changes in nutritional states and on changes in the availability of substrates which directly or indirectly influence the pattern and rate. Fluctuations in their blood and tissue concentration often alter the activity of adaptive key enzymes connected with the action of hormone inducers or repressors. The activity of enzymes participating in carbohydrate metabolism is modified if (1) the rate of enzyme synthesis is altered, (2) the inactive enzyme is converted to an active one, (3) competitive inhibitors are present, or (4) allosteric changes bring about increased access to substrates. Regulatory effects associated with carbohydrates involve *de novo* protein synthesis which is usually slow. Activation is frequently connected with phosphorylation of the enzyme protein by protein kinase in the presence of ATP or dephosphorylation by phosphatase. The regulatory effects are mainly located in the liver, and key enzymes involved in the utilization of glucose are all activated or depressed in a concerted manner. Elevated amounts of dietary carbohydrates bring about an enhanced activity of enzymes involved in glycolysis such as glucokinase, phosphofructokinase, pyruvate kinase, glycogen synthesis and lipogenesis such as fatty acid synthesis, acetyl CoA carboxylase, and in the maintenance of hexose monophosphate shunts such as glucose 6-phosphate dehydrogenase and 6-phosphogluconate dehydrogenase.[44,45] These changes are induced by insulin. Enzymes of gluconeogenesis such as glucose 6-phosphatase, fructose 1,6-diphosphatase, and pyruvate carboxylase are reduced by excess carbohydrate, the repressor being insulin. Diabetes and starvation elicit opposite effects, namely, the enzymes of glycolysis, glycogenesis, and the hexose monophosphate shunt are depressed and gluconeogenesis is induced. The inducers of the latter group of enzymes are glucocorticoids, epinephrine, and glucagon.

The maintenance of stable glucose levels in the blood is the most finely regulated of all homeostatic mechanisms in which the liver, extrahepatic tissues, and several hormones play a part.[77] The liver controls the supply of glucose and certain endocrine glands exert a fundamental effect on the utilization of carbohydrates by various hormones essential for the normal metabolism. These endocrines are the pancreatic islet cells, anterior pituitary, and adrenal cortex. Some other endocrine organs also exert an action on carbohydrate metabolism but are not required for its regulation, such as the thyroid and adrenal medulla. When blood sugar level rises to a relatively high level, kidney function influences the regulatory processes.

A. Role of the Liver

Since the liver activities represent the major source of blood sugar, this organ exercises a central role in the homeostatic mechanism.[9,25] It is directly responsible for the blood sugar level at any time. Excessive intake of sugar switches off the output of glucose from the liver. Conversely, when hyperglycemia is brought about, the alimentary intake of glucose ceases and the liver resumes its secretion into the circulation. In responses to variations in external supply and endogenous stores of carbohydrates, gluconeogenesis, glycogenesis, and glycogenolysis sets in, which includes extrahepatic processes. This is also influenced by variations in hormone levels and activity.[7,13]

Liver cells are freely permeable to glucose, whereas cells of the extrahepatic tissues are relatively impermeable. At normal blood glucose concentrations (80 to 100 mg/dℓ) the liver is the net producer of glucose. As the glucose level increases in the blood from dietary sources, hepatic output decreases, and at high levels there is a net uptake into the liver. Glucose is also synthesized in the liver from various glucogenic compounds such as some amino acids and propionate. Some precursors are products of partial glucose metabolism in certain tissues, where they are released and carried into the liver and kidney where they are transformed to glucose. Lactate, for instance, is formed in skeletal muscle and erythrocytes from glucose by oxidation, it is transformed to glucose along these processes and via circulation it becomes available for the tissues. The process represents the lactic acid cycle (Figure 4). The role of cyclic AMP in this reaction is essential as an enzyme activator. Another continuous cycle exists between adipose tissue triacylglycerol and liver or kidney. In the adipocytes, triacylglycerol is hydrolyzed to glycerol which is transferred via the

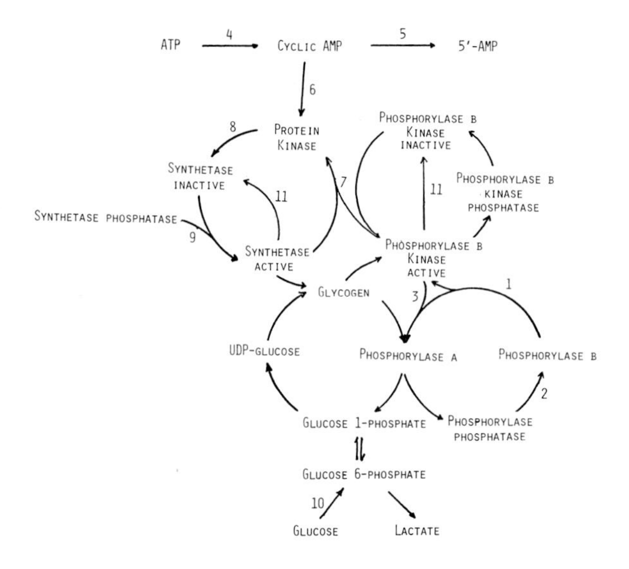

FIGURE 4. Regulation of glycogen synthesis and metabolism in muscle and other tissues. This involves the conversion of phosphorylase b to phosphorylase a (1). Phosphorylase b is active only in the presence of 5′-AMP. Phosphorylase a is a tetramer, it is hydrolyzed to the dimer by phosphorylase phosphatase (2). Two dimers of phosphorylase b may recondense to an active phosphorylase a tetramer by phosphorylase b kinase (3). These processes are related to cyclic AMP metabolism regulated by adenylate cyclase (4) and phosphodiesterase (5). Several hormones influence glycogenesis and glycogenolysis indirectly. Thyroid hormones and epinephrine or glucagon in the liver increase adenylate cyclase activity (4) and insulin enhances phosphodiesterase (5). These actions are related to the function of cyclic AMP-dependent protein kinase (6). This enzyme activates phosphorylase b kinase (7) but inactivates glycogen synthetase (8). Insulin also affects synthetase phosphatase b kinase activated by muscular contraction which is related to the release of Ca^{2+} ions (11). The role of cyclic AMP in activating the system is important.

circulation into the liver and kidney and converted to glucose. During starvation some amino acids, predominantly alanine, are transported from muscle to the liver via a glucose-alanine cycle (Figure 5). This cycle affects the net transfer of amino nitrogen from muscle to liver, and free energy transfer from liver to muscle. The energy needed for hepatic glucose synthesis is derived from fatty acid oxidation.

B. Regulatory Abnormalities in Liver Disease

The state of the liver, and particularly liver diseases, may profoundly alter glucose and glycogen metabolism. When large amounts of sugar are available and the energy requirement of all organs are fulfilled, the absorbed carbohydrate molecules including glucose, galactose, and fructose are taken up by hepatocytes and incorporated into glycogen stores; the excess is converted to fat. Conversely, during fasting glycogen is metabolized via glycogenolysis in order to maintain the normal blood glucose level and satisfy energy requirements. In man, the capacity of the hepatic glycogen depot is not great; it can only supply about 500 calories. During fasting, however, hypoglycemia does not develop since glucose and glycogen are synthesized from lactic acid and the deamination of certain amino acids and fatty acids via the formation of pyruvic acid, i.e., through the processes of gluconeogenesis and glyconeogenesis. If these processes are not adequate, hepatic fat metabolism is enhanced and ketone bodies are formed.

In man, if the liver is damaged pronounced hypoglycemia rarely occurs. Conversely, dietary hyperglycemia or moderate hypoglycemia on fasting is only apparent occasionally in patients with liver cirrhosis or hepatitis. Generally, in liver disease fairly normal glucose

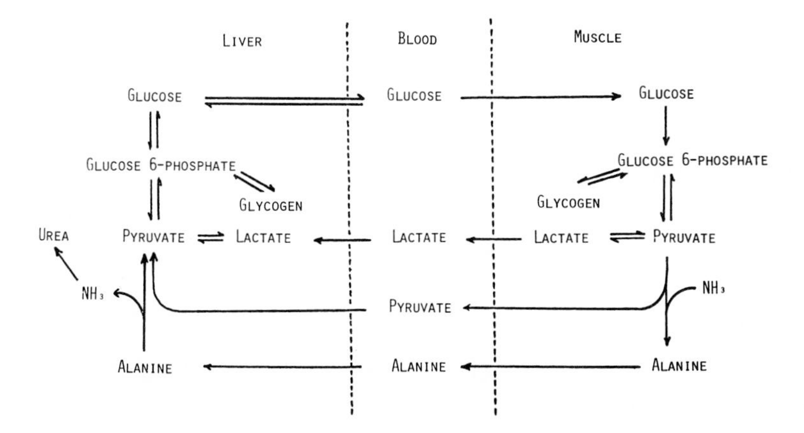

FIGURE 5. Effect of muscular and hepatic metabolism on the blood glucose level. A cycle plays part in this regulation, termed the glucose-alanine cycle. During starvation, amino acids, predominantly alanine, are transported from muscle to liver (1) yielding a net transfer of amino nitrogen. Conversely, energy is transferred from liver to muscle (2). The energy required for glucose synthesis is derived from the oxidation of fatty acids. The lactic acid or Cori cycle (3) influences this regulation.

levels are maintained in response to the metabolic demands due to the considerable reserve function of the liver and the participation of extrahepatic tissues. The latter factors can compensate for the partial loss of the hepatic contribution to the homeostatic condition. Only when these mechanisms regulating the blood sugar level are exhausted can impairment of gluconeogenesis and glyconeogenesis be seen. In these conditions the effect of hormones may also fail. In some patients with cirrhosis or diabetes the unusually inadequate response to insulin administration associated with prolonged hypoglycemia indicates a serious hepatic functional damage. Similarly, in hepatitis or cirrhosis, due to the hepatic dysfunction the incorporation of glucose into glycogen stores is incomplete and the inadequate elevations of blood sugar levels in response to epinephrine or glucagon reveal a deficient glycogen storage. In these patients, the administration of cortisone or ACTH may enhance fasting blood sugar level by stimulating gluconeogenesis.

In advanced liver disease, blood pyruvate levels are often elevated. The damaged liver does not have adequate enzyme activity to eliminate this intermediate by linking it with the citric acid cycle through CoA. The enzyme protein participating in pyruvate metabolism may be present, but the level of the coenzyme, thiamine pyrophosphate, required for function is inadequate. The phosphorylation of the coenzyme from thiamine (vitamin B_1) is catalyzed by a liver enzyme which can be impaired in hepatic disorders. Similarly, the utilization of pantothenic acid may also be deficient in liver disease. Hepatic enzymes are responsible for the synthesis of coenzyme A from pantothenic acid and other precursors, thus liver disease may influence carbohydrate metabolism through the lack of CoA synthesis. In glycogen storage diseases (Chapter 1), some enzymes responsible for the catabolism of glycogen are missing. As a result, high levels of glycogen deposit in the liver. Consequently, in this disorder low fasting blood sugar levels, enhance lipid metabolism, and impaired nitrogen retention occur.

In several liver diseases glucose tolerance tests may be abnormal; however, the cause cannot be easily interpreted since the blood sugar level is associated with the contribution of many organs to carbohydrate metabolism.[9] Fructose and galactose tolerance tests may reflect hepatic disorder since the removal of these hexoses from the blood is primarily dependent on the liver.[40,84] Particularly, galactose tolerance tests may be applicable since fructose is taken up by other organs and the results may show an inconsistency.

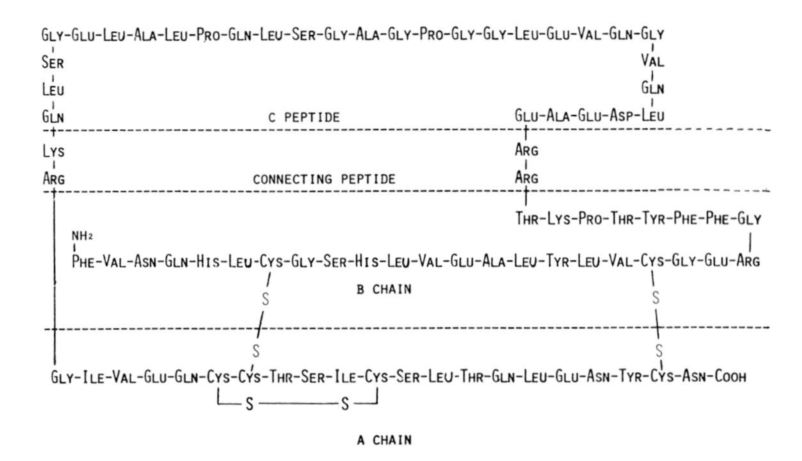

FIGURE 6. Amino acid sequence of human proinsulin. Insulin is composed of the A and B chain. A connecting peptide represents the amino acid sequence connecting the amino end of the A chain to the carboxy end of the B chain. C peptide is formed when two basic amino acids are removed from the connecting peptide.[41]

C. Role of Pancreatic Hormones

Hepatic gluconeogenesis and glucose catabolism in the tissue are sensitive to slight deviations in the normal concentration of blood sugar. The normal balance between production and utilization is maintained to a considerable extent by an existing balance between hormones derived from the pancreas, on the one hand, and from the adrenal gland and the pituitary on the other. The overall effect of insulin is to lower blood sugar, while adrenocortical and pituitary factors elevate it. These factors are antagonistic; the ratio between these hormones is of primary importance in the regulatory action and not the absolute amounts. The pancreas secretes three hormones: insulin, glucagon, and somatostatin. Each hormone is produced by individual cell type but all three are involved in glucose homeostasis.[39]

1. Insulin

This is a small peptide with a molecular weight of 6000 daltons, consisting of an A-chain of 21 amino acids and a B-chain of 30 amino acid residues.[72,73] The two chains are linked by two disulfide bridges and another internal disulfide bridge exists in the A-chain (Figure 6). The original precursor of insulin in the pancreas is a larger molecule identified as preproinsulin. This is converted to a molecule with a mass of 9000 daltons called proinsulin. Proinsulin is synthesized in the microsomal fraction of the pancreatic β-cell and transported to the Golgi apparatus where it is stored.[86] Proinsulin is converted to insulin in the secretory granules of the β-cells. This involves a proteolytic process, the double-chain molecule is shortened by the removal of a central protein or C-chain.[5,6,18,34,41] Insulin and the connecting C peptide are secreted into the blood in equal amounts. During fasting, the secretion of insulin is minimal and proinsulin release is about 15% that of insulin. In older patients, obese diabetics, pregnant diabetics and in patients with insulinomas, the circulating proinsulin is increased. Abnormalities of lipostatic processes resulting in obesity are connected with modification of plasma insulin levels. Typical plasma insulin concentrations in normal and in obese persons are presented in Figure 7.

Insulin is secreted from the pancreas into the blood as a direct response to hyperglycemia.[83] Insulin concentration in the blood runs parallel with that of glucose and its administration promptly causes hypoglycemia. Several substances can release insulin, such as free fatty acids, amino acids, ketone bodies, secretin, glucagon, or tolbutamide. Epinephrine and norepinephrine inhibit insulin release. Insulin exerts an immediate effect on muscle and

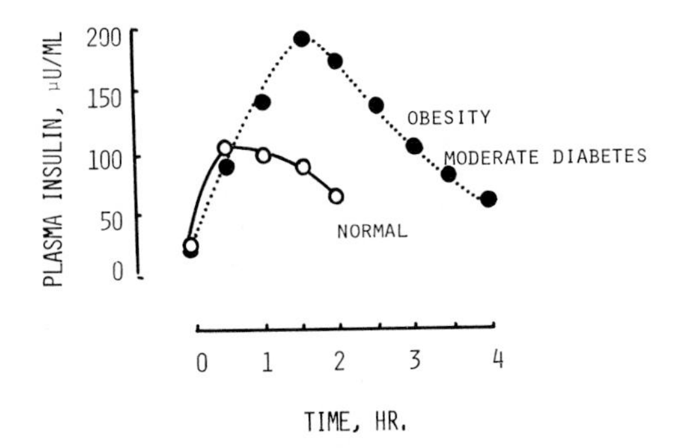

FIGURE 7. Plasma insulin levels in normal and in obese persons.

adipose tissue by increasing the rate of glucose uptake. Insulin is primarily concerned with accelerating the rate of glucose entrance into the cells of muscle and adipose tissue, either by increasing cellular permeability or by stimulating hexokinase activity which converts glucose to glucose 6-phosphate. Insulin also reduces the action of hepatic glucose 6-phosphate either directly or through other processes so that less glucose is formed from liver glycogen.

In the absence of insulin, glucose is overproduced in the liver and underutilized in the muscle and other tissues. Without insulin, hexokinase is still functioning and some glucose utilization occurs. The processes of hepatic glycogenesis and glucose utilization and also of glucose concentration in the blood are continuously exposed to disturbing actions under physiological conditions. These include intestinal absorption, physical activity and exertion, mental activity, and emotional states.[10,14] In the majority of these cases the primary effect is a rise in blood sugar. In diabetes, the entrance of glucose into the cell and cellular carbohydrate metabolizing processes are impaired; even the impairment is partly compensated by the rising blood glucose concentration. In this condition glucose processing reactions are lower than normal in the muscle and partly replaced by oxidation of keto acids and free fatty acids. The storage of glycogen in the liver and muscles is below normal levels.

2. Glucagon

This is formed by the α_2 (A) cells of the pancreas.[38,55] Plasma pancreatic glucagon is heterogeneous and includes four fractions with masses of about 2000, 3500, 9000, and 40,000 daltons. The smallest fraction may be a degradation product. The kidney is the major site of metabolism of these glucagon fractions, and in patients with renal failure glucagon levels are increased five times. The most important fraction physiologically is the 3500-dalton fraction which is stimulated to a greater extent than any other fraction and is suppressed by hyperglycemia and somatostatin.

Hypoglycemia affects the secretion of glucagon from the α-cells of the Langerhans islets of the pancreas. Glucagon activates glycogen catabolism in the liver by activating phosphorylase. The effect of glucagon is terminated by hepatic degradation. It also enhances gluconeogenesis from lactate and amino acids, but has no effect on muscle phosphorylase.[65]

In juvenile diabetics immunoreactive glucagon levels in the serum are normal, but inappropriate to glucose concentration. Their response is exaggerated to various stimuli. Determination of glucagon concentrations in the serum are important in the diagnosis of α-cell tumor of the pancreas. This tumor, glucagonoma, is associated with mild diabetes mellitus and characterized by very high levels of glucagon.[68] Very high levels of glucagon in diabetics may suggest the occurrence of this tumor.

NH_2-Ala-Gly-Cys-Lys-Asn-Phe-Phe-Trp-Lys-Thr-Phe-Thr-Ser-Cys-COOH

FIGURE 8. Amino acid sequence of somatostatin.

3. Somatostatin

This is a tetradecapeptide (Figure 8), and originally was isolated from the hypothalamus.[29,31,32,46] Somatostatin is produced in the D cells of the pancreas, which are in close proximity to the glucagon-producing A_2 cells. In diabetics with essentially no B cells, D cells form about one third of the total pancreatic islet cells, which may represent a compensatory mechanism to correct the hyperglycemia. It seems that the predominant action of somatostatin is the reduction of glucagon release. Infusions of somatostatin into insulin-requiring diabetic patients causes a decrease of elevated glucagon levels and reduces serum glucose concentrations. When insulin is withdrawn in these diabetics somatostatin prevents acetosis.

Beyond the action of somatostatin on pancreatic hormones (glucagon and insulin), it inhibits pituitary hormones (growth hormone, thyrotropin, and adrenocorticotropic hormone) and gastrointestinal hormones (gastrin and secretin). It also possesses nonendocrinologic activities such as inhibition of secretion or diminution of gastric acid secretion, pancreatic bicarbonate and enzyme release, gastric emptying time, gallbladder contraction, and acetylcholine release from peripheral nerve endings.

D. Role of Nonpancreatic Hormones

The overall effect of insulin is to decrease the blood glucose level; in contrast, pituitary and adrenocortical factors elevate it. The anterior pituitary gland secretes hormones that antagonize the effect of insulin. These hormones include the growth hormone, corticotropin, and possibly other diabetogenic components. Hypoglycemia stimulates the secretion of growth hormone. This reduces glucose uptake in several tissues such as muscle. Conversely, when growth hormone produces hyperglycemia, it stimulates the release of insulin from the pancreas and eventually leads to the exhaustion of β-cells. Part of the effect is indirect, the primary action of growth hormone is the mobilization of free fatty acids from adipose tissue which inhibit glucose utilization. Chronic administration of the growth hormone causes diabetes.

ACTH probably exerts an indirect action upon glucose utilization; primarily it accelerates the release of free fatty acids from adipose tissue. The major effect of this hormone on carbohydrate utilization is connected with the stimulation of secretion of adrenal cortical hormones.

Glucocorticoids secreted from the adrenal cortex have an important role in the metabolism of carbohydrates. Increased secretion of glucocorticoids cause gluconeogenesis by increasing protein catabolism in various tissues leading to increased hepatic uptake of amino acids and stimulation of the activity of transaminases and other enzymes involved in gluconeogenesis in the liver. Glucocorticoids also inhibit glucose breakdown in extrahepatic tissues. All actions of corticosteroids appear to be antagonistic to insulin.

The adrenal medulla product, epinephrine, increases glycogen breakdown in the muscle. The accelerated action of epinephrine on glycogenolysis is related to its ability to activate phosphorylase. When glycogen is present in the liver, the administration of epinephrine causes a heavy discharge of glucose from this organ. In the muscle, since glucose 6-phosphatase is absent, the effect of epinephrine is to increase glycolysis, resulting in excess lactate formation. Lactate diffuses into the blood, and through gluconeogenetic processes it is converted back to glycogen via the Cori cycle (Figure 9). Hypoglycemia causes a sympathetic discharge of epinephrine. The secretion of this substance stimulates glycogen breakdown, resulting in an increase of the blood glucose level.

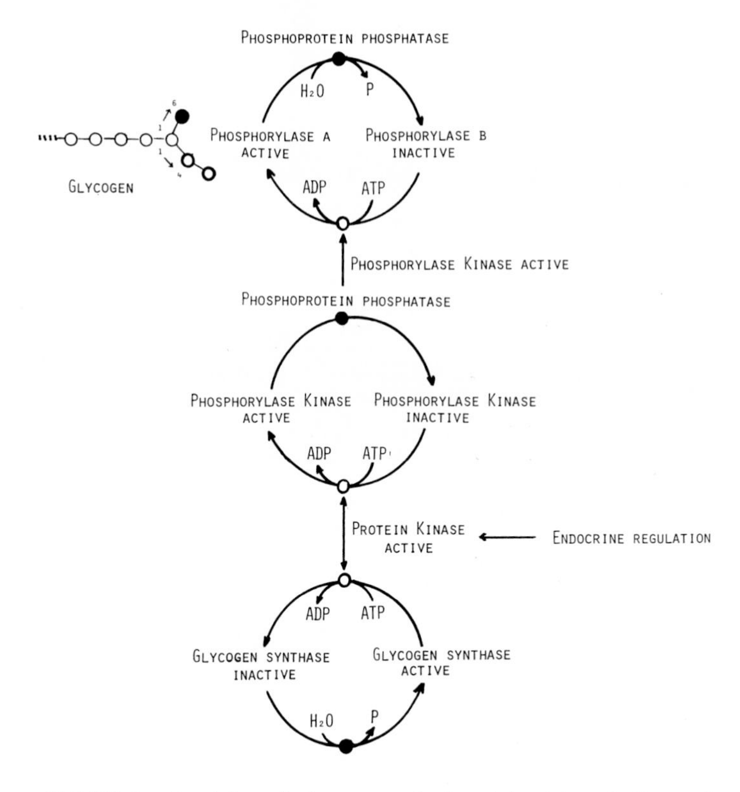

FIGURE 9. Regulation of glycogen synthesis and breakdown in liver and muscle.

Thyroid hormones also influence the blood sugar level. In hyperthyroid patients fasting blood glucose is increased; it is decreased in hypothyroid patients. The latter group has a decreased ability to utilize glucose; hyperthyroid patients, however, metabolize the glucose molecule at a normal rate. Moreover, hypothyroid individuals are much less sensitive to insulin than hyperthyroid patients or normal individuals. In experimental animals, the administration of thyroxine causes a diabetogenic action, whereas thyroidectomy inhibits the development of diabetes. The mechanism of thyroid hormone action on influencing the blood glucose level may be connected with its effect on the rate of insulin destruction in the liver or with the regulation of the response of the end organs where glucose is utilized.

E. Role of Nonhormonal Factors

Apart from hormonal influences, the process of glycogen synthesis and metabolism and glucose utilization in the liver are continually modified by several factors, even under physiological conditions.[78] These factors include glucose absorption from the intestines, emotional states, stress, physical and mental ability, and others. In the majority of these factors the primary action is connected with an elevated blood sugar level. This rise automatically suppresses the output of glucose from the liver and its metabolism in various tissues.

When glucose in the blood is raised to a relatively high level, the kidney also exerts a regulatory action. Glucose is filtered continually through the glomerulus but ordinarily it returns to the blood by reabsorption via the tubules. The process of reabsorption is associated with phosphorylation functioning in the tubular cells. This process is similar to that responsible for the intestinal absorption of sugar molecules. The capacity of the renal tubules is dependent on the enzymatic phosphorylation process, the concentration of the enzyme

Table 1
CAUSES OF HYPOGLYCEMIA

Pathological abnormalities present
- Adrenal cortical insufficiency
- Hypopituitarism
- Insulinoma
- Extrapancreatic tumors
- Massive hepatic disease

Pathological abnormalities absent
- Normal fasting plasma glucose level
- Reactive hypoglycemia
- Alimentary
- Prediabetic
- Functional

Fasting plasma glucose level low
- Drug-induced
- Salicylates, phenformin, sulfonylurea
- Alcohol-induced
- Insulin-induced

being the rate-limiting factor. The normal rate is approximately 350 mg glucose per minute. When blood sugar is raised above this level, the glomerular filtrate may contain more glucose than can be absorbed. The excess is excreted into the urine producing glucosuria.

Glucosuria occurs in normal individuals when the glucose level in the veins exceeds 170 to 180 mg/dℓ. This is the renal threshold for venous blood glucose. It shows variations with changes in the glomerular filtration rate. Since the maximum rate of tubular reabsorption is constant, this represents a more accurate measurement than the renal threshold. Defects in the renal tubules may inhibit glucose reabsorption, termed as renal glycosuria. This condition may occur even when blood glucose levels are normal and manifests as the result of disease processes or from inherited disorders.

VI. HYPOGLYCEMIA

This syndrome is characterized by low serum glucose.[24,51,52,57,61,80] It occurs in two different forms, depending on whether the hypoglycemia is acute or chronic. If the serum glucose level is quickly reduced, homeostatic mechanisms are impaired and epinephrine is released connected with sweating, trembling, weakness, and anxiety. If the decrease of serum glucose is slow, central nervous system symptoms are produced such as headache, lethargy, and irritability.

The causes of hypoglycemia are varied (Table 1). It occurs temporarily after a meal, and is called reactive hypoglycemia. In this case, fasting plasma glucose is normal and there is no anatomic lesion. Reactive hypoglycemia occurs when serum glucose levels fall below 40 mg/dℓ. The symptoms of hypoglycemia are transient. When a glucose tolerance test is carried out samples should be taken every 30 min and when the patient becomes symptomatic. The results of the 5-hr glucose tolerance test can be classified into three basic categories: (1) alimentary, (2) functional, and (3) prediabetic (Figure 10).

Alimentary hypoglycemia occurs after gastrointestinal surgery leading to an accelerated glucose absorption. In these cases, postprandial hyperglycemia is very marked and associated with a corresponding insulin release.[12,43,87] However, the rapid rise of serum glucose initiates an exaggerated insulin response resulting in a hypoglycemia. This hypoglycemia occurs 0.5 to 3 hr after eating. Functional hypoglycemia is quite common in adults; it occurs 2 to 4 hr

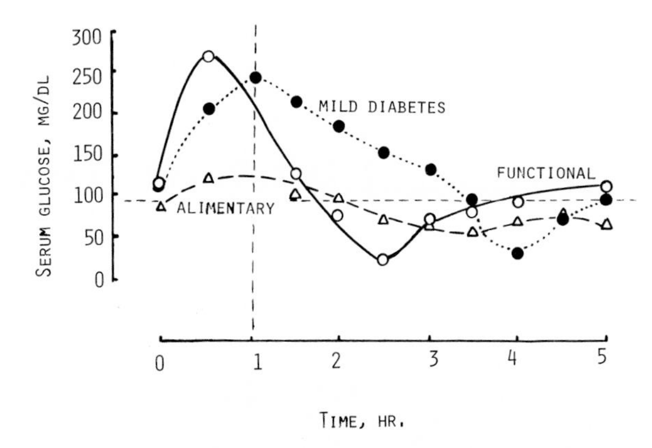

FIGURE 10. Glucose tolerance curves in various types of reactive hypoglycemia.

after meals; shortly after this time it is associated with predominant epinephrine-induced symptoms. This syndrome is commonly seen in patients with emotional problems. Diabetes mellitus usually does not develop in these two patient categories. In prediabetics or diabetics, hypoglycemia sometimes occurs in relation to a meal. In this group hypoglycemia is seen later than in the functional group and insulin response is delayed.

The causes of hypoglycemia include alcohol and other drugs. Prolonged ingestion of alcohol induces hypoglycemia when the supply of glycogen is concurrently depleted from the liver. Among the other agents, the most common causes are hypoglycemic agents such as insulin and the sulfonylureas, which account for half the drug-induced causes. Other drugs such as sulfonamides and salicylates less frequently cause hypoglycemia.[14]

Hypoglycemia may be connected with tumors such as insulinoma, adrenal cortical tumors, and gastrointestinal and hepatic carcinomas.[26,29,46,51,68] Tumor-induced hypoglycemia is generally very severe. About half of these tumors produce insulin-like substances that are associated with the development of hypoglycemia. Other diseases related to hypoglycemia are diffuse liver disease, hypopituitarism, and adrenal cortical insufficiency. Insulinoma is the most important cause of hypoglycemia and is connected with excessive secretion of insulin by pancreatic β-cell tumors. Insulinoma occurs at every age, but most commonly between 40 and 60 years. Many clinical symptoms are caused by the hypoglycemia; they have an insidious onset and can mimic a variety of neurologic or psychiatric diseases.

VII. HYPERGLYCEMIA

This condition is characterized by high concentrations of plasma glucose resulting in its excretion in the urine.[57] Chronic condition represents diabetes mellitus. Hyperglycemia may be connected with (1) total absence of insulin secretion as a consequence of a surgical removal of the pancreas; (2) periods of stress during pregnancy, dehydration, or associated with severe infection; and (3) delayed development of the pancreas in transient neonatal diabetes. Several drugs induce hyperglycemia and it may also develop secondary to endocrine diseases or antibody formation to insulin receptor (Table 2).

VIII. DIABETES MELLITUS

This is a widespread and complex disorder of metabolism.[1,2] The susceptibility to diabetes mellitus is conditioned by genetic factors. It is difficult to ascertain the overall incidence of diabetes since the onset of this disease may occur late in life and may be unrecognized for

Table 2
CAUSES OF HYPERGLYCEMIA

Primary
- Juvenile — onset diabetes mellitus (insulin-dependent)
- Maturity — onset diabetes mellitus (noninsulin-dependent)

Secondary
- Hyperglycemia resulting from pancreatic diseases
- Inflammation
- Acute pancreatitis
- Chronic pancreatitis
- Pancreatitis (connected with mumps)
- Pancreatectomy (surgical)
- Hemochromatosis (connected with infiltration)
- Tumors

Hyperglycemia resulting from major endocrine diseases
- Thyrotoxicosis
- Cushing's syndrome
- Hyperaldosteronism
- Acromegaly
- Glucagonoma
- Somatostatinoma
- Pheochromocytoma

Hyperglycemia resulting from other major diseases
- Chronic hepatic disease
- Chronic renal failure

Hyperglycemia connected with insulin receptor antibodies
- Acanthosis nigricans

Hyperglycemia connected with drug treatment
- Alloxan, dehydroascorbic acid, dithizone, streptozotocin (experimental)
- Thiazide diuretics, propranolol, phenytoin
- Oral contraceptives
- Steroids

many years. Studies carried out on various communities have shown about 1.5 to 2.0% known diabetics and revealed many unknown cases — at least double the prevalence of those initially found.

The hereditary nature of this disorder is well known. Greater frequency of diabetes mellitus occurs among close relatives of known diabetic patients than among a comparable control population. The inheritance pattern shows a simple Mendelian recessive mode of inheritance; this suggests a tendency to skip generations. However, it is important that diabetes is absent in a significant percentage of identical twin mates. Furthermore, the incidence of diabetes among the offspring of marriages between diabetic patients usually lies far below 100%.[47] This shows that the penetrance (i.e., the likelihood that a gene will receive its morphological or functional expression) of the diabetic trait is incomplete and variable. This incomplete penetrance indicates that associated environmental factors are also important in the development of diabetes. Among these environmental factors, particular importance may be attached to the environment of the fetus *in utero*. In the past years, diabetic mothers have been unable to carry pregnancies to term. Only recently have they given birth to offspring. Statistical comparisons have shown that the incidence of diabetes is greater in the offspring of diabetic mothers than those of diabetic fathers.

Diabetes mellitus is associated with a deficient secretion of insulin from the pancreas β-cells in response to glucose stimulation. This results in a disorder in the metabolism of carbohydrates, proteins, and fats. The metabolic abnormalities lead primarily to the inadequate disposal of ingested glucose in mild cases, to glucose overproduction, and dissolution of fat stores and protein reserves in more severe cases. Secondary consequences of diabetes mellitus include several serious systemic complications, such as the reduction of bone mineral content leading to bone loss and increased frequency of bone fractures. Changes in the sorbitol pathway in the peripheral nerves modify conduction velocity. Significant neurological damage is common in patients with poorly controlled diabetes, leading to diabetic neuropathy. Diabetic angiopathy, musculoskeletal disorders, and rheumatological, renal, and gastrointestinal manifestations, and other secondary effects may represent consequences of the altered metabolism of the various sites involved.[4,22,64]

Cells responding to insulin have receptor molecules in their plasma membranes to which insulin binds specifically.[13,39] The molecular mechanism of insulin action is not completely understood at this time. Insulin does not appear to enter the cells it acts upon, but rather to mediate its influence indirectly, perhaps by modulating adenyl cyclase activity in some way and ultimately by lowering the cellular cyclic-AMP content or effectiveness of action. Recent evidence also suggests that insulin may raise intracellular cyclic-GMP content and this may explain, in part, the anticyclic-AMP effects that insulin exerts in many tissues.

In some instances diabetes is connected with conditions in which cellular defects involve decreased numbers of insulin receptors in peripheral tissues, or decreased cellular responsiveness at a step subsequent to hormone-receptor interaction. Insulin resistance, a clinical syndrome represented by the failure of an individual to respond normally to endogenous secretion or exogenous administration of insulin, can lead to severe diabetes even when blood levels of the hormone are raised.[14] This resistance is caused by a target cell abnormality or the presence of insulin antagonists in the blood.

A. Endocrine Defects in Diabetes

The major endocrine defect in diabetes mellitus is the absolute or relative insulin deficiency. The nature of this defect can be studied by the production of experimental diabetes by surgical removal of all or a large portion of the pancreas or chemical destruction of all or large portions of the β-cells of the islets of Langerhans using agents such alloxan, dehydroascorbic acid, dithizone, or streptozotocin. These chemicals produce a selective necrosis of the β-cells, and inactivate the circulating insulin by the administration of antibodies effective against its own insulin. Human diabetes mellitus has been produced by (1) pancreatectomy, incident to removal of malignant tumors; (2) has been noted during and after pancreatitis; and (3) hemochromatosis, accompanied by a deposition of iron pigment in the pancreas with secondary fibrosis and loss of functioning tissue.

B. Factors Influencing Insulin Secretion

Under physiological conditions insulin secretion is stimulated by glucose. The administration of glucose brings about a rapid rise in the level of circulating insulin by entering the β-cells; glucose is then metabolized in the glycolytic pathway. There are some substances which also stimulate insulin secretion: glucagon, certain amino acids, and gastrointestinal factor, but the physiological significance of these factors is not well known. There are some other compounds which inhibit this process. Mannoheptulose completely blocks insulin release by inhibiting glucose phosphorylation in the liver. Some glucose derivatives also produce inhibition, e.g., glucosamine and 2-deoxyglucose 6-phosphate; these derivatives competitively inhibit the isomerization of glucose 6-phosphate to fructose 6-phosphate. Several naturally occurring substances influence insulin secretion. Among these are inhibitors such as adrenaline and stimulators such as adrenocorticotropin, thyrotropin, glucagon, and

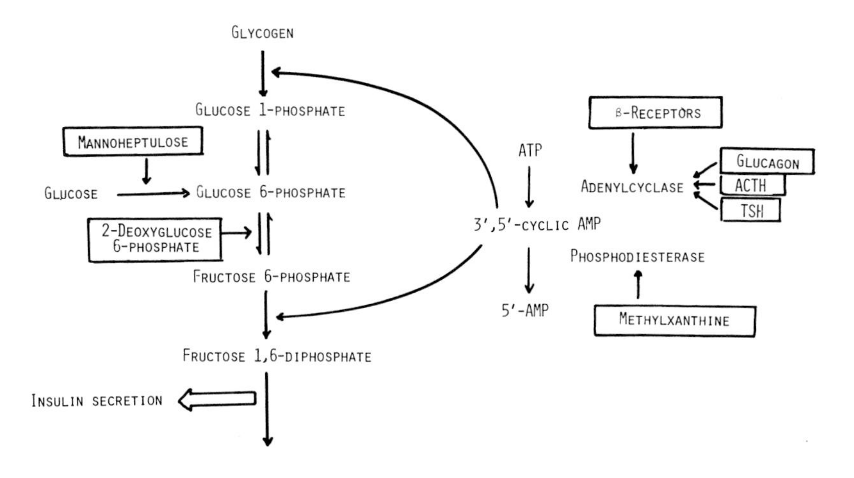

FIGURE 11. Stimulation and inhibition of insulin secretion.

cyclic AMP. The stimulatory activity of the latter hormonal factors is probably mediated through cyclic AMP. These various effects are summarized in Figure 11.

C. Metabolic Effects

The first observations about the effect of pancreatic extracts reported the lowering of the blood glucose level. Insulin primarily caused an increased rate of extraction of glucose from the blood.[8,15] It was later demonstrated that insulin accelerates the rate of glucose uptake by extrahepatic tissues other than the heart. It seems that there are insulin-responsive tissues and the primary action of insulin is to facilitate the conversion of extracellular glucose to intracellular glucose 6-phosphate (Figure 12).

Insulin directly alters hexokinase activity in the presence of inhibitory pituitary or adrenal components. It seems from some other experiments that insulin has dual effects resulting in increased penetration of glucose into the cell and correction of a decreased intracellular phosphorylating mechanism in the diabetic state. The effect on permeability is immediate, while the action on phosphorylation requires time and is sensitive to pituitary and adrenal factors. In the double role of insulin, i.e., penetrating (or transporting) and correcting phosphorylating mechanisms, there are several common components.

The primary effect of insulin is to facilitate the utilization of glucose by certain tissues, especially adipose tissue and muscle. The lack of insulin results in elevated fluid glucose levels. In the diabetic state, the hepatic enzyme structure is characteristically altered — glucose 6-phosphatase increased, glucokinase decreased — aggravating a tendency to accumulate extracellular glucose (Figure 12). The hyperglycemia tends to correct, in part, the defect in glucose utilization.

In terms of overall carbohydrate balance, a major part of the problem which faces the diabetic patient results from the occurrence of massive glucosuria as soon as the hyperglycemia surpasses the renal capacity of reabsorption for glucose. As long as insulin deficiency is not complete, compensation may occur with moderate hyperglycemia and minimal glucosuria. With increasing severity of the insulin deficiency, higher and higher concentrations of glucose are required to achieve compensation and the utilization of this increased blood glucose is to a large extent wasted by the occurrence of rapidly mounting urinary losses. The higher levels of blood glucose can be obtained only at the metabolic cost of profuse hepatic overproduction. The peripheral underutilization and hepatic overproduction of glucose are intimately related in severe states of insulin deficiency.

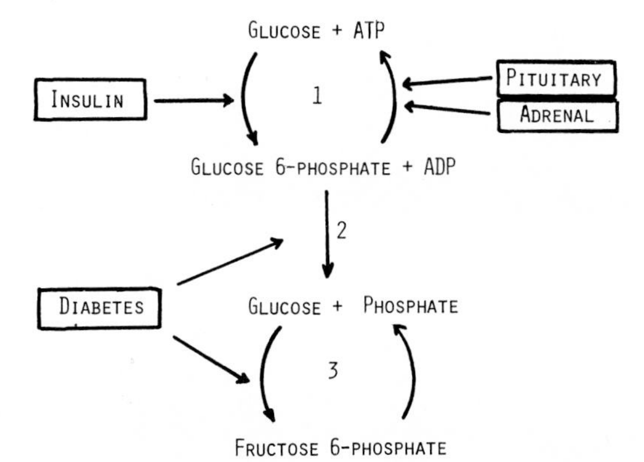

FIGURE 12. Metabolic action of insulin. Insulin increases the conversion of glucose to glucose 6-phosphate by stimulating hexokinase (1); pituitary and adrenal hormones inhibit this process. In diabetes, the consequence of the lack of insulin is that glucose 6-phosphatase (2) is increased and glucokinase (3) is decreased, resulting in an accumulation of extracellular glucose.

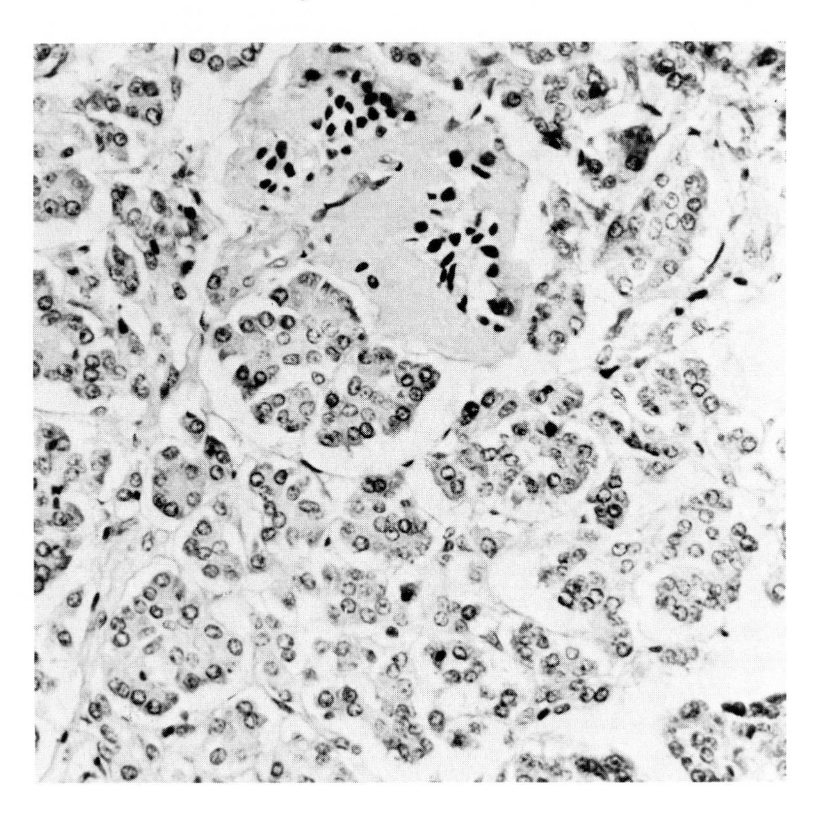

PLATE 1. Hyaline change of the Langerhans islets in diabetes mellitus. The normal structure of the islet is altered by the hyaline deposits, leaving a greatly reduced number of cells.

Diabetes mellitus is characterized by overproduction and inadequate utilization of carbohydrates. However, since there are interrelations between the metabolism of carbohydrates, components. Metabolic side effects include the accumulation of nodular deposits, hyaline and amyloid bodies in many organs (Plates 1 through 3).

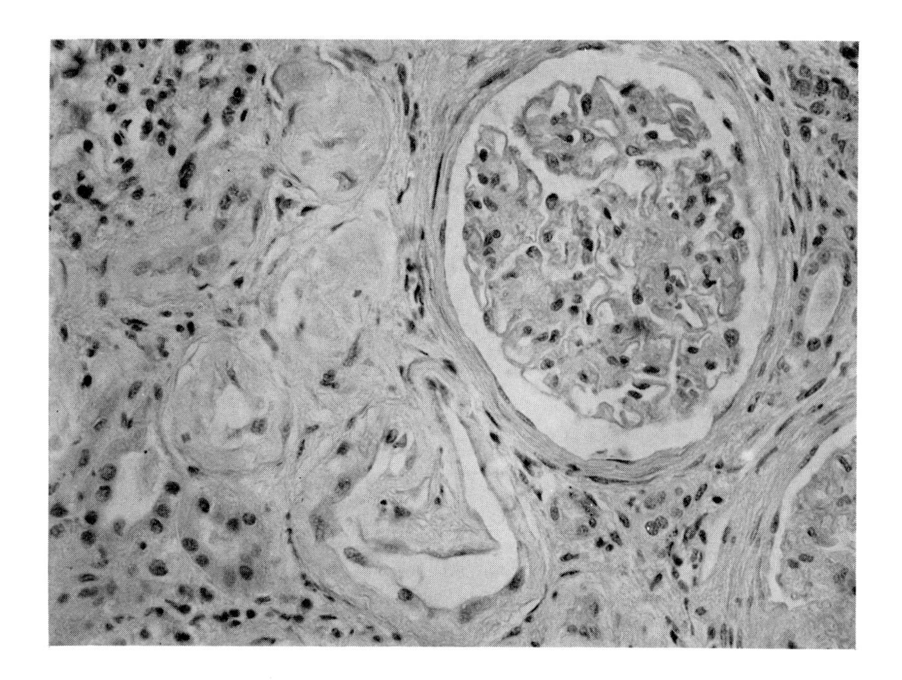

PLATE 2. Amyloid deposits in kidney. Note the thickening of the basement membrance of glomerular loops as well as the wall of afferent and efferent arterioles.

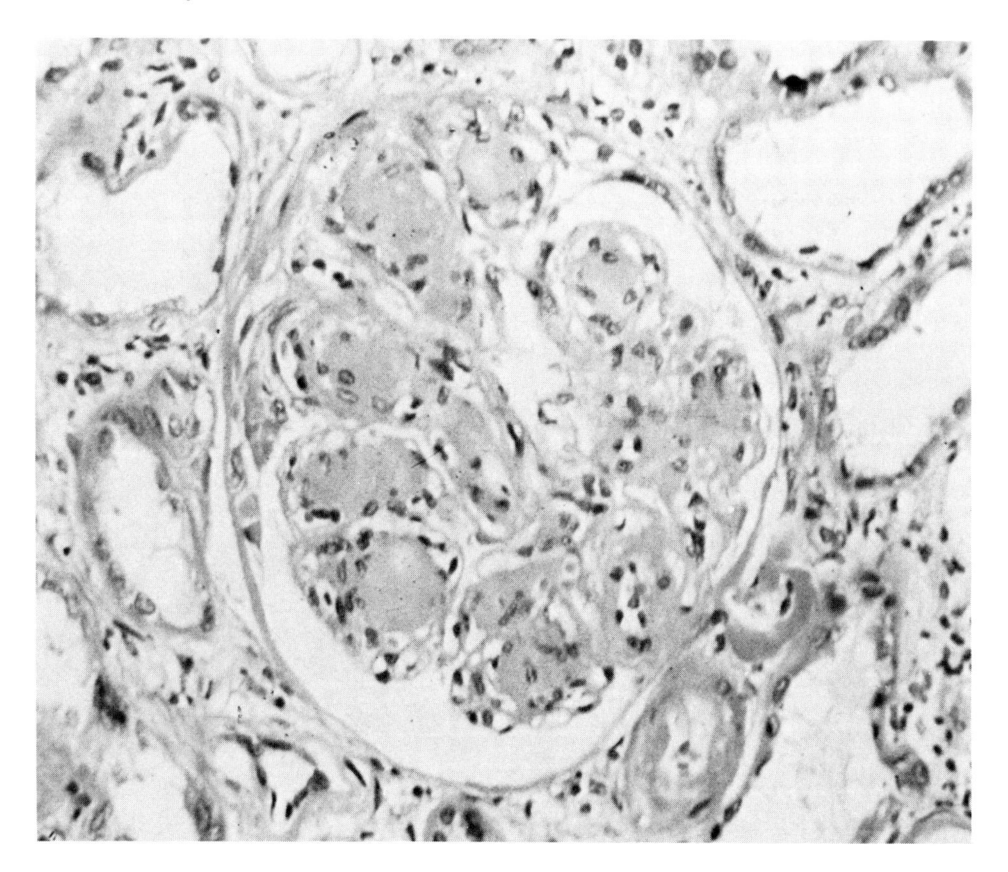

PLATE 3. Glomerular changes of Kimmerstiel-Wilson syndrome of diabetes. Note the nodular deposits altering the normal appearance of glomerular loops. Vas afferens and efferens also show diabetic changes.

D. Clinical Syndromes

The diabetic manifestations derived from hyperglycemia are termed as acute diabetic syndrome. Further manifestations develop slowly in the course of the disease and are termed the chronic diabetic syndrome. The latter frequently appears to have a vascular basis. Experimental diabetes produced in animals and measurements of the mass of the islets of Langerhans in patients suffering from diabetes suggest that the acute diabetic syndrome is the result of an absolute or relative deficiency of the hormone produced by the β-cells, i.e., a deficiency of insulin. All the symptoms and signs may be reversed by the administration of insulin.

Whether absolute or relative insulin deficiency also accounts for the symptoms of the chronic diabetic syndrome is as yet uncertain. Whether absolute or relative insulin deficiency is linked with the production of acute diabetic syndrome and the mechanism producing this deficiency is completely unknown at present. Possibilities include a biochemical lesion in the mechanisms of insulin synthesis, release and secretion, transport (destruction), and effectiveness at tissue level (receptor).

E. Course of Diabetes Mellitus

Diabetes mellitus is a genetically determined metabolic disease in which an inherited susceptibility plays an important part.[4,22,64,74-76] It is characterized by fasting hyperglycemia, atherosclerotic and microangiopathic vascular disorders, and neuropathy. Due to genetic influence, the disease has its origin at conception and exists for a prolonged period before there is a recognizable abnormality of carbohydrate metabolism. Furthermore, hyperglycemia may also manifest for years before vascular symptoms or neuropathy are recognized clinically.[20] Continuous hyperglycemia and glycosuria are the generally accepted symptoms for the definite diagnosis of diabetes. However, the great majority of clinicians and investigators agree that diabetes mellitus can be diagnosed without fasting hyperglycemia, although this is the most recognizable form of the disease. Typical neuropathic and vascular manifestations may develop in patients who have normal fasting blood glucose levels and relatively mild carbohydrate intolerance, but who are genetically predisposed to diabetes. Moreover, there is no sharp division between normal and abnormal carbohydrate metabolism in close relatives of diabetic patients.

The development of primary diabetes can be divided into four stages, based on the presence or absence of carbohydrate metabolism. These stages are prediabetes, subclinical or suspected diabetes, chemical or latent diabetes, and overt diabetes.

Prediabetes (Stage I) is characterized by a long period of time prior to the onset of identifiable diabetes mellitus (overt, chemical, or latent). This is a conceptual term which identifies an interval between fertilization of the ovum and demonstration of impaired glucose tolerance (in particular after a stress test) in an individual predisposed to diabetes on genetic grounds. This individual shows presently a normal glucose tolerance, fasting and *post cibal* blood sugar, and cortisone glucose tolerance test (GTT). Blood insulin levels after provocation may be low, normal, or elevated.

Susceptible individuals have (1) relatives with diabetes (uniovular twin of a diabetic, both parents); (2) obstetric history (large babies, over 10 lb), perinatal mortality (excluding Rh factor, incompatibility, infertility, hydramnios, excess lactation); (3) obesity (hyperinsulinemia or diminished response to insulin); and (4) vascular and neuropathological conditions (retinopathy, coronary artery disease, general arteriosclerosis). In Stage II the GTT and fasting blood sugar are normal, whereas the cortisone-GTT is abnormal. In Stage III the cortisone-GTT and the GTT, and sometimes the fasting blood sugar, are abnormal; also the postprandial blood sugar may be elevated. Overt diabetes (Stage IV) is a stage of frank, symptomatic diabetes, "juvenile" or "maturity onset" in type. Fasting hyperglycemia is often present with glycosuria.

Table 3
CHARACTERISTICS OF JUVENILE-ONSET AND MATURITY-ONSET DIABETES

	Juvenile-Onset	Maturity-Onset
Family history	Common (40%)	Less common
Age at onset	<15 years	Adult (typically 40 yr +)
Onset	Rapid	Slow
Severity	Severe	Mild
Body build	Normal or thin	<50%, obese
Ketoacidosis	Common	Rare
Stability	Unstable	Stable
Insulin therapy, sensit.	Almost all cases	Common
Sulphonylurea, sensit.	Very few (if any)	<50%
Complications	90% in 20 years	Less common, slower rate of development.
Fasting insulin (IRA)	Low	Nonobese normal or low; obese diabetic-increased.
Insulin response to glucose	Hypoinsulinemia	Increased but sluggish rise.

Secondary diabetes is associated with (1) endocrine disorders (hyperadrenalism — cortical and medullar, acromegaly, thyrotoxicosis); (2) loss of pancreatic islet tissue (inflammation, tumors) by surgical removal of 90% of the mass; and (3) stress (with temporary carbohydrate intolerance). These cases are often suspect as prediabetics.

The earliest stage of the disease is prediabetes. This stage of diabetes exists before the onset of any of the abnormalities which characterize later stages. This stage starts in a genetically predisposed individual from the time of conception until impaired glucose tolerance can be demonstrated.[20] During the prediabetic period, the GTT and cortisone-glucose tolerance tests are normal. They can be distinguished from normal controls by a delayed or decreased increase of plasma insulin level in response to glucose or amino acid stimulation.

In subclinical diabetes the fasting blood glucose level and GTT are normal under usual conditions. Among abnormal circumstances, however, diabetes may be suspected because of delayed and decreased insulin response to glucose administration and some abnormality in the cortisone-glucose tolerance test, indicating insufficient functional reserve of the islet cells of the pancreas. Stress or pregnancy may reveal the abnormality of glucose tolerance. A high percentage of women showing this impairment may develop latent or overt diabetes in later years.[66]

The latent or chemical diabetics have symptoms referable to the disease.[14] They show normal or sometimes elevated fasting blood sugar level, and the definite diagnosis of diabetes can be established by the abnormal GTT and greatly enhanced delay or significantly decreased insulin response to glucose administration.

The most advanced stage of diabetes mellitus is overt or frank diabetes. In this stage the classical symptoms of the disease are present; the fasting blood sugar level is highly elevated and there is hyperglycemia and glycosuria. The basic differences between juvenile-onset and maturity-onset diabetes are summarized in Table 3.[53]

F. Abnormalities of Calcium Homeostasis

Some reduction of the bone mass represents frequent accompanying symptoms of diabetes mellitus.[12] Osteopenia is often found in both juvenile- and adult-onset diabetes. Diabetics may have subnormal bone mineral content, and low total body calcium. Some other studies, however, failed to confirm osteopenia in diabetics, still, recent concensus favors some reduction of bone mass in diabetics below age and sex-specific normal ranges. The reduction of bone mass develops during the first 3 to 5 years of clinical symptoms and occurs in its most severe form in patients with the onset of the disease before 21 years of age. These actions may be associated with a hormonal activation of the target cells, since insulin, glucagon, or calcitonin induce an early decrease of serum calcium. The degree of the reduction of bone mass depends on the quality of diabetic control, but the association between impaired glucose homeostasis and defective calcium metabolism remains to be established.[11]

G. Magnesium Metabolism in Diabetes

Abnormal magnesium metabolism is an accompanying consequence of diabetes mellitus. Increased urinary loss of magnesium is well known, especially when the disease is poorly controlled. Losses of magnesium during diabetic ketoacidosis might represent about 2% of total body stores. Similarly, a survey of patients with various chronic illnesses has shown that diabetes is the most frequent among conditions associated with hypomagnesemia.

H. Diabetic Neuropathy

Symptomatic polyneuropathy is present in one fourth of the patients with diabetes mellitus.[23] Moreover, nearly all diabetic subjects show abnormalities in peripheral nerve function evidenced by electrophysiological measurements. These functional abnormalities representing neuropathy are the most common complications of diabetes in late phases. Many clinical, electrophysiological, histological, and biochemical data are known in the underlying pathogenic mechanism. In some cases, neuropathies associated wth diabetes mellitus are considered to be the result, at least in part, of vascular lesions.

Electrophysiological and morphological data suggest that axonal damage is the earliest manifestation of diabetic neuropathy. The predominant histologic changes in the sciatic nerves consist of axonal degeneration, shortening of the internodal lengths, and the presence of intercalated internodes; occasionally, active segmental demyelination also occurs.

The normal structure and activities of subcellular organelles and numerous enzymes are necessary for the synthesis and packaging of neurotransmitter substances as well as for the maintenance of axonal integrity and function. Various substrates and neurotransmitters are transported from the cell body through the axon by microfilament- and microtubule-mediated mechanisms. Defects in this axonal flow are associated with axonal dysfunctions. In rats, streptozotocin-induced diabetes causes a decrease of rapid axonal flow in the sciatic nerve connected with acetylcholinesterase activity and in the rate of slow flow connected with choline acetylase activity. These changes are obviated by insulin. Changes related to enzymes producing some neural protein, lipid, and carbohydrate components may reflect abnormalities. In particular, abnormally low Na^+-/K^+-ATP-ase, acetyl thiokinase, aldose reductase, and CDP-diglyceride:inositol phosphotransferase activities have been demonstrated in this experimental diabetes.

Another line of biochemical investigations in experimental animals has shown abnormalities in the axonal metabolism of myoinositol and its phospholipid derivatives. It has been suggested that these abnormalities are related to the functional disorders of diabetic nerves. The concentration of myoinositol in the peripheral nerve is about 30 times greater than that of plasma. However, the significance of the maintenance of this gradient with regard to peripheral nerve function has not been established.

I. Diabetic Ketoacidosis

This is a syndrome characterized by ketoacidosis, hyperglycemia, hyperosmolality, and dehydration.[3,49,50,59,65,70] These are connected with a relative or absolute insulin deficiency. Normally, there is a balance between the actions of insulin and the various catabolic hormones such as glucagon, thyroid hormones, cortisol, growth hormone, and catecholamines. In juvenile diabetes, due to absolute insulin deficiency an impaired balance occurs and serum

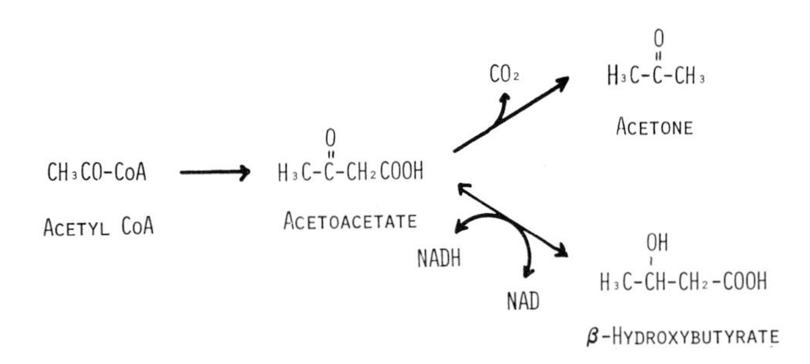

FIGURE 13. Formation of ketone bodies in diabetic ketoacidosis.

glucose concentration is elevated resulting from increased gluconeogenesis and decreased extrahepatic breakdown. These processes are connected with the activation of glucose 6-phosphatase, fructose-1,6-diphosphatase, phosphoenolpyruvate carboxylkinase, and pyruvate carboxylase (Figure 2). At the same time, excess glucagon and epinephrine enhance glycogenolysis and glucagon and cortisol promote the uptake of gluconeogenic substances such as amino acids. The developing hyperglycemia causes hyperosmolality and osmotic diuresis, leading to renal loss of glucose, water, and electrolytes.

The metabolism of free fatty acids released from adipose tissue is dependent on their plasma concentration. They form CoA derivatives and are oxidized to acetyl-CoA by the mitochondrial system. In the presence of oxaloacetate, acetyl-CoA is disposed. In diabetic ketoacidosis, acetyl-CoA is converted to acetoacetyl-CoA which is further metabolized to acetoacetate and hydroxybutyrate. Acetoacetate is decarboxylated spontaneously to acetone (Figure 13). The liver cannot utilize these ketone bodies; they enter the circulation and are excreted into the urine.

Chronic alcoholics occasionally show ketoacidosis and hyperketonemia without hyperglycemia.[48] Some of these patients are hypoglycemic and they can be distinguished from diabetic ketoacidosis patients by serum glucose levels, which are less than 200 mg/dℓ, and the lack of glucosuria.

J. Diagnosis of Diabetes Mellitus

In a prolonged state of diabetes mellitus, a diagnosis can be established easily. Mild cases and late development of the disease may cause problems in identification. Since diabetes is associated with insulin deficiency, the test applied assesses the ability of the patient's pancreas to secrete insulin as a response to some stimulation. These tests are the tolerance tests and the application of various drugs which directly stimulate insulin release.

Tests applied for the diagnosis of diabetes include: (1) fasting blood glucose level; (2) tolerance tests: the oral glucose tolerance test, the intravenous glucose tolerance test, the intravenous tolbutamide test, and the steroid (cortisone)-glucose tolerance test; and (3) hemoglobin A_{1c} content.

The diagnosis of diabetes and the assessment of the various stages depends on the proper application and interpretation of laboratory procedures.[10] There is no doubt in establishing the last stages of the disease. However, early phases of diabetes mellitus can exist without an elevation of the fasting blood sugar level; this level, therefore, should not be one of the criteria for the interpretation of the glucose tolerance test. Moreover, progression or regression from one stage to the next one show extreme variations; these may develop very slowly over many years, or may never occur, or may be rapid or explosive.

Thus, the validation of diagnostic criteria for the interpretation of the GTT must be independent of the rate of progression to overt diabetes. Moreover, there is a decrease in tolerance to glucose with advancing age in nondiabetic subjets, which may represent an age-related physiological change or may indicate an increasing incidence of the latent form of the disease. In the nonprogressive course of diabetes, the occurrence of occlusive vascular disease or neuropathy often leads to the proper diagnosis of diabetes for the first time. It is important therefore, if glucose intolerance cannot be determined, that other signs of the disease be investigated.

1. Glucose Tolerance Test

This test expresses the overall ability of the organism to maintain a balance between the various forces which influence the glucose level in the blood. GTT estimates the net result of the opposite actions represented by uptake, metabolism, and their regulation.[10,15,19,60,66] The ingestion of glucose or starch, and to a lesser degree other carbohydrates, causes a rise in the blood sugar level. The degree of increase is dependent upon the amount of glucose present in the food or formed during digestion. In the oral tolerance test, following the ingestion of 100 g glucose after an overnight fast, characteristic changes occur in the venous blood. First there is a sharp rise to a peak within 30 to 60 min. This increase is due to intestinal absorption which temporarily exhausts the capacity of the liver and other tissues to remove the excess. As blood glucose level increases the regulatory mechanism starts, overall hepatic glycogenesis increases, and glucose utilization is accelerated; insulin secretion is also raised. At the peak, glucose is removed from the blood at the same rate as it is absorbed from the intestines.

During the next period there is a sharp fall, reaching the fasting level at about 90 to 120 min. Glucose is removed from the blood faster than it enters due to stimulated utilization and completion of absorption from the intestines. The fall continues to a level slightly lower than the fasting level and the hypoglycemic phase disappears by a subsequent rise to the fasting level at 150 to 180 min. The hypoglycemic phase is a consequence of the inertia of the regulatory mechanism. Typical oral glucose tolerance curves in normal and diabetic subjects are presented in Figure 14. Liver disease and thyrotoxicosis modify the peak glucose level (Figure 15).

In the arterial blood, the changes are somewhat different from that of venous blood. The increase starts earlier and the peak may reach higher values than in venous blood. Th arterial-venous difference represents the removal of glucose from the arterial blood into the tissues and exposes the utilization of glucose for glycogen synthesis and for oxidative breakdown products, particularly by the muscles. The return to the fasting level is delayed indicating that the active removal continues further than in the venous blood. However, the two curves reach the resting level at the same time — within 150 to 180 min.

Many factors influence the GTT.[66] The tolerance is somewhat lower in old people than in young. Dietary conditions, particularly carbohydrate nutrition, influence the glucose response. In relative carbohydrate starvation the rise of blood sugar is more pronounced and the return is delayed. Conversely, high carbohydrate diet causes a reduction of the peak. Exercise before the test may cause an increase to an abnormally high level; exercise after ingesting the glucose dose may elicit a prolonged hypoglycemic phase. Emotional disturbances tend to raise blood sugar level and may also be associated with an increased period of hypoglycemia. Abnormalities are related to diseases of carbohydrate metabolism, such as diabetes mellitus, due to a deficiency of pancreatic insulin secretion; hyperthyroidism, either due to thyroid disease or temporarily induced function; increased secretion of epinephrine, adrenal cortical hyperfunction, hyperpituitarism, acidosis, asphyxia, and strenuous muscular exercise.

In diabetes, the failure of the pancreatic insulin secretion is the basic cause of the abnormality in the postabsorptive blood sugar level. The variations range between 70 to 1800 mg/dℓ. Insulin deficiency causes the high postabsorptive level, and the fasting sugar level usually runs parallel with the severity of the condition. In mild cases it may fall within normal limits. Complicating conditions, however, such as acidosis and hyperthyroidism, may also be contributing factors in maintaining the high level.

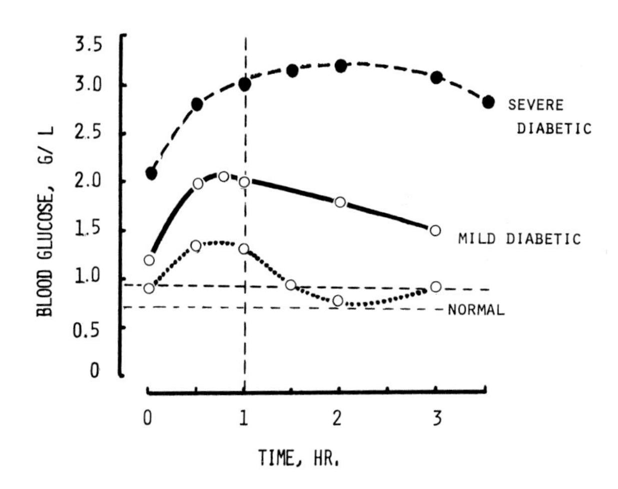

FIGURE 14. Typical oral glucose tolerance curves in various diabetic patients.

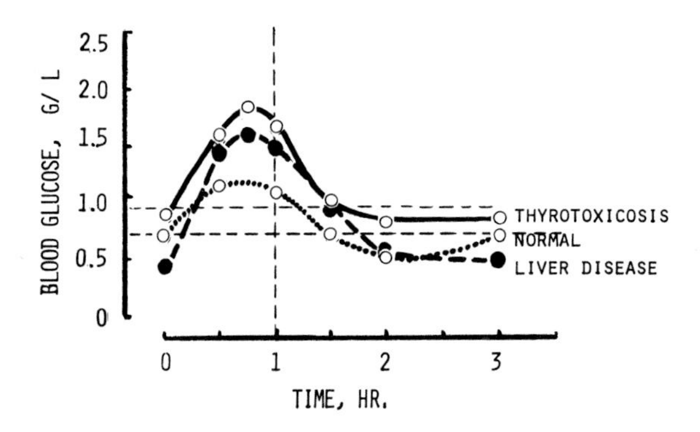

FIGURE 15. Typical oral glucose tolerance curves in various diseases.

In the case of hyperthyroidism, there is an increased hypersensitivity in the liver in the transformation of glycogen to glucose. This occurs in genuine thyroid diseases such as exophthalmic goiter or toxic adenoma, as well as in temporary conditions such as excessive administration of thyroxine. The liver hypersensitivity is probably due to sympathetic nerve impulses which are stimulated by epinephrine. Increased glycogenolysis and gluconeogenesis may also be connected with the increased tissue metabolic rate.

Increased epinephrine secretion is also an important factor in the glycogenolytic activity of the liver. Epinephrine causes an elevation of blood sugar. This may be an important factor in hyperthyroidism by the activation of the hepatic processes. Psychic hyperglycemia occurs under many circumstances. In various emotional conditions such as anxiety, fear, and anger increased amounts of epinephrine are released in the blood. Mental or cold stress or excitement elicit elevation of blood glucose as the result of the mobilization of hepatic glycogen on the account of epinephrine produced. The effect of pain and discomfort on blood sugar level is probably related to the same mechanism. Intracranial changes, such as skull fracture, concussion, hemorrhage, brain tumor, convulsive states including certain types of epilepsy, tetanus, and eclampsia bring about hyperglycemia if the liver contains adequate glycogen. If the stores are depleted, blood glucose cannot rise; it may even fall if the convulsion is prolonged.

Hyperfunction of the adrenal cortex resulting from tumor or hyperplasia is often accompanied by the development of a diminished glucose tolerance. In Cushing syndrome the hyperglycemia is due to increased hepatic gluconeogenesis. This condition is the steroid diabetes, and is associated with a resistance to insulin, absence or pronounced glycosuria, and negative nitrogen balance. In disorders of increased anterior pituitary function, hyperglycemia may set in. This occurs in acromegaly and certain types of Cushing syndrome caused by pituitary abnormality.

Acidosis counteracts the glucose-lowering effect of insulin. This is connected with an interference of glucose phosphorylation which is necessary to the uptake by the cells. This condition is particularly important in diabetes mellitus because this disease aggravates the hyperglycemic mechanism. Acidosis due to other causes such as asphyxia, excessive muscular work, fever, nephritis, or dehydration may enhance the blood sugar level. The rise may not be significant, but the association of these factors may potentiate the influence of each other.

2. *Hemoglobin Glycosylation*

Recently a new concept has emerged in the diagnosis of diabetes. Hemoglobin A_{1c} has been found to be enhanced from the 5% value in normal individuals, up to 12 to 15% in diabetic patients.[28,42] The unique structural feature of hemoglobin A_{1c} (Hb A_{1c}) is the presence of glucose linked to the α-amino terminal valine in the β-chain. Hb A_{1c} levels reflect blood sugar concentration and the degree of hyperglycemia. Thus it provides an objective, accurate assessment of the diabetic status of patients and in a prospective fashion it can be applied in diagnosing the development of metabolic consequences of the chronic disease. Other hemoglobins, Hb_{1a} and Hb_{1b}, which normally account for 1.6 and 0.8% of the total hemoglobin, respectively, are also increased in diabetics.[63]

The apparent nonenzymatic formation of Hb A_{1c} may explain some long-term systemic consequences of diabetes. In tissues with the most noticeable dysfunction, such as peripheral nerves, lens, retina, and kidney, nonenzymatic glycosylation of intracellular proteins may take place analogous to that seen within the red cells.[84] Such glycosylation may modify solubility, enzyme activity, and other functions of these proteins leading to clinical abnormalities.[62] Accumulation of sorbitol in the peripheral nerves may produce glycosylated proteins and affect neurological functions in a similar fashion. In the presence of high amounts of glucose in bovine and rat lens, proteins are nonenzymatically glycosylated on the lysine ϵ-amino group. These modified proteins possess an increased susceptibility to produce high-molecular-weight aggregates resulting in light diffraction and opacities. L-Deoxyfructose has also been detected as a glycosylating sugar molecule.

REFERENCES

1. **Alberti, K. G. M. M., Dornhorst, A., and Rowe, A. S.,** Metabolic rhythms in normal and diabetic man: studies in insulin-treated diabetes, *Isr. J. Med. Sci.*, 11, 571, 1975.
2. **Alberti, K. G. M. M. and Hockraday, T. D. R.,** Diabetic coma: a reappraisal after five years, *Clin. Endocrinol. Metabol.*, 6, 421, 1977.
3. **Arieff, A. I. and Carroll, H. J.,** Nonketotic hyperosmolar coma with hyperglycemia: clinical features, pathophysiology, renal function, acid-base balance, plasma cerebrospinal fluid equilibria and the effects of therapy in 37 cases, *Medicine*, 51, 73, 1972.
4. **Beutler, E., Matsumoto, F., Kuhl, W., Krill, A., Levy, N., Sparkes, R., and Degnan, M.,** Galactokinase deficiency in cataracts, *N. Engl. J. Med.*, 288, 1203, 1973.

5. **Block, M. B., Rosenfield, R. L., Mako, M. E., Steiner, D. F., and Rubenstein, A. H.,** Sequential changes in beta-cell function in insulin-treated diabetic patients assessed by C-peptide immunoreactivity, *N. Engl. J. Med.,* 288, 1144, 1973.
6. **Bonser, A. M. and Garcia-Webb, P.,** C-peptide measurement and its clinical usefulness: a review, *Ann. Clin. Biochem.,* 18, 200, 1981.
7. **Burnand, B., Hurni, M., and Jéquier, E.,** Insulin sensitivity and glucose disposal in young normoglycemic obese subjects, *Diabetes Metab.,* 8, 9, 1982.
8. **Burns, J. J.,** in *Metabolic Pathways,* Vol. 1, 3rd ed., Academic Press, New York, 1967, 394.
9. **Cahill, G. F., Ashmore, J., Renold, A. E., and Hastings, A. B.,** Blood glucose and the liver, *Am. J. Med.,* 26, 264, 1959.
10. **Carroll, K. F. and Nestel, P. J.,** Diurnal variation in glucose tolerance and in insulin secretion in man, *Diabetes,* 22, 333, 1973.
11. **Chadwick, V. S., Modha, K., and Dowling, R. H.,** Mechanism for hyperoxaluria in patients with ileal dysfunction, *N. Engl. J. Med.,* 289, 172, 1973.
12. **Chan, J. C.,** Acid-base and mineral disorders in children: a review, *Int. J. Pediatr. Nephrol.,* 1, 54, 1980.
13. **Chance, R. E., Root, M. A., and Galloway, J. A.,** The immunogenicity of insulin preparations, *Acta Endocrinol.,* 83(Suppl. 205), 185, 1976.
14. **Chevaux, F., Curchod, B., Felber, J. P., and Jéquier, E.,** Insulin resistance and carbohydrate oxidation in patients with chemical diabetes, *Diabetes Metab.,* 8, 105, 1982.
15. **Chiles, R. and Tzagournis, M.,** Excessive serum insulin response to oral glucose in obesity and mild diabetes: study of 501 patients, *Diabetes,* 19, 458, 1970.
16. **Cohen, R. H.,** Disorders of lactic acid metabolism, *Clin. Endocrinol. Metabol.,* 5, 613, 1976.
17. **Cohn, C. and Joseph, D.,** Effects on metabolism produced by the rate of ingestion of the diet "meal eating" versus "nibbling", *Ann. J. Clin. Nutr.,* 8, 682, 1960.
18. **Couropmitree, C., Freinkel, N., Nagel, T. C., Horwitz, D. L., Metzger, B., Rubenstein, A. H., and Hahnel, R.,** Plasma C-peptide and diagnosis of factitious hyperinsulinism: study of an insulin-dependent diabetic patient with "spontaneous" hypoglycemia, *Ann. Intern. Med.,* 82, 201, 1975.
19. **Danowski, T. S., Khurana, R. C., Nolan, S., Stephan, T., Gegick, C. G., Chac, S., and Vidalon, C.,** Insulin patterns in equivocal glucose tolerance test (chemical diabetes), *Diabetes,* 22, 808, 1973.
20. **DeFonzo, R. A.,** Glucose intolerance and aging, *Diabetes Care,* 4, 493, 1982.
21. **Dobbins, J. W. and Binder, H. J.,** Importance of the colon in enteric hyperoxaluria, *N. Engl. J. Med.,* 296, 298, 1977.
22. **Eckel, R. H., McLean, E., Albers, J. J., Cheung, M. C., and Bierman, E. L.,** Plasma lipids and microangiopathy in insulin-dependent diabetes mellitus, *Diabetes Care,* 4, 447, 1982.
23. **Ellenberg, M.,** Interplay of autonomic neuropathy and arteriosclerosis, *N.Y. State J. Med.,* 82, 917, 1982.
24. **Fajans, S. S. and Floyd, J. C.,** Fasting hypoglycemia in adults, *N. Engl. J. Med.,* 294, 766, 1976.
25. **Forbes, J. M.,** The role of the liver in the control of food intake, *Proc. Nutr. Soc.,* 4, 123, 1982.
26. **Frerichs, H. and Creutzfeld, W.,** Hypoglycemia. I. Insulin secreting tumors, *Clin. Endocrinol. Metabol.,* 5, 747, 1976.
27. **Froesch, E. R.,** Disorders of fructose metabolism, *Clin. Endocrinol. Metab.,* 5, 599, 1976.
28. **Gabbay, K. H., Hasty, K., Breslow, J. L., Ellison, R. C., Bunn, H. F., and Gallop, P. M.,** Glycosylated hemoglobins and long term blood glucose control in diabetes mellitus, *J. Clin. Endocrinol. Metabol.,* 44, 859, 1977.
29. **Ganda, O. P., Weir, G. C., Soeldner, J. S., Legg, M. A., Chick, W. L., Patel, Y. C., Ebeid, A. M., Gabbay, K. H., and Reichlin, S.,** "Somatostatinoma": a somatostatin containing tumor of the endocrine pancreas, *N. Engl. J. Med.,* 296, 963, 1977.
30. **Gelbart, D. R., Brewer, L. L., Fajardo, L. F., and Weinstein, A. B.,** Oxalosis and chronic renal failure after intestinal bypass, *Arch. Intern. Med.,* 137, 239, 1977.
31. **Gerich, J. E.,** Somatostatin, *Am. Fam. Physician,* 15, 149, 1977.
32. **Gerich, J. E.,** Somatostatin — its possible role in carbohydrate homeostasis and the treatment of diabetes mellitus, *Arc. Intern. Med.,* 137, 659, 1977.
33. **Hers, H. G., DeWulf, H., and Stalmaus, W.,** The control of glycogen metabolism in the liver, *FEBS Lett.,* 12, 73, 1970.
34. **Horwitz, D. L., Kuzuya, H., and Rubenstein, A. H.,** Circulating serum C-peptide: a brief review of diagnostic implications, *N. Engl. J. Med.,* 295, 207, 1976.
35. **Howanitz, P. J. and Howanitz, J. H.,** Disorders of carbohydrate metabolism, *Clin. Lab. Med.,* 1, 419, 1981.
36. **Hue, L. and Hers, H. G.,** Utile and futile cycles in the liver, *Biochem. Biophys. Res. Commun.,* 58, 540, 1974.
37. **Ingelfinger, F. J.,** Debates on diabetes, *N. Engl. J. Med.,* 96, 1228, 1977.
38. **Jaspan, J. B. and Rubenstein, A. H.,** Circulating glucagon-plasma profiles and metabolism in health and disease, *Diabetes,* 26, 887, 1977.

39. **Kahn, C. R., Mehyesi, K., Bar, R. S., Eastman, R. C., and Flier, J. S.,** Receptors for peptide hormones, *Ann. Intern. Med.*, 86, 205, 1977.
40. **Kaufman, V. and Froesh, E. R.,** Inhibition of phosphorylase-a by fructose-1-phosphate, α-glycerophosphate and fructose-1,6-diphosphate: explanation for fructose induced hypoglycemia in hereditary fructose intolerance and fructose-1,6-diphosphatase deficiency, *Eur. J. Clin. Invest.*, 3, 407, 1973.
41. **Kitabchi, A. E.,** Proinsulin and C-peptide: a review, *Metabolism*, 26, 547, 1977.
42. **Koenig, R. J., Peterson, C. M., Jones, R. L., Saudek, C., Lehrman, M., and Cerami, A.,** Correlation of glucose regulation and hemoglobin A_1 in diabetes mellitus, *N. Engl. J. Med.*, 295, 417, 1976.
43. **Kopf, A., Tchobroutsky, G., and Eschwege, E.,** Serial post-prandial blood glucose levels in 309 subjects with and without diabetes, *Diabetes*, 22, 834, 1973.
44. **Krebs, E. G.,** Protein kinases, in *Current Topics in Cellular Regulation*, Vol. 5, 1972, 99.
45. **Krebs, H. A. and Woodford, M.,** Fructose 1,6-diphosphatase in striated muscles, *Biochem. J.*, 94, 436, 1965.
46. **Larsson, L. I., Holst, J. J., Kuhl, C., Lundqvist, G., Hirsh, M. A., Ingemansson, S., Lindkaer-Jensen, S., Rehfeld, J. F., and Scwartz, T. W.,** Pancreatic somatostatinoma — clinical features and physiologic implications, *Lancet*, 1, 666, 1977.
47. **Lavine, R. L.,** Diabetes and pregnancy, in *New Concepts in Endocrinology and Metabolism: Hahnemann Endocrinology Symposium 1976*, Rose, L. I. and Lavine, R. L., Eds., Grune & Stratton, New York, 1977.
48. **Levy, L. J., Duga, J., Girgis, M., and Gordon, E.,** Ketoacidosis associated with alcoholism in nondiabetic subjects, *Ann. Intern. Med.*, 78, 213, 1973.
49. **McCurdy, D. K.,** Hyperosmolar hyperglycemic nonketotic diabetic coma, *Med. Clin. North Am.*, 54, 683, 1970.
50. **McSherry, E.,** Disorders of acid-base equilibrium, *Pediatr. Ann.*, 10, 302, 1981.
51. **Megyesi, K., Kahn, C. R., Roth, J., and Gorden, P.,** Hypoglycemia in association with extrapancreatic tumors: demonstration of elevated plasma NSILA-s by a new radioreceptor assay, *J. Clin. Endocrinol. Metabol.*, 38, 931, 1974.
52. **Merimee, T. J.,** Spontaneous hypoglycemia in man, *Adv. Intern. Med.*, 22, 301, 1977.
53. **Mirsky, S.,** Adult-onset diabetes, *Primary Care*, 1, 53, 1974.
54. **Morse, W. J., Sidrov, J. J., Soeldner, J. S., and Dickson, R. C.,** Observation on carbohydrate metabolism in obesity, *Metabolism*, 9, 660, 1960.
55. **Munoz-Barragan, L., Rufener, C., Srikant, C., Shannon, A., Beatens, D., and Unger, R. H.,** Immunohistologic identification of glucagon-containing cells in the human fundus, *Horm. Metabol. Res.*, 9, 37, 1977.
56. **Newsholm, E. A. and Start, C.,** *Regulation in Metabolism*, John Wiley & Sons, New York, 1973.
57. **Newmark, S. R.,** Hyperglycemia and hypoglycemia crisis, *JAMA*, 231, 185, 1975.
58. **Ng, W. G., Donnell, G. N., and Alfi, O.,** Prenatal diagnosis of galactosemia, *Lancet*, 1, 43, 1977.
59. **Oimomi, M., Yoshimura, Y., Kubota, S., Kawasaki, T., and Baba, S.,** Hemoglobin A_1 properties of diabetic and uremic patients, *Diabetes Care*, 4, 484, 1982.
60. **Olefsky, J. M. and Reaven, G. M.,** Insulin and glucose response to identical oral glucose tolerance test performed forty-eight hours apart, *Diabetes*, 23, 449, 1974.
61. **Pagliara, A. S., Karl, I. E., Haymond, M., and Kipnes, D. M.,** Hypoglycemia in infancy and childhood. II, *J. Pediatr.*, 82, 558, 1973.
62. **Perejda, A. J. and Uitto, J.,** Nonenzymatic glycosylation of collagen and other proteins: relationship to development of diabetic complications, *Coll. Relat. Res.*, 2, 81, 1982.
63. **Peterson, C. M. and Jones, R. L.,** Minor hemoglobins, diabetic "control" and diseases of post synthetic protein modification, *Ann. Intern. Med.*, 87, 489, 1977.
64. **Peterson, C. M., Jones, R. L., Koenig, R. J., Melvin, E. T., and Lehrman, M. R.,** Reversible hematologic sequelae of diabetes mellitus, *Ann. Intern. Med.*, 86, 425, 1977.
65. **Phanichphant, S., Atichartakarn, V., Nitiyanant, P., and Supasiti, T.,** Renal tubular acidosis with simultaneous lactic and keto acidoses, *Clin. Nephrol.*, 17, 319, 1982.
66. **Pozefsky, T., Colker, J. L., Langs, J. M., and Andres, R.,** The cortisone-glucose tolerance test: the influence of age on performance, *Ann. Intern. Med.*, 63, 988, 1965.
67. **Pyke, D. A.,** Genetics of diabetes, *Clin. Endocrinol. Metabol.*, 6, 285, 1977.
68. **Schein, P. S., DeLellis, R. A., Kahn, C. R., Gorden, P., and Kraft, A. R.,** Islet cell tumors: current concepts and management, *Ann. Intern. Med.*, 79, 239, 1973.
69. **Scrutton, M. C. and Utter, M. F.,** The regulation of glycolysis and gluconeogenesis in animal tissues, *Ann. Rev. Biochem.*, 37, 249, 1968.
70. **Sestoft, L., Bartels, P. D., and Folke, M.,** Pathophysiology of metabolic acidosis, *Clin. Physiol.*, 2, 51, 1982.
71. **Smith, D. A.,** Dietary considerations in diabetes mellitus, *Penn. Med.*, 85, 64, 1982.
72. **Steiner, D. F.,** Insulin today, *Diabetes*, 26, 322, 1977.

73. **Steiner, D. F. and Oyer, P. E.,** The biosynthesis of insulin and a probably precursor of insulin by a human islet cell adenoma, *Proc. Natl. Acad. Sci. U.S.A.*, 57, 473, 1967.
74. **Steiner, G.,** Diabetes and atherosclerosis: an overview, *Diabetes,* 30(Suppl. 2), 1, 1981.
75. **Stout, R. W.,** Blood glucose and atherosclerosis, *Arteriosclerosis,* 1, 227, 1981.
76. **Trivelli, L. A., Ranney, H. M., and Lai, H. T.,** Hemoglobin components in patients with diabetes mellitus, *N. Engl. J. Med.*, 284, 353, 1971.
77. **Unger, R. H.,** Glucoregulatory hormones in health and disease. A teleological model, *Diabetes,* 15, 500, 1966.
78. **Villar-Palasi, C.,** The hormonal regulation of glycogen metabolism in muscle, *Vitam. Horm.*, 26, 65, 1968.
79. **Villar-Palasi, C. and Larner, J.,** Glycogen metabolism and glycolytic enzymes, *Ann. Rev. Biochem.*, 39, 639, 1970.
80. **Vinicor, F., Faulkner, S., and Clark, C. M.,** Reactive hypoglycemia, *Hosp. Med.*, 11, 65, 1975.
81. **Weber, G.,** in *The Biological Basis of Medicine,* Vol. 2, Bittar, E. E. and Bittar, N., Eds., Academic Press, New York, 1968.
82. **Weber, G., Glazer, R. T., and Ross, R. A.,** Regulation of human and rat brain metabolism: inhibitory action of phenylalanine and phenylpyruvate on glycolysis, protein, lipid, DNA and RNA metabolism, *Adv. Enzyme Reg.*, 8, 13, 1970.
83. **Williams, R. H. and Eusinck, J. W.,** Secretion, fates and actions of insulin and related products, *Diabetes,* 15, 623, 1966.
84. **Winder, A. F., Fells, P., Jones, R. B., Kissun, R. D., and Mount, J. N.,** Galactose intolerance and the risk of cataract, *Br. J. Ophthalmol.*, 66, 438, 1982.
85. **Wood, H. G., Katz, J., and Lauden, B. R.,** Estimation of pathways of carbohydrate metabolism, *Biochem. Ztschr.*, 338, 809, 1963.
86. **Wright, P. H. and Makulu, D. R.,** Reactions of proinsulin and its derivatives with antibodies to insulin, *Proc. Soc. Exp. Biol. Med.*, 134, 1165, 1970.
87. **Yallow, R. S. and Berson, S. A.,** Dynamics of insulin secretion in hypoglycemia, *Diabetes,* 14, 341, 1965.
88. **Felig, and Wahren,** *Fed. Proc.*, 33, 1092, 1974.

FURTHER READING

O'Riordan, J. L. H., Ed., Recent Advances in Endocrinology and Metabolism, *Grune & Stratton*, New York, 1975.

Vallance-Owen, J., Ed., *Diabetes,* University Park Press, Baltimore, 1975.

Whelan, W. J., Ed., *Control of Glycogen Metabolism,* Academic Press, New York, 1968.

Dickens, F., Randle, P. J., and Whelan, W. J., Eds., *Carbohydrate Metabolism and Its Disorders,* Academic Press, New York, 1968.

Grave, Gd., Ed., *Early Detection of Potential Diabetics,* Raven Press, New York, 1979.

Katzen, H. M. and Mahler, R. I., Eds., *Advances in Modern Nutrition,* Vol. 2, John Wiley & Sons, New York, 1978.

Kryston, L. J. and Shaw, R. A., Eds., *Endocrinology and Diabetes,* Grune & Stratton, New York, 1975.

O'Riordan, J. L. H., Ed., Recent Advances in Endocrinology and Metabolism, *Grune & Stratton*, New York, 1975.

Vallance-Owen, J., Ed., *Diabetes,* University Park Press, Baltimore, 1975.

Whelan, W. J., Ed., *Control of Glycogen Metabolism,* Academic Press, New York, 1968.

Chapter 5

ABNORMALITIES OF NUCLEIC ACID AND PURINE OR PYRIMIDINE SYNTHESIS AND METABOLISM

I. INTRODUCTION

Nucleoproteins are characterized by the presence of nucleic acid prosthetic groups usually attached to a basic protein such as histone or protamine. They constitute a large part of the nuclear material of the cell. Nucleic acids are essential components of life — they participate in the molecular mechanisms of vital processes: how genetic information is stored, replicated, and transcribed. All living cells contain nucleoprotein and some of the simplest systems seem to be built almost entirely from nucleoprotein, such as viruses. Chromatin is largely composed of nucleoproteins, indicating that these compounds are involved in cell division and the transmission of hereditary factors. Any abnormality in the mechanism of nucleoprotein formation is, therefore, followed by an alteration in cell growth and reproduction. Abnormalities in nucleoprotein synthesis are connected with folic acid or vitamin B_{12} deficiency, radiation-induced mutations, or the radiomimetic effects of many chemicals such as nitrogen mustards. These abnormalities do not only affect chromosome structure and gene action, but also suppress mitosis.

The building stones of nucleic acid are the mononucleotides. These molecules also play an essential role in intermediary metabolism and energy transformation as part of coenzymes. Mononucleotide-derived coenzymes serve in many oxidation-reduction reactions and in the metabolism of many cellular components such as sugar derivatives, amines, and several small- and large-molecular-weight organic acids. The various purine and pyrimidine bases occurring in nucleotides are derived by appropriate substitution on the ring structure of the parent substances (Figure 1). There are three main pyrimidine bases: cytosine, thymine, and uracil and two purine bases: adenine and guanine. When a purine or pyrimidine base is attached to a pentose molecule by glycosidic linkage and a phosphate is added, nucleotide is formed. There are two types of nucleotides: containing either ribose or 2-deoxyribose; these are the ribonucleic or deoxyribonucleic acids. When a nucleotide is treated with an appropriate enzyme, nucleotidase, the phosphoric acid is removed and a nucleoside is formed.

The biosynthesis of ribo- and deoxyribonucleotides is an essential process and these compounds are direct precursors of the nucleotide coenzymes, DNA and RNA. In the production of mononucleotides the pathway of pyrimidine and purine synthesis is vital. Almost all living organisms can synthesize these bases from very simple precursors. In man, several inborn errors of metabolism cause defects in the regulation of the synthesis and catabolism of purine nucleotides leading to disease, such as xanthine oxidase deficiency, nucleoside phosphorylase, adenosine deaminase immune deficiency, and Lesch-Nyhan syndrome.[11,19-21,33,65,67] Genetic abnormality connected with pyrimidine deficiency is responsible for the hereditary orotic aciduria.[25,30,87,88] Excessive quantities of uric acid in the body can produce gout with various manifestations such as acute and chronic arthritis,[2,5,9,13,26,27,41,54,83,89,90,102,103] hyperuricemia, deposition of sodium urate tophi in tissues, and nephropathy.[15,34,63,73,96,101] Hypouricemia also occurs in about 1% of the population.

II. SYNTHESIS AND CATABOLISM OF PYRIMIDINES

Nucleoproteins are degraded to nucleic acids and further to purine and pyrimidine bases constituting nucleic acids. Intake of preformed purines and pyrimidines from the diet is not essential since the human organism is capable of producing them. These compounds are

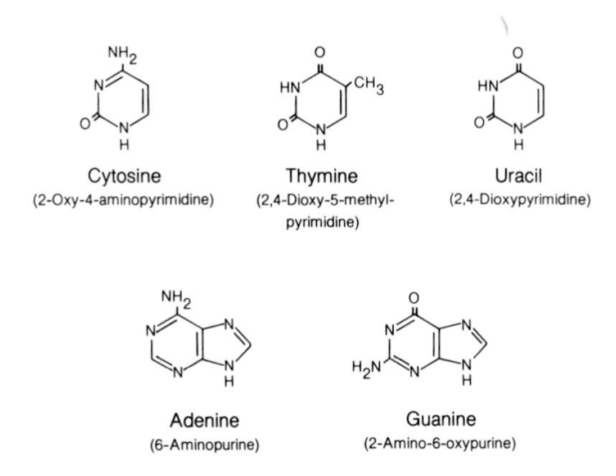

FIGURE 1. Structure of various pyrimidine and purine bases.

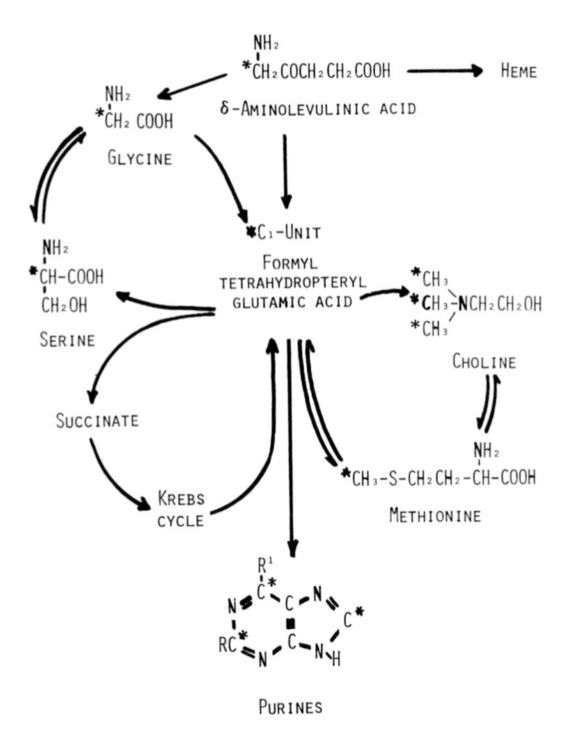

FIGURE 2. Schematic presentation of the incorporation of one carbon unit into purines and other body constituents.

largely derived from endogenous components. The nitrogens of purine bases come mainly from various amino acids such as glycine, aspartic acid, and glutamine broken down to ammonium salts;[75,80] the carbon atoms of the skeleton are derived from respiratory carbon dioxide, formate, serine, and δ-amino-levulinic acid through various transformations (Figure 2). Pyrimidines are formed from carbamoylaspartic acid, which is an intermediate compound in the urea cycle. Formate and serine contribute further carbon atoms to the biosynthetic process, and vitamin B_{12} is involved directly in this mechanism. The regulation of pyrimidine and purine synthesis is controlled by feedback factors influencing the activity of enzymes taking part in these processes.

There are two alternate pathways in the synthesis of pyrimidine nucleotides: *de novo* synthesis by a route exclusive of orotic acid synthesis (Figure 3), and the salvage pathway

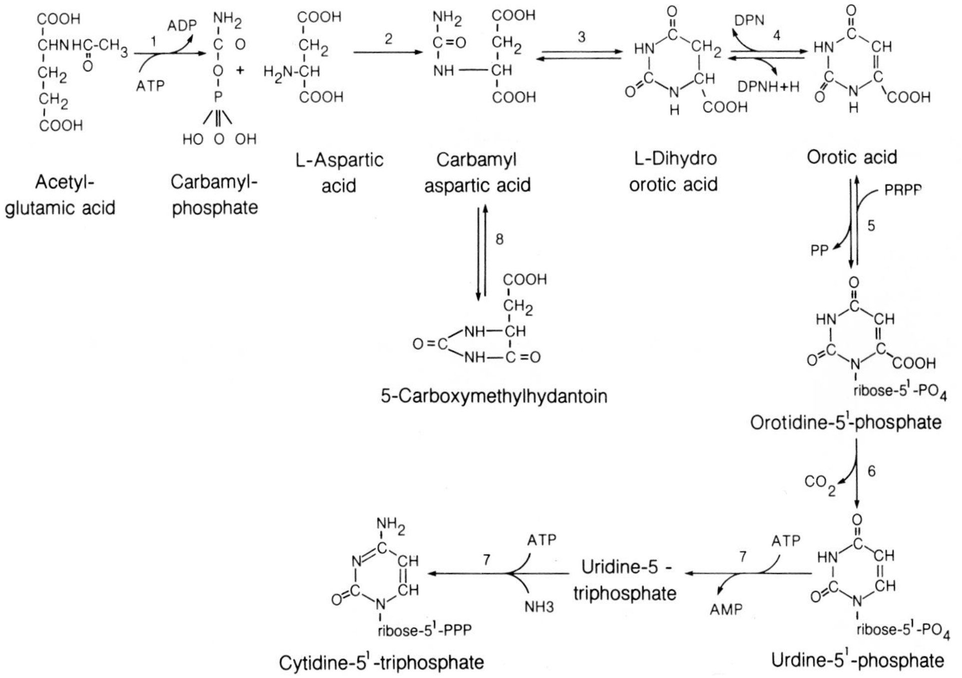

FIGURE 3. Pathway of pyrimidine nucleotide synthesis. From acetyl-glutamic acid in presence of ATP, carbamyl phosphate synthetase (1) produces carbamyl phosphate which is converted in the presence of aspartic acid by aspartate transcarbamylase (2) to carbamyl aspartic acid. This is then reversibly transformed to L-dihydroorotic acid by dihydroorotase (3) and further oxidized to orotic acid in presence of NAD by dihydroorotic dehydrogenase (4). Orotic acid and 5-phosphoribosyl-1-pyrophosphate form orotidine-5′-phosphate catalyzed by orotidylic 1-pyrophosphorylase (5). Decarboxylation by orotidylic acid decarboxylase (6) leads to uridine-5′-phosphate and kinase, and transphosphorylation reactions (7) produce further uridine and cytidine nucleotides. Side reaction catalyzed by 5-carboxymethylhydantoinase (8) leads to the formation of 5-carboxymethylhydantoin.

by the utilization of preformed pyrimidine bases taken from dietary sources or from nucleotide turnover (Figure 4). The simple precursors of the *de novo* synthesis are carbon dioxide, ammonia, and β-alanine. Uracil is the precursor of uridine 5′-phosphate in the salvage pathway. The coordination of these pathways is probably related to a feedback mechanism and to changes in the primary structure of participating enzymes through association-dissociation processes into subunits or to changes in conformation.[52] During metabolism, pyrimidines are converted to end products which are removed from the organism. The breakdown of uracil leads to the formation of carbamyl-β-alanine and β-alanine; metabolism of thymine or dihydrothymine results in the appearance of β-ureidoisobutyric and β-aminoisobutyric acid (Figure 5).

III. SYNTHESIS AND CATABOLISM OF PURINES

Feeding experiments in animals using radiolabeled precursors and subsequent degradation studies revealed that various purines have common precursors in their biosynthesis.[47,49,104] Small-precursor molecules such as carbon dioxide, formate, glycine, or ammonia are added stepwise when the purine ring is formed (Figure 6). The precursors are amply supplied by any diet and therefore the *de novo* purine biosynthesis is largely independent from dietary purine sources. Cofactors such as folic acid are, however, essential in this process. Some vitamin deficiencies may cause severe impairment in cellular purine production.

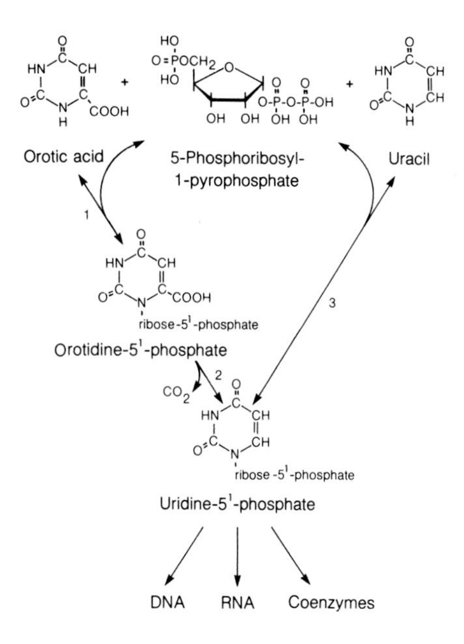

FIGURE 4. Pathways of salvage synthesis of pyrimidine nucleotides. Orotidine-5′-phosphate is formed by the reversible pyrophosphorolysis between orotic acid 5-phosphoribosyl-1-pyrophosphate catalyzed by orotidylic pyrophosphorylase (1), then decarboxylation by orotidylic decarboxylase (2) leads to uridine-5′-phosphate. This nucleotide may be formed from uracil and 5-phosphoribosyl-1-pyrophosphate (3), a reaction analogous to that catalyzed by orotidylic pyrophosphorylase.

Cytosine —($-NH_3$, 4)→ Uracil —($+2H$, 1)→ Dihydrouracil —($+H_2O$, 2)→ β-Ureidopropionic acid (Carbamyl β-alanine) —($+H_2O-CO_2-NH_3$, 3)→ β-Alanine

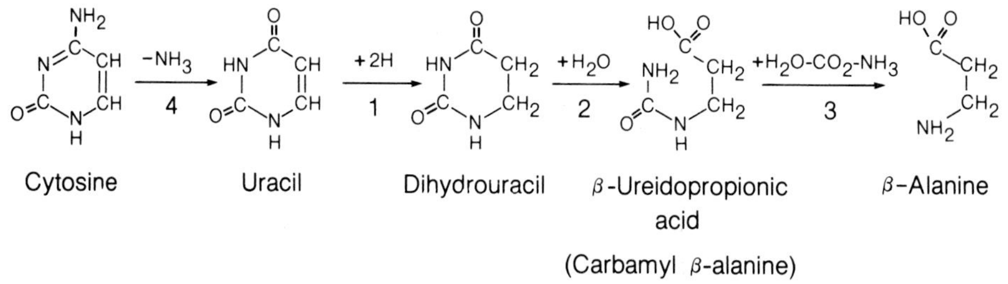

FIGURE 5. Metabolic breakdown of pyrimidine nucleotides. The pathway involves the reduction of the pyrimidine ring to a dihydropyrimidine derivative catalyzed by a dehydrogenase (1) followed by the opening of the ring by hydrolysis (2) and further decomposition (3) with the loss of ammonia and CO_2. Deamination (4) and other processes provide an interaction between various nucleotides in catabolic pathways.

Specifically, in the biosynthesis of purines and pyrimidines and subsequently in the formation of ATP, coenzymes and nucleic acids phosphoribosyl pyrophosphate play an essential role.[36,74] The formation of these molecules is dependent on two factors: the availability of ribose-5-phosphate precursor, and phosphoribosyl pyrophosphate synthetase enzyme. The proportion of these factors is different in various tissues, representing various degrees of regulation in the synthesis of purines and related compounds. The regulation of purine metabolism is also associated with feedback mechanism;[17,18,29,45,65] adenosine monophosphate and guanosine monophosphate exert inhibition on several enzymes (Figure 7).

The salvage pathway contributes to a great extent to purine metabolism.[23] In this pathway, free purines such as adenine, hypoxanthine, or guanine originate from dietary nucleoproteins by catabolisms; they are incorporated into tissue nucleic acids catalyzed by hepatic enzymes.

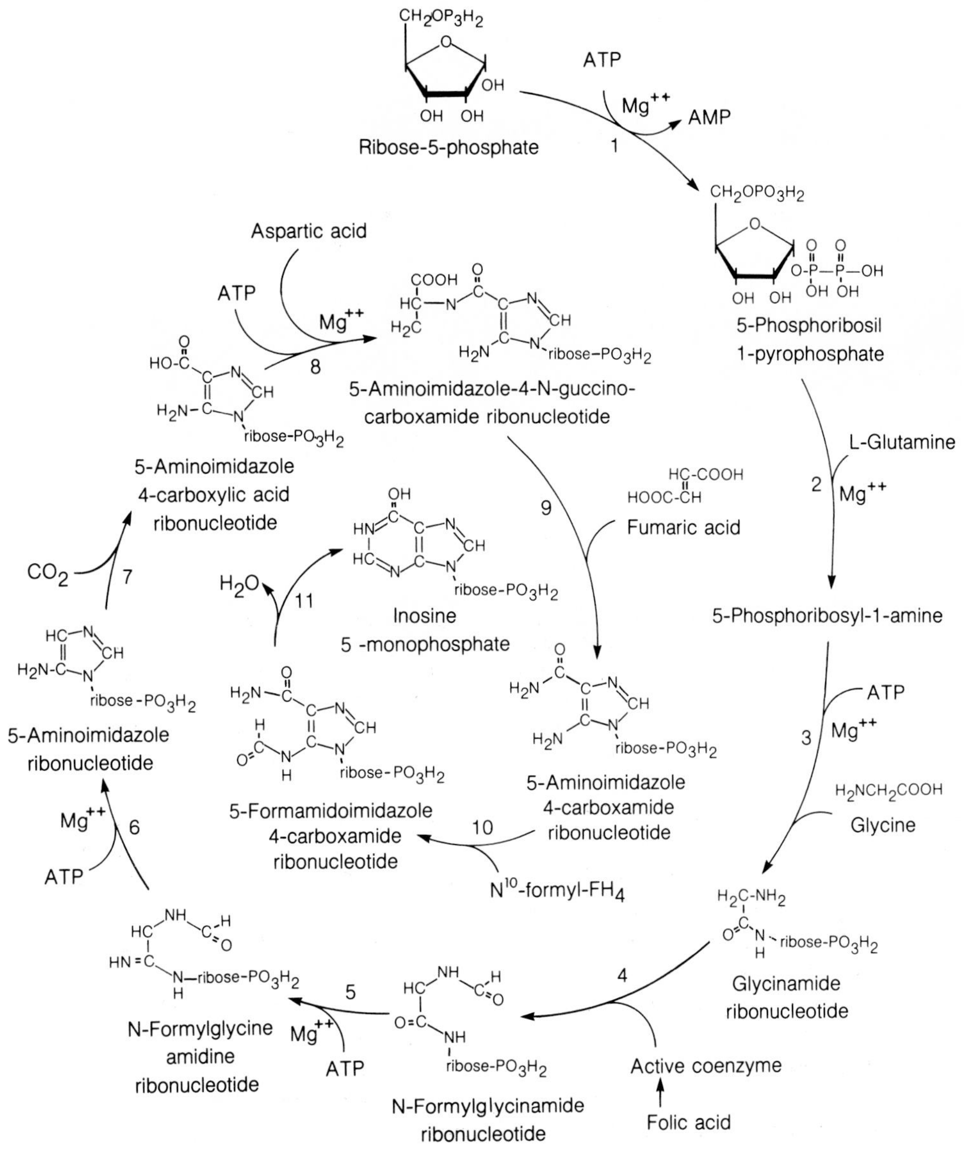

$CH_2OP_3H_2$
Ribose-5-phosphate
ATP
Mg^{++}
AMP
1
$CH_2OPO_3H_2$
5-Phosphoribosil
1-pyrophosphate
L-Glutamine
2 Mg^{++}
5-Phosphoribosyl-1-amine
ATP
3 Mg^{++}
H_2NCH_2COOH
Glycine
$H_2C\text{-}NH_2$
Glycinamide
ribonucleotide
4
Active coenzyme
Folic acid
N-Formylglycinamide
ribonucleotide
5
Mg^{++}
ATP
N-Formylglycine
amidine
ribonucleotide
Mg^{++} 6
ATP
5-Aminoimidazole
ribonucleotide
CO_2 7
5-Aminoimidazole
4-carboxylic acid
ribonucleotide
Aspartic acid
ATP
Mg^{++}
8
5-Aminoimidazole-4-N-guccino-
carboxamide ribonucleotide
9
Fumaric acid
5-Aminoimidazole
4-carboxamide
ribonucleotide
10
N^{10}-formyl-FH_4
5-Formamidoimidazole
4-carboxamide
ribonucleotide
H_2O 11
Inosine
5 -monophosphate

FIGURE 6. Biosynthesis of purine nucleotides. Sequences start from ribose 5-phosphate; ribose-5-phosphate pyrophosphokinase (1) converts it to 5-phosphoribosyl-1-pyrophosphate. This reacts with glutamine catalyzed by phosphoribosyl pyrophosphate aminotransferase (2). 5-Phosphoribosyl-1-amine is converted in the presence of glycine to glycinamide ribonucleotide catalyzed by glycinamide ribonucleotide synthetase (3). This reaction provides four atoms in the imidazole ring of the purine base. The imidazole ring is completed by formylation of the amino group by glycinamide ribonucleotide formyltransferase (4) in presence of folic acid, to yield formylglycinamide ribonucleotide. From glutamine, another nitrogen of the purine ring is derived catalyzed by formylglycinamidine ribonucleotide synthetase (5). The imidazole ring is then closed and 5-aminoimidazole ribonucleotide is formed by aminoimidazole ribonucleotide carboxylase (7) and further converted to 5-aminoimidazole-4-*N*-succinocarboxamide ribonucleotide by synthetase enzyme (8) in presence of aspartic acid. 5-Aminoimidazole-4-*N*-succinocarboxamide is catabolized to 5-aminoimidazole-4-carboxamide ribonucleotide and fumaric acid by adenyl succinase (9). In the next step, 5-amino-4-imidazole carboxamide ribonucleotide is converted to 5-formamido-4-imidazole carboxamide ribonucleotide by formyl transferase (10) using N^{10}-formyltetrahydrofolic acid, and finally inosine-5′-monophosphate is formed by ring closure catalyzed by isosinidase (11). Inosine 5′-monophosphate is the intermediate for the synthesis of adenosine 5′-monophosphate and guanosine 5′-monophosphate (see Figure 7).

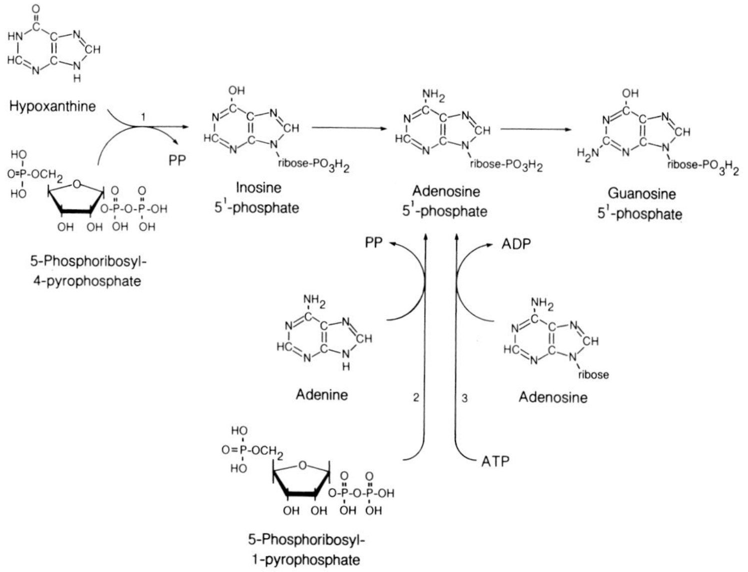

FIGURE 7. Pathways of salvage synthesis of purine nucleotides. Hypoxanthine reacts with 5-phosphoribosyl-1-pyrophosphate to yield inosine-5′-monophosphate and pyrophosphate catalyzed by pyrophosphorylase (1). Inosine 5′-monophosphate is an intermediate in the synthesis of adenosine 5′-monophosphate and guanosine 5′-monophosphate. Adenosine 5′-monophosphate is also formed from adenine by pyrophosphorylase (2) or from adenosine by kinase (3) action.

In children suffering from Lesch-Nyhan syndrome the daily uric acid excretion in the urine is around 47 mg/kg body weight.[45] In normal children this is about 10 mg/kg; the difference represents the loss of the amount of uric acid which is normally retained for the resynthesis of purine derivatives through the salvage pathway.

The mechanism regulating the balance between the *de novo* and salvage synthesis is not known. In some tissues the level of various enzymes regulating both the *de novo* and salvage pathways is adequate. Other tissues only contain the salvage processes, such as leukocytes and bone marrow. The efficiency of nucleotide and nucleic acid production and the quantity produced in these organs are, therefore, dependent entirely on the available exogenous purine intake. The main organ of purine biosynthesis is the liver to a minor extent muscle contributes to the endogenous purine pool. In the liver, purine nucleotides are dephosphorylated and nucleosides are transformed to bases which are transported to various other tissues. Further conversion of these bases, particularly in erythrocytes, results in the formation of hypoxanthine and xanthine.[16] These are released from the red blood cells and distributed in other tissues where they are used for the synthesis of nucleotides by salvage reaction. Many pathways exist in these processes for the reconstruction of the nucleotide levels in normal tissues (Figure 7).[81]

In the degradation process of polynucleotides DNA or RNA, two types of mononucleotides are formed — 3′- and 5′-mononucleotides, depending on the nature of the enzymes involved. Different enzyme mechanisms regulate dephosphorylation, deamination, and the hydrolysis of the N-glycosidic bond in various derivatives. Furthermore, there are many processes of interconversion between various purine nucleotides (Figure 8), but finally, by oxidative processes, the metabolism of purines leads to uric acid (Figure 9). Purine metabolism is also connected with the formation of histidine, pteridines, and flavins. Histidine is an essential amino acid; riboflavin and folic acid belong to the vitamin group and has to be taken from the diet rather than synthesized *de novo*.

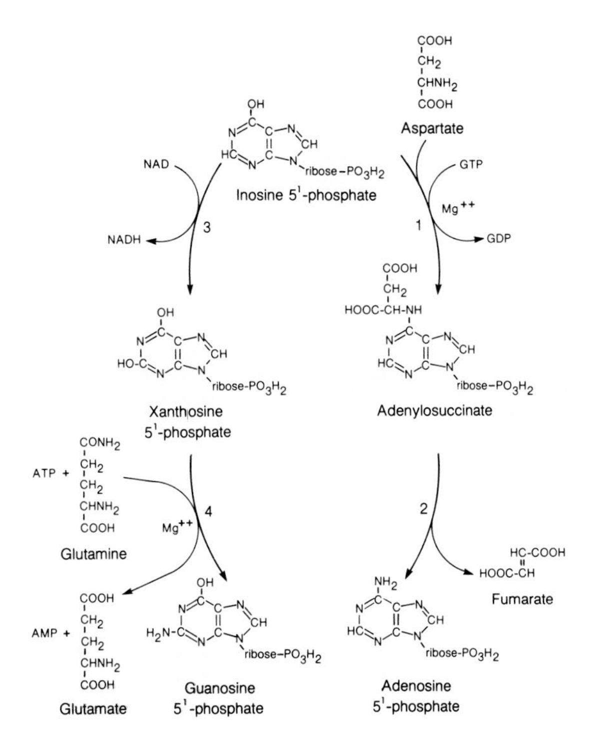

FIGURE 8. Interconversion between purine nucleotides. Inosine 5′-monophosphate reacts with aspartic acid catalyzed by adenylosuccinic synthetase (1). Adenylosuccinic acid is further converted to adenosine-5′-monophosphate and fumaric acid catalyzed by adenylosuccinate lyase (2). Inosine 5′-monophosphate is also transformed to guanosine 5′-monophosphate in two steps: oxidation to xanthosine 5′-monophosphate by inosinic dehydrogenase (3) followed by the conversion to guanosine 5′-monophosphate in the presence of ATP and glutamine catalyzed by guanosine 5′-monophosphate synthetase (4).

In general terms, diseases of purine and pyrimidine metabolism are connected with marked alterations in the rate of control of pyrimidine or purine synthesis.[17] Some of them are associated with inherited disorder in metabolism, others with the impairment of elimination. Disturbances of uric acid metabolism in Lesch-Nyhan syndrome lead to excessive uric acid production, while in gout, uric acid levels are elevated in the blood and urine due to decreased elimination.[33] Secondary metabolic pathways leading to further transformation of orotic acid is blocked in orotic aciduria due to enzyme defects and results in increased excretion.

IV. BLOOD URIC ACID LEVEL

Uric acid formed during the metabolism of endogenous or exogenous nucleic acids is excreted unchanged.[4,32,35,77] In the blood, uric acid is not bound to proteins; it is distributed between serum and red blood cells. The concentration of uric acid in other body fluids is the same as in blood. Uric acid filters through the glomerulus freely, but very efficiently; about 90% is reabsorbed through the tubules. When renal disease is apparent, interference with glomerular filtration is connected with elevated blood levels.[34] In contrast, if kidney disease causes defective tubular reabsorption, low serum levels of uric acid are produced.

In human blood, the normal uric acid concentration is low in children and it rises after puberty to adult levels;[8] it is greater in men (150 to 450 mg/ℓ) than in women (130 to 400 mg/ℓ). The higher level is probably associated with a lower clearance of urate in men than in women. The uric acid concentration in the blood may also be controlled by hormones.

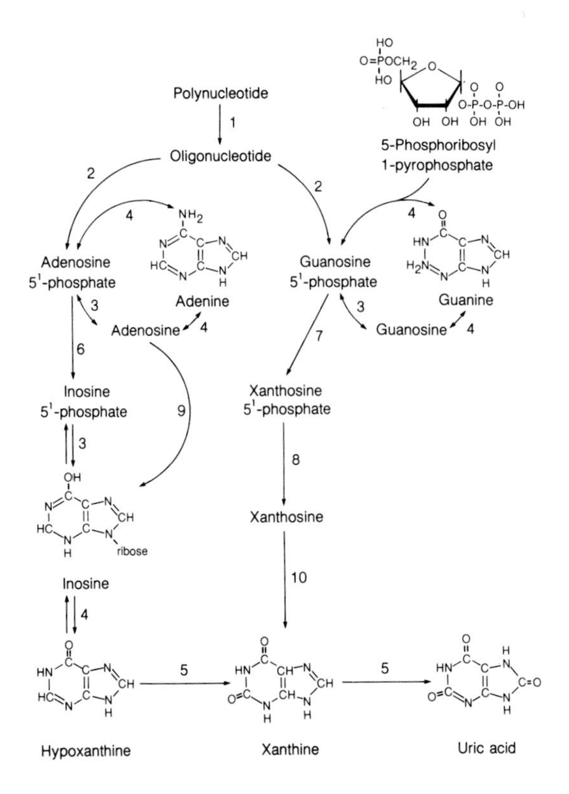

FIGURE 9. Enzyme defects in gout. Enzymatic breakdown of polynucleotides occurs through the action of various nucleases (1). Oligonucleotides are further metabolized by phosphodiesterases (2) to yield 3'- and 5'-mononucleotides such as adenosine 5'-monophosphate or guanosine 5'-monophosphate. These are cleaved by specific nucleoside 5'-phosphatases (3) or by nonspecific phosphatases resulting in corresponding purine nucleoside and phosphate. The purine nucleoside is further converted to free purine base and ribose or ribose 1-phosphate catalyzed by nucleoside phosphorylase (4). Hypoxanthine is then oxidized by xanthine oxidase (5) to yield xanthine, or further to uric acid. In addition to these general reactions, adenosine 5'-monophosphate is converted to inosine 5'-monophosphate by adenylic deaminase (6). Guanosine 5'-monophosphate is directly deaminated by guanase (7) to yield xanthosine 5'-monophosphate, which is further converted to xanthine by xanthosine phosphorylase (8). In contrast to gout, lack of adenosine deaminase (9) or nucleoside phosphorylase (4) leads to hypouremia. Both latter disorders are connected with immune deficiency.

In men, higher rates of 17-ketosteroid secretions show a close correlation with higher blood uric acid levels. After menopause the hormonal control is reduced in women and consequently blood uric acid level is increased.

The amount of uric acid in the blood is influenced by many other factors.[10,22,26,68] It is elevated in kidney disease and in any abnormal conditions when rapid tissue metabolism occurs. Various conditions such as leukemia, hemolytic and pernicious anemia, massive infarct, eclampsia, or polycythemia are associated with elevated serum uric acid levels. These types of secondary hyperuricemia are seldom associated with permanently increased values characteristic of the clinical gout.[36,44,98]

V. GOUT

A. Development of the Disease

This disease is presumably due to an inherited metabolic error, characterized by elevated

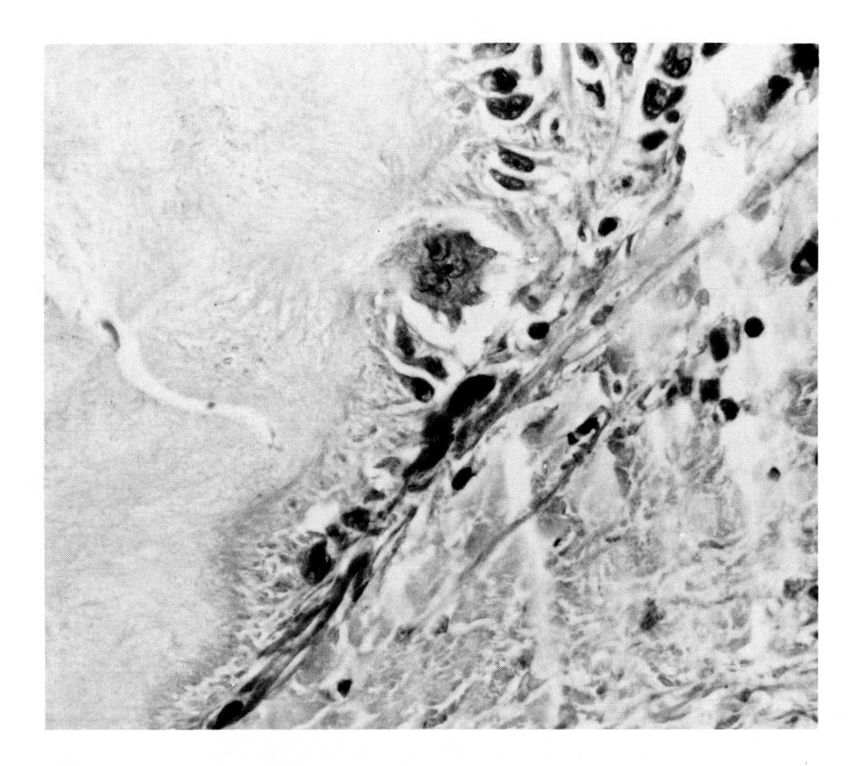

PLATE 1. Gout. Central deposits of urate crystals surrounded by a granulomatous reaction composed of histiocytes and giant cells of foreign body type.

levels of sodium urate in the blood and uric acid in the urine.[2,14,41,89,90,102,103] Three types of gout are known: asymptomatic hyperuricemia, acute, and chronic gouty arthritis.[13,27,57,99] The disease may be asymptomatic at the start, or for a long while, but in some individuals clinical manifestations and recurrent attacks occur resembling those of acute arthritis. In asymptomatic hyperuricemia the blood uric acid level is raised, but arthritic symptoms are not yet apparent. This stage of the disease starts at puberty in the gouty male patient, while in female patients recurrent attacks are frequently delayed until the menopause. In men with the genetic trait for gout, sometimes hyperuricemia remains the only sign of disease and there may be any further abnormality or downhill progress. Women carrying the trait usually do not have hyperuricemia and gout only appears several years after menopause, if it occurs at all. The attacks are usually triggered by the precipitation of sodium urate crystals in the joint and neighboring tissues.[79] (See Plate 1.)

In primary gout or acute gouty arthritis the arthritic conditions may become chronic; more urate crystal deposits (tophi) are formed around the joints and cartilages.[100] These tophi can be characterized by histological methods. There are no marked changes of plasma uric acid levels preceding an acute attack of gouty arthritis, and no change after the attack; urinary excretions may be elevated.[94] The rise is probably mediated by the uricosuric action of corticosteroids which are produced during the inflammatory process of the gout. Increased formation of uric acid is also a consequence of the enhanced leukocyte turnover and leukocytosis is associated with inflammation. Thus, the miscible pool of uric acid in gouty patients is much larger than in normal individuals. The excess is mainly stored in the tophi and stored urate is in equilibrium with the urate present in body fluids. The administration of uricosuric drugs can decrease the urate content of the storage without a significant reduction in the blood. This can be explained by constant intermixing between parts of the pool.

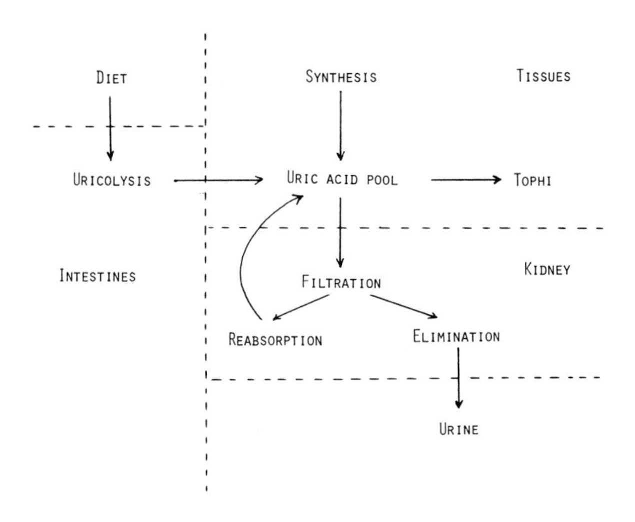

FIGURE 10. Role of the kidney in uric acid metabolism.

The kidney is often involved in the clinical manifestation of gout (Figure 10).[96] Renal stones, partially or entirely containing uric acid, may be formed at any stage of the gout.[5,15,37,93,94,101,106] Very rarely, the kidney injury is associated with the symptomless period of hyperuricemia.[22,95] In primary gout, renal conditions usually develop involving particularly the tubules and the interstitial tissue. In these structures the accumulation of crystalline urate often produces renal lithiasis. Vascular and glomerular damage of the kidney and hypertension are also common in gout.

Chronic gouty arthritis may develop as a residue and continuation of an acute attack, but it may manifest independently in a previously uninvolved joint.[105] The time interval from the initial attack to the start of the chronic symptoms is variable, ranging from 3 to over 40 years. In chronic conditions acute attacks may occur involving the chronically affected sites, but in the late phase they may completely disappear. In addition, renal dysfunction is present in many patients, and the occurrence of urate crystals in the medulla may be connected with vascular or pyelonephritic changes.[96] The incidence of kidney stones is much greater in this condition than in either the acute phase or the asymptomatic disease.

Increased urate levels in the blood may occur in other diseases where there is an abnormally increased turnover of nucleic acids.[43] In some hematopoietic diseases, mainly in chronic and acute leukemia, polycythemia, chronic hemolytic anemia, or myeloid metaplasia, occasional gouty attacks may also occur.[3,38,44,76] Several diseases are connected with impairment of purine biosynthesis and as a consequence of this defect hyperuricemia may develop, such as in nongouty type arthritis,[27,35,43,82,83] atherosclerosis,[8,40,58,69] diabetes mellitus,[10,66,78] and certain types of glycogen storage disease. [1,28,46,53]Secondary gout occurs in subjects maintained on a total fast for long periods. In this case, the development of the gouty condition is probably due to the increased nucleic acid catabolism and the presence of excessive amounts of breakdown products.[61]

B. Mechanism of Pathogenesis

There are various mechanisms elucidating the pathogenesis of gout.[54,55,92] The majority of cases seem to be associated with an inherited defect of purine metabolism, leading to an excessive degree of glycine transformation to uric acid (Figure 6).[52] In gouty patients greater amounts of glycine are incorporated into purine synthesis than into the other pathways utilizing this amino acid. Purine synthesis is regulated at several steps; the rate-limiting reaction in this mechanism is the combination of glutamine with phosphoribosyl pyrophosphate to form phosphoribosylamine which is catalyzed by phosphoribosyl pyrophosphate-aminotransferase. The regulation of this reaction may represent the most important event in the synthesis of purines from precursors.[50,91] The rate of phosphoribosylamine synthesis, in turn, is controlled by the concentrations of various substrates such as glutamine and phosphoribosyl pyrophosphate.[7,80] Any metabolic event resulting in enhanced phosphoribosyl pyrophosphate leads to an increased purine synthesis. The level of intracellular glutamine and the production of ribose which brings about an enlargement of the tissue phosphoribosyl pyrophosphate pool can also serve the rate-limiting role in purine biosynthesis. Many patients showing significant overproduction of uric acid have a phosphoribosyl aminotransferase deficiency.[51,56,65] The decreased enzyme level may play an important role in the biochemical mechanism of inherited gout.

Other conditions influencing intermediates of this rate-limiting step also affect uric acid production, In Von Gierke's disease, glycogen storage disease Cori type I (Chapter 1), the accumulated glucose 6-phosphate can be transformed to pentose phosphate and hyperuricemia can develop.[1,28,53] Similarly, patients with an enhanced glutathione reductase activity have shown elevated blood urate levels. This enzyme catalyzes glutathione reduction; it requires NADPH, which is primarily produced by the pentose phosphate pathway. When glutathione reductase is enhanced, pentose phosphate biosynthesis is accelerated in order to maintain the supply of NADPH. Consequently, the increased pentose phosphate level will stimulate phosphoribosyl pyrophosphate biosynthesis.

The activity of phosphoribosyl pyrophosphate-aminotransferase can be blocked by adenosine di- and triphosphate and less effectively by guanosine mono- and diphosphate or inosine monophosphate, thereby suggesting an important feedback inhibition in the regulation of purine biosynthesis. Adenosine monophosphate and guanosine monophosphate also inhibit several other enzymes in the purine biosynthesis pathway. Moreover, in the salvage pathway the incorporation of free purines into nucleotides is catalyzed by various hepatic enzymes such as adenine phosphoribosyl transferase and hypoxanthine-guanine phosphoribosyl transferase. Both enzymes require phosphoribosyl pyrophosphate as the ribose phosphate precursor. Through the formation of this compound and in the presence of adequate substrate levels, these enzymes therefore influence uric acid production via feedback mechanisms.

The various uricosuric drugs lower the levels of sodium urate by diminishing the renal tubular reabsopriton and thus increasing excretion. Allopurinol is an inhibitor of xanthine oxidase (Figure 11), acting as an antimetabolite to hypoxanthine and xanthine.[12,34,63,85] This drug blocks the available receptor sites for the natural substrate on the enzyme molecule. Allopurinol decreases uric acid production, but also increases the levels of hypoxanthine and xanthine in the blood and urine.[24] These compounds, however, are more soluble than uric acid so that the complications related to high blood and tissue uric acid level are abolished or at least diminished. Furthermore, allopurinol, by decreasing the formation of uric acid, reduces the production of stones in the kidney.

Many patients with primary gout have a lower urate binding capacity in the plasma than normal subjects, which may be responsible for the tendency of these patients to precipitate and deposit sodium urate crystals in affected tissues.[12,66] These crystals can be recovered from the synovial fluid obtained from the joints of these patients. The inflammatory process and the pain accompanying acute attacks of gout are associated with their presence in the affected tissues. The response is attributed to the size and the needle shape of the urate crystals. Acute gout cannot be produced by the oral, intravenous, or subcutaneous administration of uric acid into patients. Similarly, amorphous particles provoke no or little inflammation.

Although the involvement of the kidney in the manifestation of acute and chronic gouty arthritis is established, it is believed that the renal involvement is secondary since gout is an inherited defect of purine metabolism.[5,15] The mechanisms responsible for uric acid

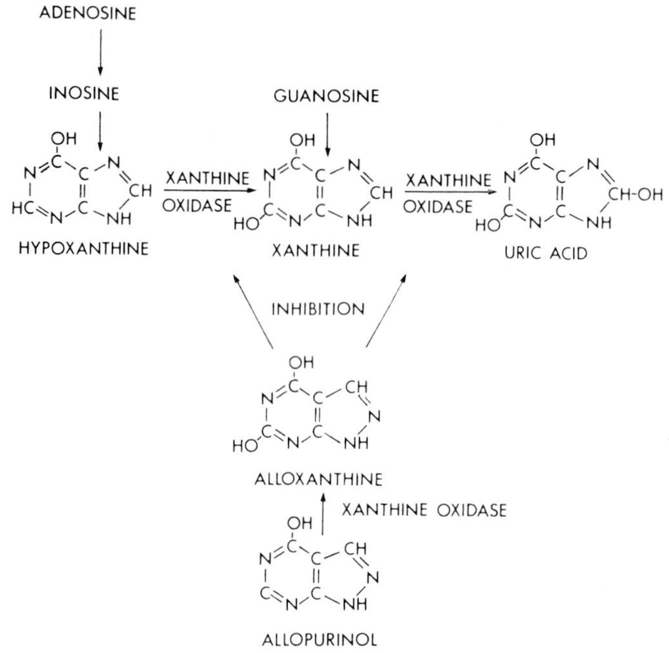

ADENOSINE

INOSINE

GUANOSINE

HYPOXANTHINE

XANTHINE OXIDASE

XANTHINE

XANTHINE OXIDASE

URIC ACID

INHIBITION

ALLOXANTHINE

XANTHINE OXIDASE

ALLOPURINOL

FIGURE 11. Mechanism of allopurinol inhibition of hypoxanthine metabolism.

elimination are normal but inadequate due to overload. There are, however, observations that in some cases the primary defect may be in the kidney, involving lack of secretion of uric acid associated with impaired enzymic transport by the renal tubules. Although other functions of the kidney are normal, the consequence of this deficiency is a lowered excretion of urate. This type of primary renal gout may be associated with progressive kidney failure incident to glomerulonephritis or to some other destructive process. Impairment of kidney function and the decline of glomerular filtration occurs frequently as complications in later stages.

An acute attack of gout is usually connected with the precipitation of uric acid crystals. The solubility of urate decreases with decreasing pH. In normal conditions, increased anaerobic glycolysis results in an elevated lactic acid formation, and lactic acid accumulation is the more likely factor in acidification of tissues. During muscular activity lactic acid accumulates in the muscle and in the synovial fluid before it is transferred into the blood. This lowers the pH of the tissue and, therefore, a pH gradient is generated between tissues and blood. Urate salts are soluble at higher pH, but when the pH is lowered they precipitate. Precipitation of urate crystals in the synovial fluid initiates an inflammatory reaction accompanied by lymphocyte accumulation. Since in lymphocytes the rate of anaerobic glycolysis is increased, lymphocyte accumulation triggers a vicious circle which further potentiates the precipitation of urate crystals.

The accumulation of uric acid can also derive from overproduction or decreased excretion of purines due to increased tubular reabsorption. Inhibition of transfer through the epithelial membrane offers another explanation for the increased urate concentration. Changes in urinary uric acid content are related to blood levels. Moreover, the fact that uric acid excretion in the saliva of patients with gout is lower than in normal individuals indicates that an inhibitor of urate carrier may be present in the organism of these patients.

The actual accumulation of uric acid may be related to reduced kidney function.[5,73] These compounds and many other catabolites of purine metabolism are normally removed by glomerular filtration. It is suggested that precipitation of sodium urate in crystal form occurs from supersaturated body fluids during the inflammatory response which involves local leukocytosis and increased production of lactate, thus lowering the local pH. A further drop in pH results in more crystal formation. The effect of chronic alcoholism potentiating hyperuricemia may be connected with this mechanism.[61,86] Ethanol is metabolized by alcohol dehydrogenase which uses NADPH cofactor. The reduction of this cofactor shifts the conversion of pyruvate to lactate. The levels of hyperlacticacidemia are adequate to suppress the excretion of uric acid through the kidney and thus enhances the possibility of crystal formation.

VI. OROTIC ACIDURIA

This is a rare genetic disease, associated with the excessive urinary excretion of orotic acid.[25,30,84,97] The clinical symptoms consist of retarded development and growth and megaloblastic anemia which is not responsive to the normal hematological treatment. The biochemical defect is attributable to the decreased activity of orotidylic pyrophosphorylase and orotidylic acid decarboxylase in the erythrocytes (Figure 3). Due to these enzyme defects the metabolic control mechanism also becomes faulty and orotic acid overproduction occurs. The low plasma level of orotic acid excludes the involvement of a renal tubular defect causing excessive urinary excretion.

Similar conditions manifest following the administration of the antineoplastic compound 6-azauridine. This compound acts as an antagonist to orotidylic acid decarboxylase and produces a transient oritic aciduria in otherwise normal individuals. When these people were given uridine and cytidine or the corresponding nucleotides, the excess excretion of orotic acid retuned to normal. The pyrimidine nucleosides or nucleotides abolish orotic acid synthesis, acting as feedback inhibitors. When patients suffering from orotic aciduria were treated with a mixture of yeast cytidylic and uridylic acids, these nucleotides caused a marked reduction of orotic acid excretion. Supplying the end products of pyrimidine synthesis, the disease condition can be reversed. Due to the response of the reticuloendothelial system, hemoglobin is inceased to normal levels. Erythrocytes, however, become normal and megaloblasts disappear from the bone marrow. Retardation of the development and other signs of the disease also show reversibility.

VII. XANTHINURIA

This disorder is characterized by an excessive urinary excretion of xanthine replacing uric acid. Oxypurine levels are increased in the serum due to a high renal clearance compensatory action to the elevated plasma oxypurine level and they do not indicate a primary kidney abnormality. In contrast to the increased oxypurine, uric acid levels are low in the urine and serum. The defect in purine metabolism is related to a deficiency in xanthine oxidase activity, normally found in the liver and intestines. In these patients enzyme levels are usually low in both organs. If these patients are kept on a purine-restricted diet, uric acid is virtually absent in the plasma and urine. In some cases xanthine stones occur in the kidney.

VIII. LESCH-NYHAN SYNDROME

This is an inherited disease associated with hyperuricemia and uricosuria. (References 6, 11, 31, 33, 39, 42, 45, 70—72, and 87.) Affected children are normal at birth and during early infancy. Signs appear at the age of 3 to 4 months and characteristic neurological symptoms, spasticity, and mental retardation become apparent. Uric acid levels are very high in the blood and spinal fluid.[59] Biochemical changes reflect the massive overproduction of uric acid, and extremely excessive amounts are present in the urine. These abnormalities

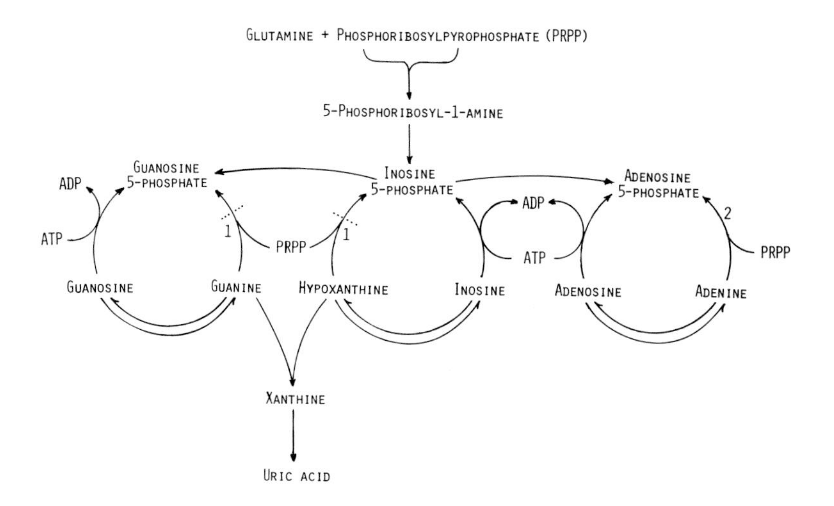

FIGURE 12. Enzyme defects in Lesch-Nyhan syndrome. Characteristic changes include increased excretion of uric acid due to decreased activity of hypoxanthine guanine phosphoribosyl transferase (1) activity resulting in elevated levels of 4-phosphoribosyl-1-pyrophosphate. Adenine phosphoribosyl transferase (2) activity is increased.

are associated with decreased activity, sometimes almost total lack, of hypoxanthine-guanine phosphoribosyl, transferase, increased activity of adenine phosphoribosyl transferase, and increased tissue levels of phosphoribosyl pyrophosphate (Figure 12). Abnormal enzyme activities are fairly widely distributed in various tissues including liver, brain, red blood cells, and skin fibroblasts.[64]

Phosphoribosyl pyrophosphate is also elevated in many tissues.[39,51] These changes indicate that in this disease the salvage pathway of nucleic acid synthesis is deficient, consequently, the total pool of degradation products from both endogenous and exogenous sources is elevated.[62] In some cases of Lesch-Nyhan syndrome, plasma dopamine-β-hydroxylase is increased. This enzyme catalyzes the conversion of dopamine to norepinephrine. Enzyme damage occurs, especially when these patients are exposed to stress that generates acute stimulation of the sympathetic nervous system. The mechanism between enhanced dopa hydroxylase and defective response to sympathetic stimulation is not known.[60]

IX. HYPOURICEMIA

Low serum urate levels (less than 120 μmol/ℓ) are found in 1% of the population.[48] In some cases the hypouricemia is transient, in some cases it is connected with genetic disorders, and in others it can be considered as secondary. Xanthine oxidase deficiency is very rare. It is due to the inactive or missing enzyme. Serum uric acid levels are usually below 60 μmol/ℓ. Excretion of hypoxanthine and xanthine in the urine is high. Connected with the increased levels of precursors, xanthine deposits in the muscle may sometimes occur.

Fanconi syndrome, renal tubular defect of urate reabsorption, some neoplastic diseases, advanced hepatic disease, and ingestion of some drug such as salicylates, uricosurics, and allopurinol may cause hypouricemia.

X. NUCLEOSIDE PHOSPHORYLASE AND ADENOSINE DEAMINASE DEFICIENCY

Nucleoside phosphorylase converts adenosine, guanosine, and inosine to adenine, guanine, and hypoxanthine, respectively. Adenosine deaminase catalyzes the transformation of adenosine to inosine (Figure 9). The activity of these enzymes is reduced in rare congenital diseases, leading to decreased uric acid production. Furthermore, these deficiencies cause severe functions of B- and T-cell functions resulting in immune deficiency diseases. 5′-Nucleotidase deficiency has also been reported to be connected with immune deficiency in some patients.

REFERENCES

1. **Alepa, F. P., Howell, R. R., Klinenberg, J. R., and Seegmiller, J. E.,** Relationship between glycogen storage disease and tophaceous gout, *Am. J. Med.*, 42, 58, 1967.
2. **Auscher, C., de Gery, A., and Pasquier, C.,** Xanthinuria, lithiasis and gout in the same family, *Adv. Exp. Med. Biol.*, 76A, 405, 1977.
3. **Ayvazian, J. H.,** Xanthinuria and hemochromatosis, *N. Engl. J. Med.*, 270, 18, 1964.
4. **Balis, M. E.,** Uric acid metabolism in man, *Adv. Clin. Chem.*, 18, 213, 1976.
5. **Barlow, K. A. and Beilin, L. J.,** Renal disease in primary gout, *Q. J. Med.*, 37, 79, 1968.
6. **Beardmore, T. D., Fox, I. H., and Kelley, W. N.,** Effect of allopurinol on pyrimidine metabolism in the Lesch-Nyhan syndrome, *Lancet*, 2, 830, 1970.
7. **Becker, M. A., Lossman, M. J., Itkin, P., and Simkin, P. A.,** Gout with superactive phosphoribosylphosphate synthetase due to increased enzyme catalytic rate, *J. Lab. Clin. Med.*, 99, 495, 1982.
8. **Benedek, T. G.,** Correlations of serum uric acid and lipid concentrations in normal, gouty, and atherosclerotic men, *Ann. Intern. Med.*, 66, 851, 1967.
9. **Bendersky, G.,** Etiology of hyperuricemia, *Ann. Clin. Lab. Sci.*, 5, 456, 1975.
10. **Berkowitz, D.,** Gout, hyperlipidemia and diabetes relationships, *JAMA*, 197, 117, 1966.
11. **Burkhardt, W. C., Jackson, J. F., Clement, E. G., and Sherline, D. M.,** Prenatal diagnosis of the Lesch-Nyhan syndrome, *J. Reprod. Med.*, 21, 169, 1978.
12. **Bluestone, R., Kippen, I., and Klinenberg, J. R.,** Effect of drugs on urate binding to plasma proteins, *Br. Med. J.*, 4, 590, 1969.
13. **Bollet, A. J.,** Diagnostic and therapeuatic aid in gout and hyperuricemia, *Med. Times*, 109, 23, 1982.
14. **Boss, G. R. and Seegmiller, J. E.,** Hyperuricemia and gout. Classification, complications and management, *N. Engl. J. Med.*, 300, 1459, 1979.
15. **Cameron, J. S.,** Uric acid, gout and the kidney, *J. Clin. Pathol.*, 39, 1245, 1981.
16. **Carcassi, A., Marcolongo, R., Marinello, E., Riario-Sforza, G., and Boggiano, C.,** Liver xanthine oxidase in gouty patients, *Arthritis Rheum.*, 12, 17, 1969.
17. **Clifford, A. J., Steine, L., and Castles, J. J.,** Activities of purine pathway enzymes in gouty human fibroblasts, *Clin. Chim. Acta*, 74, 255, 1977.
18. **Darlington, L. G., Slack, J., and Scott, J. T.,** Family study of lipid and purine levels in gouty patients, *Ann. Rheum. Dis.*, 41, 253, 1982.
19. **DeMars, R., Sato, G., Felix, J. S., and Benke, P.,** Lesch-Nyhan mutation: prenatal detection with amniotic fluid cells, *Science*, 164, 1303, 1969.
20. **Dent, C. E. and Philpot, G. R.,** Xanthinuria, an inborn error of metabolism, *Lancet*, 1, 182, 1954.
21. **Dickinson, C. J. and Smellie, J. M.,** Xanthinuria, *Br. Med. J. S.*, 1217, 1959.
22. **Dreifuss, F. E., Newcombe, D. S., Shapiro, S. L., and Sheppard, G. L.,** X-linked primary hyperuricemia, *J. Ment. Def. Res.*, 12, 100, 1968.
23. **Edwards, N. L., Recker, D. P., and Fox, I. H.,** Hypoxanthine salvage in man: its importance in urate overproduction in the Lesch-Nyhan syndrome, *Adv. Exp. Med. Biol.*, 122, 301, 1980.
24. **Elion, G. B., Kovensky, A., and Hitchings, G.,** Metabolic studies of allopurinol: an inhibitor of xanthine oxidase, *Biochem. Pharmacol.*, 15, 863, 1966.
25. **Fallon, H. J., Smith, L. H., Graham, J. B., and Burnett, C. H.,** A genetic study of hereditary orotic aciduria, *N. Engl. J. Med.*, 270, 878, 1964.
26. **Fessel, W. J.,** Hyperuricemia in health and disease, *Semin. Arthritis Rheum.*, 1, 275, 1972.
27. **Fessel, W. J.,** Distinguishing gout from other types of arthritis, *Postgrad. Med.*, 63, 134, 1978.
28. **Fine, R. N., Strauss, J., and Donnell, G. N.,** Hyperuricemia in glycogen storage disease Type I, *Am. J. Dis. Child.*, 112, 572, 1966.
29. **Fox, I. H. and Kelley, W. N.,** Phosphoribosylpyrophosphate in man: biochemical and clinical significance, *Ann. Intern. Med.*, 74, 424, 1971.

30. **Fox, R. M., O'Sullivan, W. J., and Firkin, B. G.,** Orotic aciduria. Differing enzyme patterns, *Am. J. Med.*, 47, 332, 1969.
31. **Garther, S. M., Scott, R. C., Goldstein, J. R., Campbell, B., and Sparkes, R.,** Lesch-Nyhan syndrome: rapid detection of heterozygotes by the use of hair follicles, *Science*, 172, 572, 1971.
32. **Geiderman, J. M. and Dawson, W. J.,** Diagnostic athrocentesis: indications and method, *Postgrad. Med.*, 65, 109, 1979.
33. **Ghadimi, H., Bhalla, C. K., and Kirschenbaum, D. M.,** The significance of the deficiency state in Lesch-Nyhan disease, *Acta Paediatr. Scand.*, 59, 233, 1970.
34. **Gibson, T., Highton, J., Simmonds, H. A., and Potter, C. F.,** Hypertension, renal function and gout, *Postgrad. Med. J.*, 55 (Suppl. 3), 21, 1979.
35. **Grayzel, A. I., Liddle, L., and Seegmiller, J. E.,** Diagnostic significance of hyperuricemia in arthritis, *N. Engl. J. Med.*, 265, 763, 1961.
36. **Greene, M. L. and Seegmiller, J. E.,** Elevated erythrocyte phosphoribosylpyrophosphate in X-linked uric aciduria, *J. Clin. Invest.*, 48, 32, 1969.
37. **Gutman, A. B. and Yu, T. F.,** Uric acid nephrolithiasis, *Am. J. Med.*, 45, 756, 1968.
38. **Gutman, A. B. and Yu, T. F.,** Secondary gout, *Ann. Intern. Med.*, 56, 675, 1962.
39. **Halley, D. and Heukels-Dully, M. J.,** Rapid prenatal diagnosis of the Lesch-Nyhan syndrome, *J. Med. Genet.*, 14, 100, 1977.
40. **Hansen, O. E.,** Hyperuricemia, gout and atherosclerosis, *Am. Heart J.*, 72, 570, 1966.
41. **Healy, L. A. and Hall, A. P.,** The epidemiology of hyperuricemia, *Bull. Rheum. Dis.*, 20, 600, 1970.
42. **Hoefnagel, D.,** Seminars on the Lesch-Nyhan syndrome, *Fed. Proc.*, 27, 1045, 1967.
43. **Hughes, G. R., Barnes, C. G., and Mason, R. M.,** Bony ankylosis in gout, *Ann. Rheum. Dis.*, 27, 67, 1968.
44. **Huguley, C. M., Bain, J. A., Rivers, S., and Scoggins, R.,** Refractory megaloblastic anemia associated with excretion of orotic acid, *Blood*, 14, 615, 1959.
45. **Itiaba, K., Melancon, S. B., Dallaire, L., and Crawhall, J. C.,** Adenine phosphoribosyl transferase deficiency in association with sub-normal hypoxanthine phosphoribosyl transferase in families of Lesch-Nyhan patients, *Biochem. Med.*, 19, 252, 1978.
46. **Jakovcic, S. and Sorensen, L. B.,** Studies on uric acid metabolism in glycogen storage disease associated with gouty arthritis, *Arthritis Rheum.*, 10, 129, 1967.
47. **Kamoun, P., Chanard, C., Brami, M., and Funch-Brentans, J. L.,** Purine biosynthesis *de novo* by lymphocytes in gout, *Clin. Sci. Biol. Med.*, 54, 595, 1978.
48. **Kay, N. E. and Gottlieb, A. J.,** Hypouricemia in Hodgkin's disease, *Cancer*, 32, 1508, 1973.
49. **Kelley, W. N., Fox, I. H., and Wyngaarden, J. B.,** Essential role of phosphoribosylpyrophosphate in regulation of purine biosynthesis, *Clin. Res.*, 18, 457, 1970.
50. **Kelley, W. N., Greene, M. L., Rosenbloom, F. M., Henderson, J. F., and Seegmiller, J. E.,** Hypoxanthine-guanine phosphoribosyl transferase deficiency in gout, *Ann. Intern. Med.*, 70, 155, 1969.
51. **Kelley, W. N. and Meade, J.,** Studies on hypoxanthine-guanine phosphoribosyl transferase in fibroblasts from patients with the Lesch-Nyhan syndrome, *J. Biol. Chem.*, 246, 2953, 1971.
52. **Kelley, W. N., Rosenbloom, F. M., Henderson, J. F., and Seegmiller, J. E.,** A specific enzyme defect in gout associated with overproduction of uric acid, *Proc. Natl. Acad. Sci. U.S.A.*, 57, 1735, 1967.
53. **Kelley, W. N., Rosenbloom, F. M., Seegmiller, J. E., and Howell, R. R.,** Excessive production of uric acid in type I glycogen storage disease, *J. Pediatr.*, 72, 488, 1968.
54. **Khachadurian, A. K.,** Hyperuricemia and gout: an update, *Am. Fam. Physician*, 24, 143, 1981.
55. **Klinenberg, J. R.,** Hyperuricemia and gout, *Med. Clin. North Am.*, 61, 299, 1977.
56. **Klinenberg, J. R., Compion, D. S., and Olsen, R. W.,** A relationship btween free urate, protein-bound urate and gout, *Adv. Exp. Med. Biol.*, 76B, 159, 1977.
57. **Reynolds, M. D.,** Gout and hyperuricemia associated with sickle-cell anemia, *Semin. Arthritis Rheum.*, 12, 404, 1983.
58. **Kohn, P. M. and Prozan, G. B.,** Hyperuricemia — relationship to hypercholesterolemia and acute myocardial infarction, *JAMA*, 170, 1909, 1959.
59. **Lesch, M. and Nyhan, W. L.,** A familial disorder of uric acid metabolism and central nervous function, *Am. J. Med.*, 36, 561, 1964.
60. **Lloyd, K. G., Hornykiewicz, O., Davidson, L., Shannak, K., Farley, I., and Goldstein, M.,** Biochemical evidence of dysfuncton of brain neurotransmitters in the Lesch-Nyhan syndrome, *N. Engl. J. Med.*, 305, 1106, 1981.
61. **Maclachlan, M. J. and Rodman, G. P.,** Effects of food, fast and alcohol on serum uric acid and acute attacks of gout, *Am. J. Med.*, 42, 38, 1967.
62. **Martinez-Ramon, A. and Grisolia, S.,** Increased incorporation of aspartate and decreased incorporation of orotate in fibroblasts from Lesch-Nyhan patients, *Biochem. Biophys. Res. Commun.*, 96, 1011, 1980.

63. **McPhaul, J.,** Hyperuricemia and urate excretion in renal disease, *Metabolism,* 17, 430, 1968.
64. **Merrill, C. R., Goldman, D., and Ebert, M.,** Protein variations associated with Lesch-Nyhan syndrome, *Proc. Natl. Acad. Sci. U.S.A.,* 78, 6471, 1981.
65. **Migeon, B. R.,** X-linked hypoxanthine-guanine phosphoribosyltransferase deficiency, *Biochem. Genet.,* 4, 377, 1970.
66. **Mikkelsen, W. M.,** The possible association of hyperuricemia and/or gout with diabetes mellitus, *Arthritis Rheum.,* 8, 853, 1965.
67. **Nadler, H. L.,** Prenatal detection of genetic defects, *J. Pediatr.,* 74, 132, 1969.
68. **Naff, G. B. and Byers, P. H.,** Possible implication of complement in acute gout, *J. Clin. Invest.,* 46, 1099, 1967.
69. **Novak, A., Knesl, E., and Müller, M. M.,** Fatty acid composition of plasma lipid fractions in gout, *Adv. Exp. Med. Biol.,* 76A, 563, 1977.
70. **Nyhan, W. L.,** The Lesch-Nyhan syndrome, *Adv. Nephrol.,* 3, 59, 1974.
71. **Nyhan, W. L.,** The Lesch-Nyhan syndrome, *Dev. Med. Child. Neurol.,* 20, 376, 1978.
72. **Nyhan, W. L., Bakay, B., Connor, J. D., Marks, J. F., and Keele, D. K.,** Hemizygous expression of glucose 6-phosphate dehydrogenase in erythrocytes for the Lesch-Nyhan syndrome, *Proc. Natl. Acad. Sci. U.S.A.,* 65, 214, 1970.
73. **Ostberg, Y.,** Renal urate deposits in chronic renal insufficiency, *Acta Med. Scand.,* 183, 197, 1968.
74. **Page, T., Bakay, B., and Nyhan, W. L.,** An improved procedure for detection of hypoxanthine guanine phosphoribosyl transferase heterozygotes, *Clin. Chem.,* 28, 1181, 1982.
75. **Pagliara, A. S. and Goodman, A. D.,** Elevation of plasma glutamate in gout, its possible role in the pathogenesis of hyperuricemia, *N. Engl. J. Med.,* 281, 767, 1969.
76. **Paik, C. H., Alavi, I., Dunea, G., and Weiner, L.,** Thalassemia and gouty arthritis, *JAMA,* 213, 296, 1970.
77. **Paulus, H. E., Coutts, A., Calabro, J. T., and Klinenberg, J. R.,** Clinical significance of hyperuricemia in routinely hospitalized men, *JAMA,* 211, 270, 1970.
78. **Perheentupa, J. and Raivio, K. O.,** Fructose-induced hyperuricemia, *Lancet,* 2, 528, 1967.
79. **Phelps, P.,** Appearance of chemotactic activity following intraarticular injection of sodium urate crystals, *J. Lab. Clin. Med.,* 76, 622, 1970.
80. **Raivio, K. O. and Seegmiller, J. E.,** Role of glutamine in purine synthesis and interconversion, *Clin. Res.,* 19, 161, 1971.
81. **Rajan, K. T.,** Lysosomes and gout, *Nature (London),* 210, 959, 1966.
82. **Reginato, M. A., Valenzuela, R. F., Martinez, C. V., Passano, G., and Daza, K. S.,** Polyarticular and familial chondrocalcinosis, *Arthritis Rheum.,* 13, 157, 1970.
83. **Reich, M. L.,** Arthritis: avoiding diagnostic pitfalls, *Geriatrics,* 46, 53, 1982.
84. **Rogers, L. E. and Porter, F. S.,** Hereditary orotic aciduria. A urinary screening test, *Pediatrics,* 42, 423, 1968.
85. **Rundles, R. W., Wyngaarden, J. B., Hitchings, G. H., and Elion, G. B.,** Drugs and uric acid, *Ann. Rev. Pharmacol.,* 9, 345, 1969.
86. **Saker, B. M., Tofler, O. B., Burvil, M. J., and Reilly, K. A.,** Alcohol consumption and gout, *Med. J. Aust.,* 1, 1212, 1967.
87. **Shin-Buehring, Y. S., Osang, M., Wirtz, A., Haas, B., Rahm, P., and Schaub, J.,** Prenatal diagnosis of Lesch-Nyhan syndrome and some characteristics, *Pediatr. Res.,* 14, 825, 1980.
88. **Smith, L. H.,** Hereditary orotic aciduria-pyrimidine auxotrophism in man, *Am. J. Med.,* 38, 1, 1965.
89. **Sorensen, L. B.,** Hyperuricemia and gout, *Adv. Intern. Med.,* 15, 177, 1969.
90. **Sorensen, L. B.,** The pathogenesis of gout, *Arch. Intern. Med.,* 109, 379, 1962.
91. **Sperling, O., Frank, M., Ophir, R., Lieberman, U. A., Adam, A., and deVries, A.,** Partial deficiency of hypoxanthine-guanine phosphoribosyl-transferase associated with gout and uric acid lithiasis, *Rev. Eur. Stud. Clin. Biol.,* 15, 942, 1970.
92. **Spilman, E. L.,** Uric acid synthesis in the nongouty and gouty human, *Fed. Proc.,* 13, 302, 1954.
93. **Steele, T. H.,** Control of uric acid excretion, *N. Engl. J. Med.,* 284, 1193, 1971.
94. **Steele, T. H.,** Renal excretion of uric acid, *Arthritis Rheum.,* 18, 793, 1975.
95. **Steele, T. H. and Rieselback, R. E.,** The contribution of residual nephrons within the chronically diseased kidney to urate homeostasis in man, *Am. J. Med.,* 43, 876, 1967.
96. **Steele, T. H. and Rieselback, R. E.,** The renal mechanism for urate homeostasis in normal man, *Am. J. Med.,* 43, 868, 1967.
97. **Tubergen, D. G., Krooth, R. S., and Heyn, R. M.,** Hereditary orotic aciduria with normal growth and development, *Am. J. Dis. Child.,* 118, 864, 1964.
98. **Wall, B. A.,** Acute gout and systemic lupus erythematosus, *J. Rheumatol.,* 9, 305, 1982.
99. **Wallace, S. L., Robinson, H., Masi, A. T., Decker, J. L., and McCarty, D. J.,** Selected data on primary gout, *Bull. Rheum. Dis.,* 28, 992, 1978.

100. **Weissmann, G.,** Lysosomes and joint disease, *Arthritis Rheum.*, 9, 834, 1966.
101. **Woeber, K. A., Ricca, L., and Hills, A. G.,** Pathogenesis of uric acid urolithiasis, *Clin. Res.*, 10, 45, 1962.
102. **Wyngaarden, J. B.,** Pathophysiology of hyperuricemia in primary gout, *Trans. Am. Clin. Climat. Assoc.*, 81, 161, 1969.
103. **Wyngaarden, J. B. and Jones, O. W.,** The pathogenesis of gout, *Med. Clin. North Am.*, 45, 1241, 1961.
104. **Yip, L. C., Yu, T. F., and Balis, M. E.,** Aspect of purine metabolic aberrations associated with uric acid overproduction and gout, *Adv. Exp. Med. Biol.*, 122, 307, 1980.
105. **Yu, T. F.,** Some unusual features of gouty arthritis in women, *Semin. Arthritis Rheum.*, 6, 247, 1977.
106. **Yu, T. F. and Gutman, A. B.,** Uric acid nephrolithiasis in gout. Predisposing factors, *Ann. Intern. Med.*, 67, 1133, 1967.

FURTHER READING

Davidson, J. N., *The Biochemistry of Nucleic Acids*, 4th ed., John Wiley & Sons, New York, 1969.

Gutman, A. B., Bondy, P. K., and Rosenberg, L. E., *Metabolic Control and Disease*, 8th ed., W. B. Saunders, Philadelphia, 1980.

Harbers, E., Domagk, G. F., and Müller, W., *Introduction to Nucleic Acids. Chemistry, Biochemistry and Functions*, Reinhold, New York, 1968.

Jacob, S. T., Ed., *Enzymes of Nucleic Acid Synthesis and Modification*, CRC Press, Boca Raton, Fla., 1983.

Martin, S. J., *The Biochemistry of Viruses*, Cambridge University Press, Cambridge, 1978.

Wyngaarden, J. B. and Kelley, W. N., *Gout and Hyperuricemia*, Grune & Stratton, New York, 1976.

INDEX

A

B

D

E

F

G

H

I

J

K

M

N

R

S

T

U

V

W

X

Y

Z